American Institute of Chemical Engineers

Topical Conference on Applications of Micro-Reactor Engineering 2006

Held at the 2006 AIChE Spring National Meeting

April 23-27, 2006
Orlando, Florida, USA

Printed from e-media with permission by:

Curran Associates, Inc.
57 Morehouse Lane
Red Hook, NY 12571
www.proceedings.com

ISBN: 978-1-60423-525-8

Some format issues inherent in the e-media version may also appear in this print version.

American Institute of Chemical Engineers

Topical Conference on Applications of Micro-Reactor Engineering
2006

TABLE OF CONTENTS

Session 71 - Micro Process Engineering and Intensified Process Systems Poster

CoChair: Roshan Jachuck

Chair: David Brenchley

Chair: Jean-Christophe Eloy

Session 139 - Microtechnology and Process Analytics

CoChair: Ray Chrisman

Chair: Melvin Koch

Session 140 - Particles, Emulsions and Dispersions II

CoChair: Kazuhiro Mae

Chair: Michael Koehler

Session 180 - Fine Chemicals and Pharmaceuticals Production in Micro-Systems

CoChair: YW Kwon

Chair: P. Watts

Preface for Topical "Applications of Microstructured Reactors"

In 2002, the journal Chemical Engineering Progress (January issue, p. 315) quoted that microstructured reactors are just at the upheaval of the "S"-curve where product maturity is converted into societal gain, tightly following the Lab-on-a-chip systems. Now after about 4 more years, we have indeed come much closer to the applications, and this is what our Topical will be about. Evidence for that are the growing number of companies in the field, that virtually any larger R&D laboratory in chemical industry is equipped with the technology, and that more and more pilot / production plants are built and announced. Concerning scientific impact, there is hardly any conference without a session on the topic, and papers and special issues thereabout gain high attention. From a niche position, the technology is moving in the center of chemical engineering, being recognized as a complementary tool.

Thus, the Topical was structured according to such classical application disciplines, e.g. into mixing / heat transfer, polymerizations, fine chemistry, catalytic processes, and dispersions /particles. In addition, the sessions on process control and cost / eco efficiency / exergy analyses stand for the industrial use and for process and plant engineering. It is thus just logical that the umbrella for this event is provided by the Process Development Group within AIChE which established just a year ago the position of a new group entitled "Micro Process Engineering".

It turned out that this Topical is the largest at this AIChE Spring Meeting in Orlando, even though there are many, many other conferences on microstructured reactors. More than 90 papers were submitted, resulting in more than 50 oral presentations and about 20 posters. Accordingly, we expect about 100 attendees to the sessions. The scientific reporting of the conference results will be done via three special issues in two chemical engineering journals. Catalysis Today will cover the catalytic investigations, to be issued probably in December 2006. Chemical Engineering & Technology will provide a place for papers on all other topics, with two special issues in November 2006 and February 2007. Selection of the best papers for this purpose will be made during the Topical to ensure and acknowledge high scientific quality.

The AIChE Annual Conference in San Francisco is just around the corner. Five sessions on microfluidics or microstructured reactors have been announced there, now moving the technology within the discrete application topicals. This is the next logical step in making microstructured reactors a routine tool in academia and chemical industry. Still, there will be an specific Topical on industrial uses of microstructured reactors and process intensification, where the technology itself will be more in the focus.

We are looking much forward to the AIChE meeting in Orlando and to welcome researchers from all over the world.

Kind regards,

Volker Hessel
Chair Micro Process Engineering
IMM, Mainz, Germany
TU/e Eindhoven, The Netherlands

Kurt Vanden Bussche
Co-chair Micro Process Engineering
UOP LLC, Des Plaines, IL

23a The Impact of Microtechnologies in Chemical and Pharmaceutical Production Processes

Barbara Pieters[1], Géraldine Andrieux[1], Volker Hessel[2], Holger Löwe[2], Jean-Christophe Eloy[1], and **Jean F. Jenck**[3]. (1) Yole Développement, 45 rue Sainte-Genevieve, Lyon, F-69006, France, (2) Institut für Mikrotechnik Mainz GmbH, Carl-Zeiss-Straße 18-20, Mainz, D-55129, Germany, (3) ENKI Innovation, 3 chemin des Balmes, Sainte Foy, F-69110, France
Confronted to high economical stakes in a very high competitive environment, the chemical and pharmaceutical industries are looking for innovative solutions: the challenge is not only to provide service to our consumer society, but to do so with markedly lower reliance on materials, energy, labour and waste. If technology was in the past viewed as a cause of environmental degradation, it is now recognized for its key contribution to sustainability by decoupling economic growth from environmental impact. Through the application of 'green' chemistry and engineering, the industry has already implemented innovations in the field of renewable resources, eco-efficient products, energy efficiency, waste reduction and reuse, inherently safer processes [1]. New R&D areas, such as 'white' biotechnologies, catalytic processes, 'green' solvents, alternative energies, can make a significant contribution to improved sustainability. A key topic is Process intensification (PI), where the motivation is "doing more with less" [2,3]. First designed as a strategy to reduce the size of plants, PI takes a broader dimension as a method to give full potential to physico-chemical transformation by minimizing diffusional limitations. It is about adapting the equipment rather than the physico-chemical transformation to existing and often unsuited apparatus. In this context, microtechnologies show wide possibilities to replace large, expensive, inefficient equipment by smaller, more compact, safer, better controlled (therefore more economic) installations. They also allow creating new reaction conditions leading to development of new molecules. In R&D, microtechnologies have shown their value in high-throughput screening [4,5] by notably reducing the analysis cost through lower volumes used and faster analysis [6]. The current trend is now to transfer such innovation into production. Through flow process, the reaction conditions created in R&D could thus be applied directly for production purpose. We will thus discuss how microtechnologies could impact the chemical processing [7-10]. The advantages will be presented for each market segment targeted [10]. We will analyse the balance between technological advantages and technological hurdles for their implementation in chemical processing. Learning from some preliminary attempts to introduce the technology into processing, we will try to highlight the main challenges to overcome today to succeed. After this technological overview, we will present an economical analysis to highlight what could be the benefits over the investment. As an archetypical example, a 'novel-chemistry' process, tailored by IMM for microreaction engineering, will be benchmarked to its conventional counterpart. A cost analysis will be given for the high-pressure and high-temperature variant ('high-p,T', [11]) of the Michael addition [12] and Kolbe-Schmitt [13] synthesis, where an enormous increase in productivity is observed which dramatically change, e.g. in the space-time yields. It will be shown how this impacts the operational costs (OPEX) on one side, the investment costs (CAPEX) on the other side. Based on our analysis, we will

conclude on the expected evolution of this field from a technological, economical and industrial point of view.

[1] Jenck, J.F., Agterberg, F., Droescher, M.J.; "Products and processes for a sustainable chemical industry: a review of achievements and prospects", Green Chem. 6 (2004), 544-556 [2] a) Ramshaw, C.; "Process Intensification and Green Chemistry", Green Chem. 1 (1999), G15-G17 b) Aoune, A., Ramshaw, C.; "Process Intensification: Heat and Mass Transfer Characteristics of Liquid Films on Rotating Discs." Int. J. Heat Mass Transfer, 42 (1999) 2543-2556. [3] Stankiewicz, A. I., Moulijn, J. A.; "Process intensification: transforming chemical engineering", Chemical Engineering Progress 1 (2000) 22-34. [4] Andrieux, G., Pieters, B., Labat C., Eloy J.C. ; "Emerging Market for Microfluidics Applications", EMMA, Yole Développement report (2004) [5] Ducrée, J., Zengerle, R., Newmann, J., Ernst, H., Riedmüller, K., Messner, S., Andrieux, G., Provence, M., Eloy, J.C.; "Flow Map – Microfluidics Roadmap for Life Sciences", Flow Map consortium & EC (2004) [6] Murphy, W., Volpe, A. F., Weinberg, H.; "High-throughput approaches to catalyst discovery", Current Opinion in Chemical Biology 7 (2003) 427-433. [7] Hessel, V., Löwe, H., Müller, A., Kolb, G.; Chemical Micro Process Engineering - Processing and Plants, Wiley-VCH, Weinheim (2005). [8] Hessel, V., Hardt, S., Löwe, H.; Chemical Micro Process Engineering - Fundamentals, Modelling and Reactions, Wiley-VCH, Weinheim (2004). [9] Stange, T., Kiesewalter, S., Russow, K., Hessel, V., Provence, M., Balsalobre, C., Boulon, P., Mounier, E.; "PAMIR – Potential and applications for MicroReaction technology, a market survey", EC Competitive and sustainable growth programme (1998 – 2002) [10] Pieters, B., Andrieux G., Eloy, J.C.; "Microtechnologies for process intensification", Yole Développement market analysis (2005) [11] Hessel, V., Löb, P., Löwe, H.; "Development of reactors to enable chemistry rather than subduing chemistry around the reactor - potentials of microstructured reactors for organic synthesis", Curr. Org. Chem. 9, 8 (2005) 765-787. [12] Löwe, H., Hessel, V., Löb, P., Hubbard, S.; "Addition of secondary amines to &alpha&beta-unsaturated carbonyl compounds and nitriles by using microstructured reactors", Org. Proc. Res. Dev. (2005) submitted. [13] Hessel, V., Hofmann, C., Löb, P., Löhndorf, J., Löwe, H., Ziogas, A.; "Aqueous Kolbe-Schmitt synthesis using resorcinol in a microreactor laboratory rig under high-p,T conditions", Org. Proc. Res. Dev. 9, 4 (2005) 479-489.

23b Micro Reaction Technology and Molecular Industries – Changing the Paradigm of Plant Engineering Cost Efficiency to Innovation Value Responsiveness

Thomas Schwalbe and Keith Simons. Acclavis LLP, 84 Browne Street #3, Brookline, Ma 02446, MA 02246
Chemical Industries in the established industrial regions are under existential pressure to change. Output is practically stagnant and bound to relocate to China and India. There, suppliers face 5 and 3 % of established regions' labor cost respectively. As long as there is excess supply, these regions can always beat their competition on cost. Innovation and the intellectual capital are of course more important. The number of Chinese students enrolled in chemistry and engineering has doubled to 4 million in the time period from 1998 to 2001, representing 33 % of all students enrolled in these disciplines among the regions USA, Japan, Europe, India and China. Thus, intellectual output productivity of chemical processing innovation becomes critical.

Among the established regions, Japan enjoys a unique opportunity. Growing exports to China have led to temporary growth and significant profitability increase. Since chemical industry strives to drastically reduce bank lending, there is a challenge to change the way in which the industry works, in particular to save on plant cost and inventory and to enhance profitability. Microreaction Technology (MRT), Process Intensification (PI) and Reaction Process Design (RPD) are enabling technologies. They promise continuous innovation for better products at lower cost. More importantly, they enable a compression of the supply chain to end customers by providing new chemistry on specification and on demand. The technologies permit an industrial transformation in a similar manner to what has been called the revolution of automotive industries. The latter has been predominantly driven by Japanese firms. Beyond technology it brought processes to increase new product sales, shorten time to market, allow for Just-in-Time manufacture and permit continuous improvement.

Microreaction technology (MRT), PI and RPD are developing and need to be measured against their contribution to this transformation process. MRT and PI are based on continuous manufacturing, which is always more cost effective than batch manufacturing. The switch from batch to continuous manufacturing yields cost, quality and environmental benefits. Labor cost are lower, throughput time is reduced by 98 % and many non-value adding steps are avoided. In a Genzyme case study, the company had shown how a switch of their Havershill Renagel™ manufacturing operation to continuous processing reduced their plant capital investment by 40 %.

To catch up with the underlying technology, MRT, PI and RPD will be briefly described. More importantly, their implementation status will be benchmarked against their potential to drive the necessary industry transformation. Several examples of high value transformations from MRT will be referenced and the current level of new chemistry implementation addressed.

Measures and metrics of short term business success will be juxtaposed to the strategic goal of industrial sustainability.

Web Page: www.acclavis.com

23c Basf's Eco-Efficiency Analysis Method - a Tool to Assess Micro-Reactor Technologies

Charlene Wall, Corporate Safety & Ecology, BASF Corporation, 100 Campus Drive, MS F-402, Florham Park, NJ 07932

In order to truly understand advantages of new technologies such as micro-reactors, quantitative tools are necessary. Ideally, these tools should assess life-cycle impacts of new versus existing technologies, and provide the opportunity to optimize new technologies. BASF's Eco-efficiency Analysis Method is such a tool. This presentation will describe the analysis method, and include a case study in which different processes were compared. The intent is to provoke consideration of the application of this type of tool in the development of micro-reactor technologies. BASF's eco-efficiency analysis methodology was developed in 1996 and to date over 240 analyses have been completed. It is based upon life-cycle analysis principles of ISO 14040, with additional enhancements that allow the results to be used as a concise decision-making tool. First, a product or technology application is selected, and the current or future alternatives are identified. Next, six environmental categories are evaluated: Raw materials consumption, Energy consumption, Land use, Air, water and solid waste emissions, Toxic potential of the substances employed and released and risk potential. Life-cycle data are compiled for all of the categories, a weighting scheme is used to aggregate the results, and they are normalized in order to generate the ecological fingerprint. The fingerprint depicts the relative impacts of the alternatives in each of the environmental categories. The total environmental impact is then combined with an economic analysis that addresses costs over the product life-cycle. The Eco-efficiency Portfolio depicts relative environmental impacts on one axis, and the economic impacts on the second axis. The most eco-efficient products will provide both environmental benefits, and be most likely to succeed in the marketplace based on favorable cost aspects. BASF has used Eco-efficiency Analysis for products used in industries ranging from construction, to automotive, to human and animal nutrition. Applications of eco-efficiency analysis include: Strategic planning – determining which products or technologies should be pursued, Marketing – demonstrating in the marketplace which technologies are the most environmentally sound, Research and development – evaluating product portfolios and identify areas for optimization, Stakeholder and government dialogue – sharing knowledge externally about which technologies are the best and why, and demonstrating commitment to sustainable development

23d Microreactor Technology: a Revolution for the Fine Chemical and Pharmaceutical Industry?

Niklaus Bieler, Laurent Ducry, and Dominique M. Roberge. Exclusive Synthesis, Process R&D, Lonza Ltd., Walliser Werke, Visp, Switzerland
50% of reactions in the fine chemical / pharmaceutical industry could benefit from a continuous process based mainly on microreactor technology. However, the frequent presence of a solid phase still hinders the widespread application of such a technology as a multi-purpose solution. For small scale and pilot productions, speed in process R&D as well as avoidance of scale-up issues are the main drivers. On the other hand, for large scale productions, a gain in yield and safety are the main motivation for micoreactor technology. The gain in yield must be significant in order to cope with the increase in capital expenditure associated with the development of a new technology.

Potential Evaluation of Micro Reaction Technology – A breakthrough of this technology?

Dr. D. Schmalz*, Dr. M. Häberl*, Dr. N. Oldenburg*, Dr. M. Grund*, Prof. Dr. habil. U. Kunz§

* Central Process Development

Merck KGaA

Frankfurter Straße 250

D-64293 Darmstadt

E-Mail: dirk.schmalz@merck.de

§ Institute of Chemical Process Engineering

Clausthal University of Technology

Leibnizstraße 17

D-38678 Clausthal

Introduction

The fine chemicals and pharmaceutical industry (FCPI) has been characterized for decades by batch processes in laboratorys as well as production plants. The rational for the use of this technology can be focused in short life cycles and relatively small outputs in these sections. Thus conventional, dedicated and often continuously working plants do not appear economically applicable. Batch wise working plants possess a comparatively high flexibility (capacity -, assortment and structure flexibility), which is a basic condition for the employment in the FCPI because of frequently changing products.

Nevertheless micro reaction technology (MRT), a new continuous technology, has been in the focus of interest in the chemical industry for the last decade. Economic advantages are often mentioned yet but no quantification has been given yet [1][2].

So far in no publications a method has been developed to quantify the economic benefits of MRT. In order to be able to number possible savings, first a reference value of current costs of process development and production must be determined. So far no economic characteristic numbers are available. The maximum of attainable savings depends on the unit operations, which are technically feasible and meaningfully applicable. Beyond that the total process must be considered for an evaluation. Recent examples mentioned in literature are only related to th synthesis and do not include the whole process chain. Process chains in the FCPI yet possess a high portion of downstream operations, which must be included in an evaluation. Resource savings, like reduced amounts of raw materials and an improvement in yield by the employment of the MRT can not be evaluated in general. It is essential to look at each example process individually. Finally a concludingstatement about economy needs an evaluation with economic methods.

Potential theory

In order to rate the efficiency of new technologies in process industry, methods, which concern themselves with the evaluation of alternative energies, can be consulted [3][4]. Estimations of potentials are also used for the evaluation of environmental products [5]. These evaluations of potential can not be generalized because they are too case specificl. In addition the material aspect, which plays a large role in the process industry, is not sufficiently considered. Therefore no generally accepted rules are available.

Transferred to the existing question the potential is defined as the efficiency of the MRT and thus the possible reductions of costs regarding the process development and production compared to bath wise working operations. The potential considered It is differentiated between the theoretical, technical, material and economical potential.

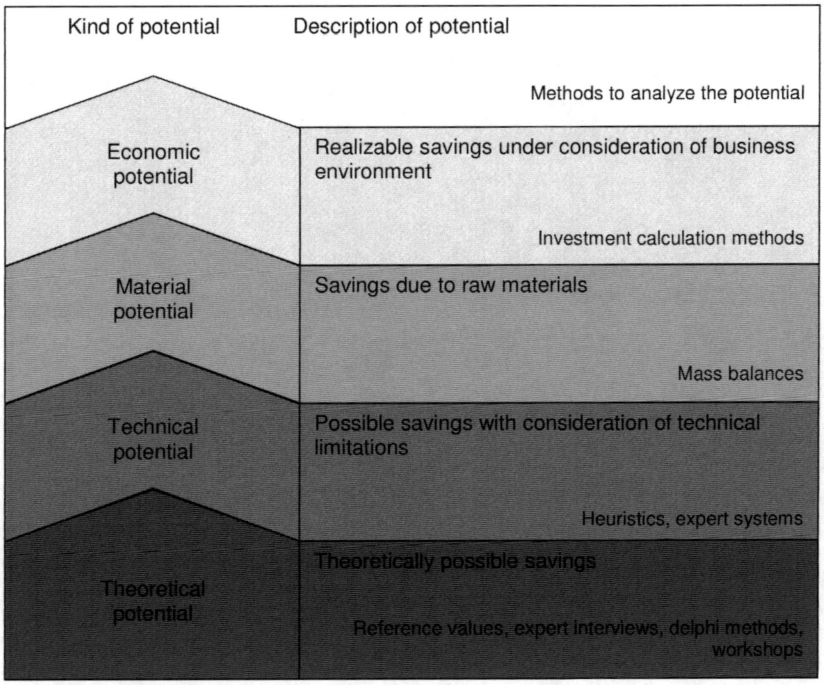

Figure 1: Kind of potentials, its descriptions and methods to analyze the potential [5]

The theoretical potential describes the maximally attainable savings with the use of the MRT. The technical potential describes the possible savings considering existing unit operations. The material potential is quantified as the savings of raw materials in relation to the batch technology The economic potential sets the savings calculated with the material potential in relation towards economic key indicators.

Specifics of fine chemistry and pharmaceutical industry

The five most important characteristics of the FCPI are represented in figure 2. At first the very high specific product price up to several hundred thousand Euros can be named here. A second important size is the number of chemical stages, which must be accomplished from the first raw material to the actual product. In the petrochemistry very few chemical stages have to be accomplished, the number of stages rises over the individual sections to 10-20 needed stages in the pharmaceutical industry. The third characteristic is the specific waste material. In the petrochemistry there is the specific waste material smaller than 1 kg per kg value product. This specific relation in fine chemistry and pharmaceutical industry can raise up 1000 kg/kg. The development time of products ranges between several years in the petrochemistry and up to 15 years in the pharmaceutical industry. Last but not least the production amounts in the FCPI lies in the region of several hundred kilograms up so several tons wheres in the petrochemistry several million tons.

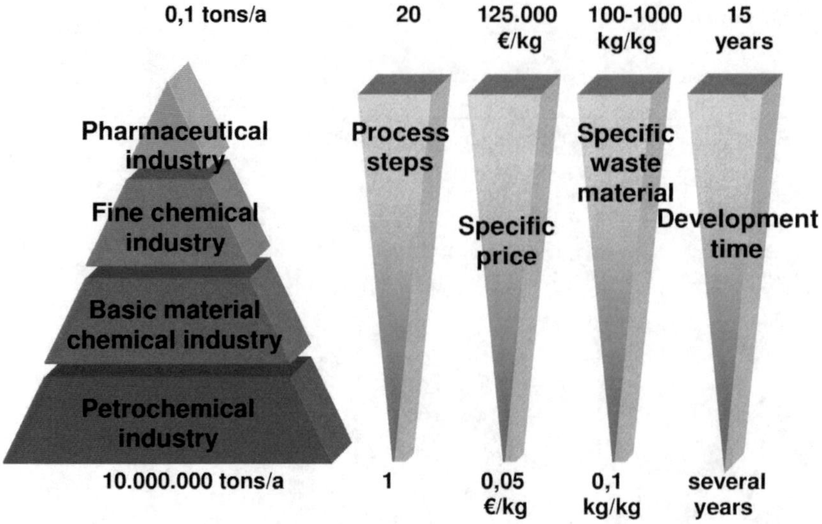

Figure 2: Classification of the FCPI on the basis of characteristic variables

Reaction portfolio

The determination of the theoretical potential is achieved by using a typical and randomized portfolio for the FCPI. In this reaction portfolio contains 62 reactions from FCPI which are clustered in a first step regarding following criteria:

- Reaction type (Grignard reaction, Wittig reaction, etc.)
- Homologous reactions (e. g. nearly the same raw materials)
- Similar sequences of process steps (e. g. reaction, hydrolysis, etc.)
- Similar technical and logistical demands (e. g. volume, steel grade, etc.)

62 reactions were clustered into 11 reference processes. With the determination of one reference process a conclusion by analogy was made for all reactions. In the existing reaction portfolio the main reaction type are Hydrogenations and isomerisations (23%), Grignard reactions (20%) and lithiations (35%) as well as 4 reactions with smaller amounts.

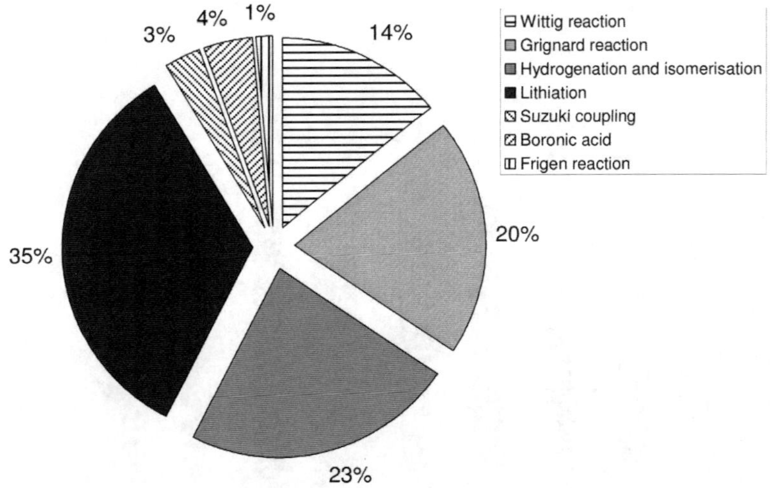

Figure 3: Mass fraction of the considered reactions

Theoretical potential

The possible savings by the employment of the micro reaction technology is to be understood as the theoretical potential. It is assumed that there are no restrictions regarding technical, material and economic restrictions. Thus the theoretical potential represents an upper limit of the reduction of costs potential for the process development and production compared to the classic batch wise process development and production.

Table 1: Costs of capital and running costs

Cost of capital	Running costs
Depreciation	Personnel costs
Building	Maintenance costs
Micro reaction plant	Building
Imputed interest	Micro reaction plant
Building	Costs of analytics
Micro reaction plant	Other costs

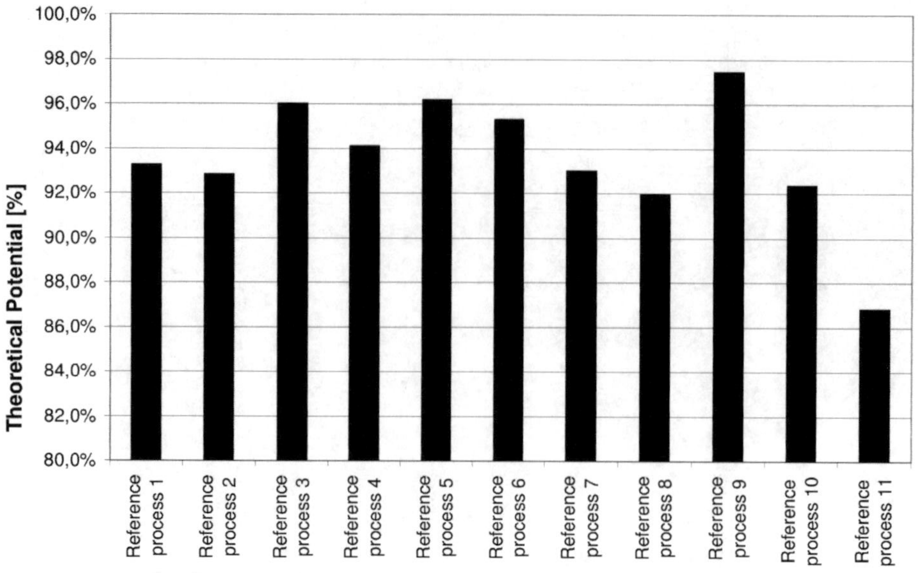

Figure 4: Theoretical potential in dependence on the reference process

The theoretical potential of the considered portfolio is very high over all phases of the process development. Looking of the process development in the laboratory the savings are based on the quadruple number of experiments that can be done. Due to the almost not existing scale-up factor from laboratory to pilot plant validation experiments can be reduced to an absolute minimum. Therefore additional savings can be realized. In the production phase the savings are most significant. This is due to reduced costs of investment a well as personal costs.

Technical potential

The technical potential describes the fraction of the theoretical potential considering the technical feasibility. It represents the first subset of the theoretical potential. The technical potential considers restrictions in using MRT. These restrictions can be determined e.g. with expert systems and heuristics [6][7]. On this basis the following obstacles can exists for the MRT [7]:

• Suitable unit operations for the execution of the process can not provided (e.g. cristallisations)

• Suitable parameters (concentration, etc.), with which the raw materials, products and by-products are in solution, are missing.

• Due to process intensification decomposition reactions are not avoidable

• There are no builing materials available.

Therefore only reactions steps and aqueous downstream processes, which are presently feasible with MRT, are considered in the evaluation. Savings in the laboratory are estimated <between 20% and 30%. The potential in the pilot plant scale is about 85% of the total costs.

In the production, which exhibits 95% of the overall saving, savings between 12% and 85% in comperision to the batch productin can be observed.

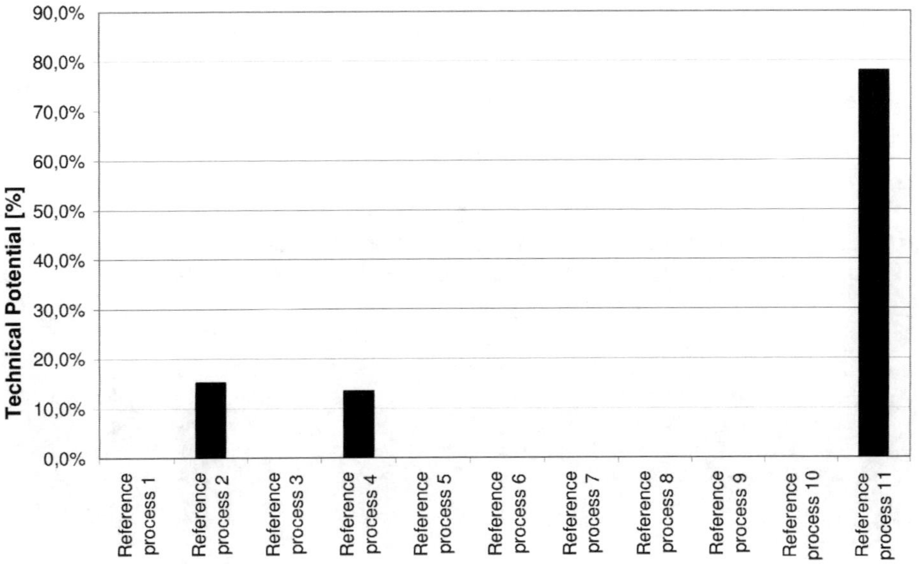

Figure 5: Technical potential in relation to the reference process

Material potential

In the context of the material potential the material balance from the realized experiments in the potential is considered. The changed material costs and costs of waste streams are considered. The investigations done show a slightly improved conversion regarding the

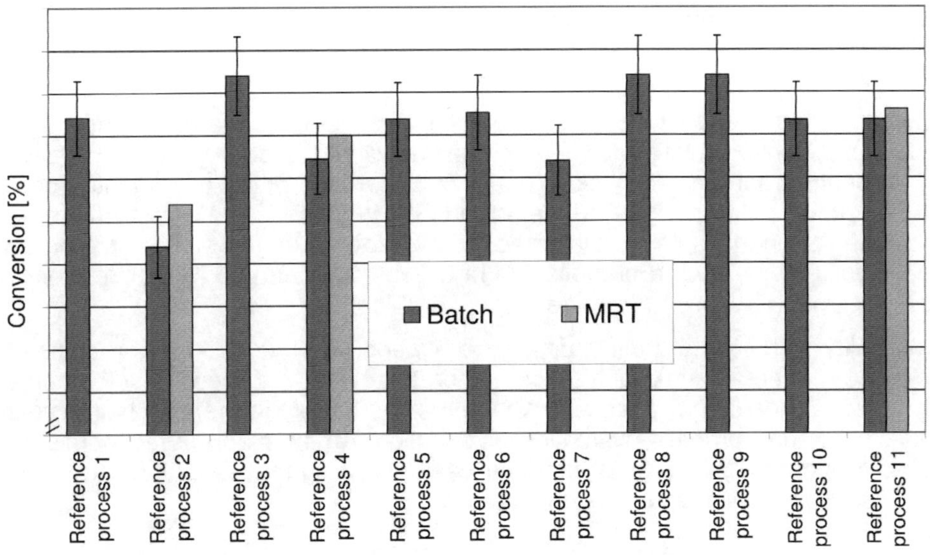

Figure 6: Material potential in dependence on the reference process

key components. In order to confirm these savings further investigations must be accomplished to exclude possible coincidental errors. For this reason it is assumed the material potential corresponds to the technical potential.

Economic potential

The economic potential describes the potential which can be realized under the valid boundary conditions in the FCPI. For the determination of the economical potential investment calculation methods are used. With these methods the questions is answered, when it is economically meaningful to insert alternative technologies under the given boundary conditions.

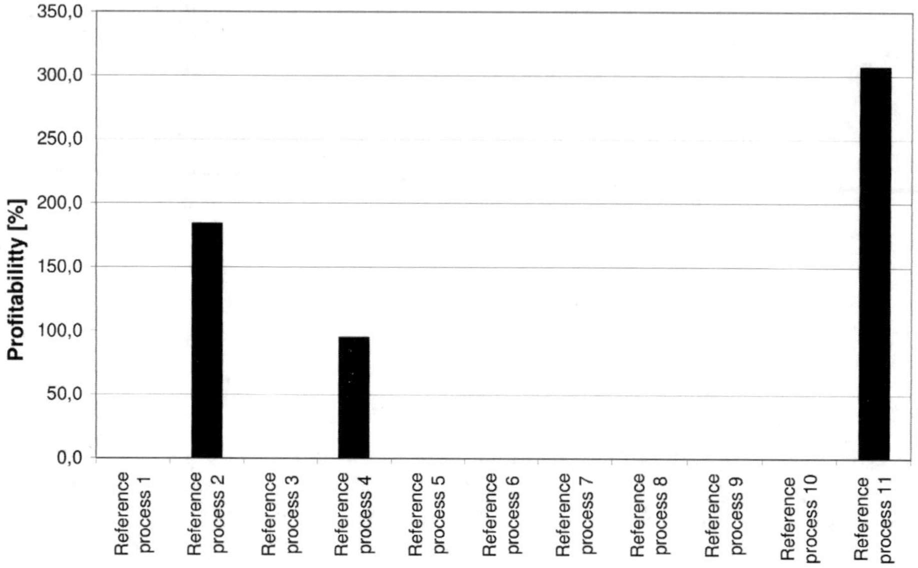

Figure 7: Profitability in dependence on the reference process

A criteria for the evaluation of a technology is the profitability. This describes the relationship of savings by the technology in relation to the used capital. With the profitability the relative advantage of a technology can be evaluated. For the consideration of liquidity-oriented considerations with the amortization period a further criteria is introduced. The amortization period is defined as the assigned capital in relation to the average return flows. The profitability of the use of MRT for the considered portfolio lies between 100% and 300% per reference process. The payback period is less than 1 year.

A minimum profitability of 15% and a minimum payback period of 3 years form the basis of the evaluation. In order to be able to reach these numbers, at least 3 process developments must be accomplished in the laboratory scale. The economic benefit in the pilot plant scale can be realized by only one process evaluation. Production in new plants is economically feasible with outputs around 10 tons to 30 tons per year. By using these key indicators the advantage of the use of the MRT can be clearly quantified.

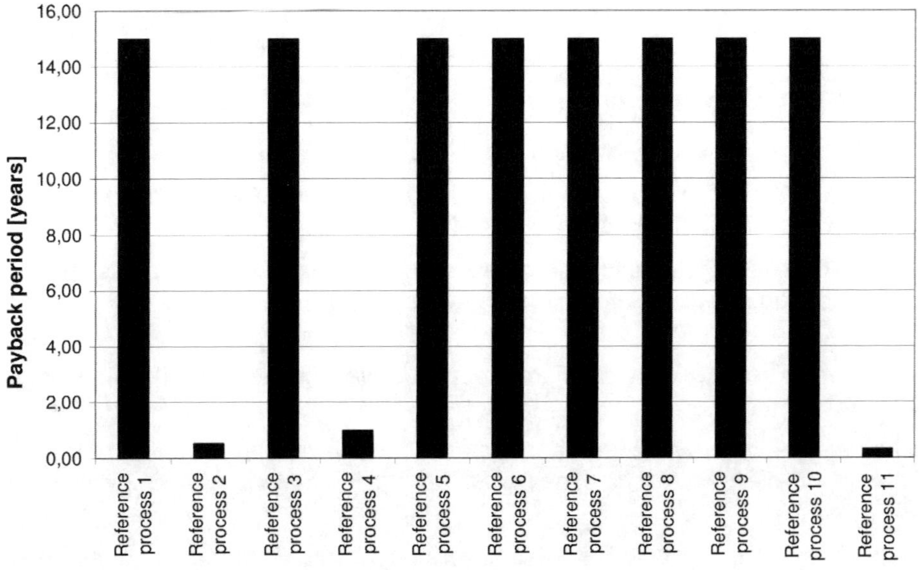

Figure 8: Payback period in relation to the reference process

Conclusion

For the evaluation of the MRT a method using different types of potentials was provided and applied systematically to the process development and production in the FCPI. A differentiation between the theoretical, technical, material and economical potential was developt. The method was applied on a real industrial portfolio consists of more thean 60 relevant processes.

The theoretical potential represents the upper limit of possible savings with a MRT system. It was shown that the theoretical potential of the considered portfolio amounts to 94% of the total costs, whereby largest savings lie in the context of production.

In the investigation of the technical potential technical restrictions were regarded. It could be shown that certain reaction steps and the aqueous processing with MRT system can be accomplished. Downstream processes must be accomplished however further in the batch procedure. The technically realizable potential is about 7% of the total costs, whereby 95% of savings are expected in production.

The material potential is identical to the technical potential, since no significant improvements in the chemical conversion could be recognized.

A minimum profitability of 15% and a minimum payback period of 3 years form the benchmark of the evaluation. In order to be able to reach these numbers, at least 3 process developments must be accomplished in the laboratory scale. Around the advantages in the pilot scale economically only one evaluation will needed. Production is economically feasible with outputs around 10 tons to 30 tons per year. With these key indicators the advantage of the use of the MRT and the method could be proven.

Literature

[1] C. Boswell, Chem. Market Rep. **2004**, 266, No. 11

[2] M. Matlosz, IMRET 7, Lausanne, Switzerland **2003**

[3] K. Wagner, S. Wittkopf, Der Energieholzmarkt Bayern, Bayerische Landesanstalt für Wald und Forstwirtschaft (Hrsg.) **2000**

[4] P. Aleman, Y. Orlando, Potentialbestimmung und Wirtschaftlichkeitsanalyse der regenerativen Energiequellen in Kolumbien, PhD-Work, Darmstadt University of Technology **1995**

[5] K. Hungerbühler, Mitteilungsblatt der Fachgruppe Umweltchemie- und Ökotoxikologie der GDCh **2000**, Nr. 3

[6] T. Stief, Numerische Untersuchungen des dynamischen Verhaltens von Mikrostrukturen im Hinblick auf die periodische Reaktionsführung, PhD-Work, University of Dortmund **2001**

[7] D. Schmalz, Eine systematische Potentialbewertung für die Einführung neuer Technologien in der Prozessindustrie am Beispiel der Mikroreaktionstechnik, PhD-Work, Clausthal University of Technology and Merck KGaA **2005**

Economic Aspects of Microreactors in Production Scale Use

Dirk Kirschneck, Walter Linhart and Günter Tekautz
Microinnova KEG, Reininghausstraße 13a, 8020 Graz, Austria
tel: +43 316 908606-0, mail: office@microinnova.com, web: www.microinnova.com

A lot of research has been done on micro reaction technology. Some companies started to use this technology on production scale level. Microinnova transfers micro reaction technology into production scale. Two Microinnova developed applications with throughputs higher than one ton per hour have been started up recently or will be started up in the next weeks. A complete pharmaceutical plant has reached the decision stage. Scale-up experiments have already been done successfully. The examples have proofed technical feasibility. Some of the success stories have been published recently. The next step is to evaluate the economic impact.

Micro reaction technology is not automatically an economic push. This shows the first known production scale example of micro reaction technology. One of the eighteen steps has been successfully carried out in a micro reactor. This technical successful implementation does not turn the product into an economical product. The product was shut down several years later.

Now first results from production scale implementations done by Microinnova can be presented. The economic impact of two main aims: energy savings and debottlenecking of a production application will be demonstrated. The economic impact of two other projects will be shown. The first project cuts out a process step, the second one shows the yield increase reached by micro reaction technology.

Life cycle assessment of microreaction technology versus conventional batch technology

Dana Kralisch, Guenter Kreisel, *Institute for Technical Chemistry and Environmental Chemistry, Friedrich-Schiller-University Jena, Germany*

The pros and cons of microreaction technology were investigated using a life cycle assessment (LCA) method. Life cycle assessment is currently the best way of evaluating environmental impacts of alternative products and processes. The evaluation is based on their use of energy and materials and takes into account the entire life cycle of processes or products from "cradle to grave".

In this LCA the ecological advantages associated with the transfer of a chemical synthesis from a macro-scale discontinuous batch process to a continuous micro-scale set-up were researched. The two-step production of *m*-anisaldehyde from *m*-bromo anisole was taken as an example reaction (figure 1). This lithium-organic synthesis, due to its highly exothermic behavior, can only be carried out under significant safety precautions and high cooling energy consumption in conventional batch reaction equipment. When transferred to a continuous micro-scale set-up, the increased surface-to-volume ratio allows for an improved control of the heat supply and –removal, and the reaction can be conducted isothermally.

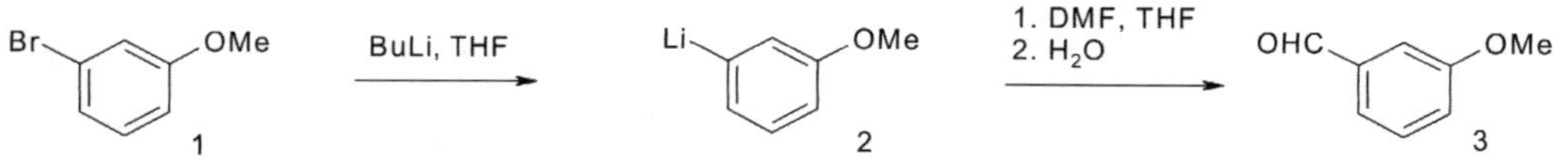

Figure 1: Model reaction

We analyzed two scales of producing *m*-anisaldehyde, the laboratory scale and the industrial scale, to get in-depth information about possible ecological advantages of microreaction technology.

On the one hand the reaction was carried out in a discontinuous 10 L double-wall reactor as well as in the continuous Cytos® Lab System[1]. While running this reaction in the continuous micro-scale set-up we observed an increase in yield up to 28 %. This can be attributed to the short residence time of the instable lithium-containing intermediate (2) (a few seconds) before being submitted to stage II of the reaction.

[1] D. Kralisch, G. Kreisel. *Chem. Ing. Tech.*, **2005**, *77, (6)*, 784-791

The short residence time of the intermediate (relative to its rate of decomposition) leads to a higher yield when compared to the laboratory scale batch process. In this latter instance, several hours are necessary for the dosage of the n-butyllithium, depending on the batch size[2]. The results were presented on the AIChE Spring Meeting 2005[3].

One the other hand we recently compared the synthesis of *m*-anisaldehyde in a 400 L vessel, cooled down to 193 K by evaporating liquid nitrogen, and in the continuous Cytos® Pilot System at T = 273 K. Because of the efficient cooling system of the batch process no yield increase can be expected for the isothermal microreaction mode. Nevertheless, the results of this second step of comparison clearly indicate significant ecological advantages achieved with the continuous synthesis in the microreactor for the industrial scale, too. Two selected results, expressed as environmental impact potentials regarding global warming (GWP) and human toxicity (HTP), are illustrated at figure 2 and 3. They are referred to a functional unit (FE) of 1 t *m*-anisaldehyde as basis of comparison.

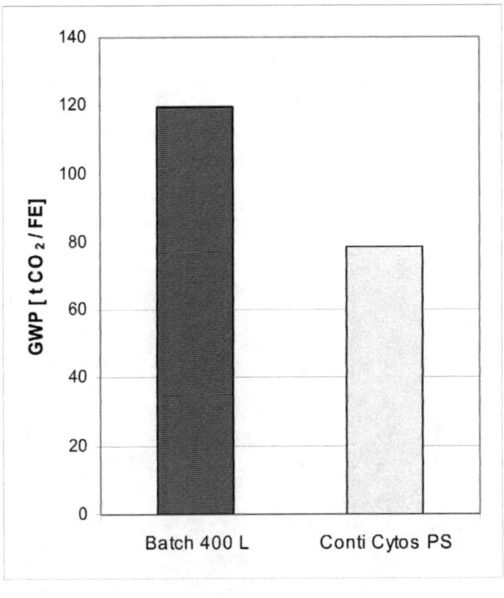

Figure 2: Global warming potential

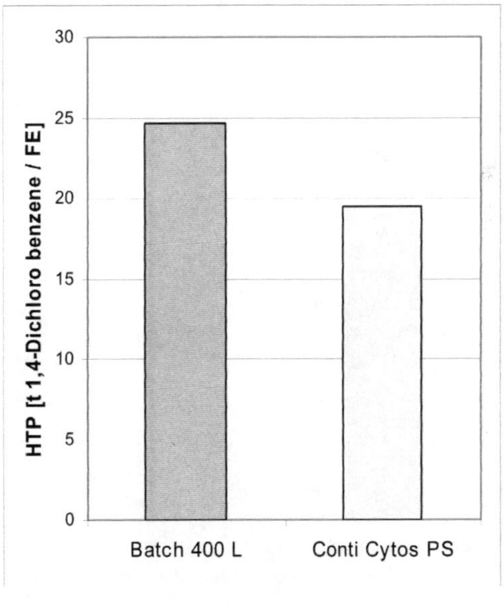

Figure 3: Human toxicity potential

[2] T. Schwalbe, V. Autze, M. Hohmann, W. Stirner; Org. Proc. Res. & Dev. **2004**, 8, 440-454
[3] D. Kralisch, G. Kreisel, A. Stark, "Assessment of the ecological potential of microreaction technology" Spring 2005 AIChE National Meeting. New York, NY: AIChE. 2005.

This outcome can be related to the increase of the reaction temperature by 80 K and thus the avoidance of a cryogenic cooling system. Compared to these saving potential, the additional energy and material demand during the fabrication step of the microstructured reactors only plays a minor role. Against this background we have found lower environmental burdens for all considered impact categories in case of the continuous micro-scale set-up. Thus our results point out, that microreaction technology can be a promising way to more ecological sustainability of production processes.

The financial support of the Deutsche Bundesstiftung Umwelt (DBU), Osnabrück, Germany, is gratefully acknowledged (ref. no. 19 348). Additionally the authors thank CPC Systems GmbH and Bayer AG for fruitful team-work in this cooperation as well as for providing the data for the life cycle assessment.

27a An Overview of Microchannel Reactor Technology and Its Applications

Yong Wang, Process Science and Engineering Division, Pacific Northwest National Laboratory, P.O. Box 999 / K8-93, Richland, WA 99352
Microchannel reactors consist of stacks of closely spaced thin plates designed to form microchannels with specific dimensions, regularity, and connectivity in the spaces between the plates. The advancement of this novel reaction technology requires the close link between reactor design and optimization of catalyst composition and microstructures. The inherent and unique characteristics include efficient mass and heat transfer and precise control of the hydrodynamic environment, which are often desired for distributed, mobile and intensified chemical processing. In this talk, the recent advancements in this technology and its applications will be reviewed.

Paper 27b

Two-Phase Pipe Side Process Model for Fired Heaters

C.B. Panchal **, S. A. Lottes, and M. Petrick
Argonne National Laboratory
9700 S Cass Avenue
Argonne, IL 60439
CPanchal@anl.gov

** To whom correspondence should be sent

Key Words: fired heater, crude heater, two-phase pipe flow, petroleum fouling, API530

Prepared for Presentation at the 2006 National Meeting, Orlando, FL, April 23-27, 2006

Two-Phase Pipe Side Process Model for Fired Heaters

C.B. Panchal, S,A. Lottes, and M. Petrick

Argonne National Laboratory

Argonne, IL 60439

ABSTRACT

The objective of this study was to develop multiphase heat transfer and pipe flow analytical methodologies that are applicable to large diameter pipes and U-bends found in refinery heaters. The technical approach was implemented in three phases. Phase I focused on identifying and evaluating multiphase correlations to predict process side flow regimes, pressure drop, and the heat transfer coefficient applicable to fired heaters. This involved an extensive literature survey, systematic evaluation of literature correlations for applicability to process heaters, and selection of the most promising correlations that could be used in a validated pipe-side process model. In an on-going effort a threshold fouling/coking model is being incorporated into the process model.

INTRODUCTION

The process heater is a work horse of refining and petrochemical industries. A typical refinery consumes about 445,000 Btu/barrel of crude processed, most of which is provided by process heaters. The recent trend of high utilization is forcing the refining industry to operate process heaters close to, or beyond, design capacities. A long-term effort was initiated to develop a pipe-flow process model, which includes multiphase flow regime predictions, heat transfer, and pressure drop in straight sections of the heater pipes as well as in U-Bends. The ultimate goal is to integrate the pipe-side performance model with a mechanistic coking model and combustion side heat flux distribution to predict localized coking and evaluate threshold coking conditions. This study was initiated in support of an on-going DOE project whose objective was to develop a new generation of refinery heaters. The technical approach was implemented in two phases. The objective of Phase I effort was to identify and evaluate multiphase literature correlations to predict process side flow regimes, pressure drops, and heat transfer coefficients. This involved an extensive literature survey, systematic evaluation of literature correlations for applicability to refinery heaters, and selection of the most promising correlations. The Phase II tasks included 1) the development of a process model; 2) evaluation of the literature correlations to select those applicable to fired heaters; and 3) comparative analysis of the process model with the API-530 correlation. In the next phase a fouling/coking model is being incorporated into the process model to determine the localized coking rate and threshold coking conditions. This process model developed was applied to a generic crude heater to demonstrate its capabilities.

EVALUATION OF LITERATURE CORRELATIONS

A broad spectrum of multiphase flows are encountered within the process side tubes in process heaters as the fluid is vaporized. The multiphase flows that occur are the result of the heater design and operating parameters. The flow regime (characterized by the distribution of vapor and liquid) changes dramatically along the pipe length and thus directly impacts the heat transfer and pressure drop. In order to design and safely operate the fired heaters appropriate heat transfer and pressure drop models/correlations must be utilized.

A critical review of the literature data and prediction correlations was performed to identify literature correlations that could be applicable for fired heaters. The bases for selecting correlations from this initial screening analysis were as follows:

- Most frequently cited in the literature;
- Recent literature reviews and comparison with the literature data; and
- Based on phenomenological models which incorporate a) physical properties, b) pipe size, c) orientation (horizontal vs. vertical up and down flows; d) individual flow regimes.

A comparative analysis was performed to determine applicability of these correlations to process conditions typical of refinery fired heaters. In general, the literature correlations have been developed with experimental data obtained under test conditions that were outside the parameter space that is typical for refinery heater conditions. A key objective of the analysis was to identify correlations that are based on mechanistic models and/or a wide range of experimental databases to minimize the prediction uncertainty in applying to refinery heaters.

The outcome of the comparative analysis was a list of recommended correlations that can be used to analyze fired heaters; they are presented in Table 1 for straight sections and in Table 2 for U-Bends. These correlations were subsequently incorporated into the performance prediction model for analyzing process heaters.

Table 1 Recommended Correlations for Straight Sections

	Horizontal Flows	Vertical Flows	
		UP Flows	Down Flows
Flow regimes	Taitel-Dukler (1976)	Speddings et al. (1998) Hewitt-Robert (1969)	Speddings et al. (1980)
Pressure drop	Chisholm (1973)	Slug/churn Flows – Friedel (1976) Annular - Speddings et al. (1998)	Slug/churn Flows – Friedel (1976) Annular - Speddings et al (1998)
Heat transfer coefficient	Slug flow – Shah (1976) Annular – Chen (1966)	Annular – Steiner&Taborek (1992) Slug/Churn – Wadekar (1990)	Annular – Numerich-Claus (1991) Slug/Annular - Wadekar (1990)
Void fraction		Spedding-Hand (1995)	Spedding-Hand (1995)

Table 2 Recommended Correlations for U-Bends

	Horizontal Bends	Vertical Bends
Flow regimes		Chen et al. (1983)
Pressure drop	Chisholm (1980)	Chisholm (1980)
Heat transfer coefficient	Farukhi (1973) Moshfeghian (1978)	Annular – Farukhi (1973); Moshfeghian (1978) Stratified – Katan (1998)

PROCESS MODEL

The technical approach was to develop a pipe-side process model and couple it with heat flux distribution of the combustion space, which could be obtained in different ways, such as from a CFD analysis or from empirical design methods. This coupled model will be able to predict local process parameters as well as the overall performance of the fired heater. The process model predicts the performance of both the convective section, including the steam coil, and the radiative section. The performance analysis of the convective section is analyzed first, taking process conditions at the cross-over point as input parameters. Point-wise integration is followed for each row to predict process conditions at coil inlet and the heat rate of the superheated steam section. The process model predicts the stack-gas temperature and can be compared with the measured value. Due to cross-flow arrangement of process and steam pipes with respect to flue-gas flow, temperature of the stack is expected to be non-uniform. Depending on location of temperature sensor, the measured stack gas temperature may or may not be representative of a mixed-average value. The process model also predicts stack-gas pressure drop across the convective section. The bridgewall temperature can be derived from the combustion model. Alternatively, if accurate measurement of stack-gas temperature is available, it can be used as an input parameter. An iteration procedure is then used to calculate the bridgewall temperature.

Process Model for Radiative Section

The radiative section process model predicts flow-regime based heat transfer coefficients and pressure drops. This approach is based on an assumption of uniform flow distribution in all passes. However, if maldistribution of flow is suspected, the process model can be applied to individual passes separately. The model was unidirectionally coupled with the specified heat-flux distribution to calculate process conditions along each pipe of a given pass. The computed parameters include flow rates of vapor and liquid phases, axial pressure profile, and pipe wall temperatures. The process model computes local conditions at each step and selects correlations based on flow regimes. The forward-march computation starts at the crossover point for one of the passes and ends at coil-out point. The Fortran program has been written with separate subroutines for calculating flow regimes, pressure drops, and heat transfer coefficients based on local process conditions. The computer model is structured so that, if desired, it can be readily incorporated into a comprehensive fully coupled Fired Heater model. The results from the Fortran program are transferred into an Excel™ program for post processing the results.

Performance Model for Convective Section

In order to simulate the overall performance of the fired heater, it was necessary to develop a performance model for the convective section. In addition to process side calculations, the performance model predicts the flue-gas side heat transfer coefficient and pressure drop. Analysis of the cross-flow between flue gas and process fluid requires a 2-dimensional calculation method. A forward-march computation method is employed starting from the bridgewall point where the flue gas enters the convective section. The computation of local heat flux and pressure drop is carried out for each row along with the flue-gas temperature and pressure drop. At the end of this computation, the coil-in process temperature is compared with the measured value and iteration is performed to match these values. If the stack-gas temperature is used as the input parameter, the iteration is on the bridgewall temperature.

Because no assumptions are made for calculating the local heat flux, this first principle performance prediction method provides an independent check on the heat duty in the convective section and hence for the whole heater. Fluid process conditions, such as temperature, vapor quality, and pressure, at crossover are generally not always measured and thus must be computed; these values are needed as input parameters for the radiative section model.

Coking/Fouling Model

Panchal et al (1997) developed a threshold fouling model applicable to a preheat train. The model was developed for single-phase flows; however, it is applicable to two phase flows provided local fluid dynamics can be calculated (Kuru and Panchal, 1997). This fouling model was incorporated into this process heater model. The rate of fouling is expressed in terms of generation of precursor in the boundary layer and its removal before firm adhesion of deposits on the wall occurs.

$$dR_f/dt = \alpha \, Re^{\beta} \, Pr^{-0.33} \exp\left(-E/RT_f\right) - \lambda \, \tau_s$$

where the Reynolds number, Re and Prandtl number, Pr are calculated for properties at the bulk and film temperature T_f (average of bulk and metal wall temperatures), respectively. The Reynolds number in this equation was based on the two-phase flow viscosity. The wall shear stress, τ_s is calculated for the liquid flow at the wall surface. Constants α, β, λ and the activation energy E are determined from laboratory and/or field data. When the equation is equated to zero, it provides threshold fouling/coking conditions between the in-side tube metal temperature and the wall shear stress.

CRUDE HEATER

The two-phase process model was applied to a crude heater for which overall operating conditions have been reported by Lawler (1979). In a previous study, the effects of preheat train fouling on energy consumption were analyzed for this refinery (Panchal and HuangFu, 2000). Operating conditions used in this model are summarized in Table 3. Heater dimensions were not given by Lawler (1979). A heater with four passes, 8-inch pipe diameter (7.625 inside diameter), and 30 vertical pipes per pass was modeled in this analysis. With this heater configuration the mass velocity was 278 lb/s ft^2. Physical and thermodynamic properties of typical crudes with API gravity in the range of 30 and 35 were used.

Table 3 Process parameters of the crude heater for the process model

Process Parameters	Input	Calculated
Process Fluid	Crude	
Crude Flow Rate, lb/hr	1,271,325	
Mass Flow Rate, lb/s ft^2 (for assumed pipe size)	278	
Vapor Quality at Coil Inlet	0.05	
at Crossover		0.15
at Coil Outlet		0.47
Coil Inlet Pressure, psia		406
Pressure at Crossover, psia		178
Coil Outlet Pressure, psia (assumed)	30	
Temperatures, °F		
Coil In (for clean preheat train conditions)	560	
Crossver		597
Coil Outlet		709
Bridgewall (assumed)	1,600	
Exit Flue Gas		569
Heat Rate Process, Million Btu/hr		
Convective Section		55
Radiative Section		146
Total		201
Average Radiation Heat Flux, Btu/hr ft^2		14,074

Flow Regimes

The process model predicts flow regimes at each point along the pipe length and selects appropriate correlations. The flow regime in the convective section is generally slug flow due to low vapor qualities. In the radiative section, the flow regime in up flow pipes starts with churn flow at low vapor qualities and then changes to annular flow as vapor qualities increases. However, the annular flow may not be fully developed and liquid slugs at some frequency would be expected. The flow regime in down flow pipe for most of the heater remains annular/slug, which is combination of annular flow with systematic slug flow at a regular frequency. The flow regime in U-bends of the radiative section at low vapor qualities is either separated or semi-annular, which could produce dry out on inner surface of U-bends and downstream regions. Later, the flow regime becomes annular, which would reduce potential dry out of the heated surface. An important fact to remember is that the transition of flow regimes could occur with change in process conditions, especially mass velocity and local vapor qualities. Therefore, evaluating the impact of possible change in flow regimes is always recommended.

Heat Transfer Coefficient and TMTs

The local heat transfer coefficients predicted by the process model and API-530 correlation differ significantly. Figure 1 shows that the heat transfer coefficient predicted by the process model increases as the vapor quality increases from crossover to coil outlet; however, no significant variation in the coefficient is predicted by API-530. Also, the heat transfer coefficients for up flow and down flow pipes are different, especially at lower vapor qualities. Note that API-530 correlation is independent of flow direction, up flow or down flow.

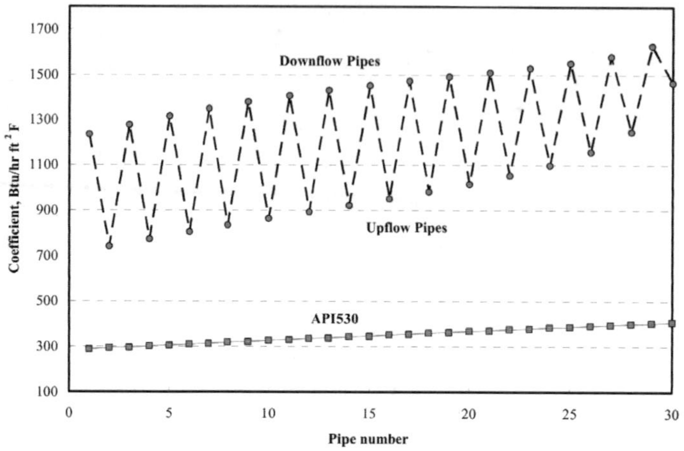

Figure 1. Heat transfer coefficient predicted by the process model and API-530.

Figure 2 shows TMTs based on circumferentially averaged heat fluxes for hot-end (Pipe#30 up flow) pipe with a higher vapor quality of 45% and cold-end pipe (Pipe#1 down flow) with a lower vapor quality of 15%. In general API-530 predicts higher TMTs at all heat fluxes, and the over prediction increases in the region of high heat fluxes. The maximum TMT predicted by API-530 correlation for Pipe#30 is 822°F as compared to 777°F by the process model. The difference between TMT predicted by the two methods at the point of TMT measurement would be even greater when circumferential variations of heat flux are taken into account. Note that the API-530 does not differentiate TMTs in up flow or down flow, while the process model shows a difference in expected TMTs for up flow and down flow pipes. Depending on flow regimes, and transition that may occur during change in process conditions, the maximum TMT may or may not occur at the measurement location or even on the pipe equipped with the TMT sensor.

Coking Rate

Constants used for analyzing fouling of the preheat train (Panchal and HuangFu, 2000) were used in this analysis. Figure 3 shows the coking rate ((hr ft^2 F/Btu)/day) for cold-end, intermediate, and hot-end pipes of the heater. The coking rate for the cold-end pipe is nearly ¼ the fouling rate calculated for the hot-end of the preheat train, which reflects the effects of high two-phase flow wall shear stress in the process heater as compared to single-phase flow shear stress in the preheat train exchanger on the local fouling rate. The coking rate increases with increase in bulk-flow and in-side tube metal temperatures; however, the increasing wall shear stress as the vapor quality increases has opposite effects. The coking rate accelerates towards the hot-end pipes of the heater. It should be noted that the coking rate is calculated based on circumferentially averaged heat flux and in-side tube metal temperatures. The coking rate would be much higher locally for circumferential variation of heat flux combined with two-phase maldistribution of liquid and vapor phases. Locations of individual pipes with respect to the burner configurations would further enhance this effect of localized coking. This process model is a first step towards predicting localized coking and taking corrective steps before significant coking occurs.

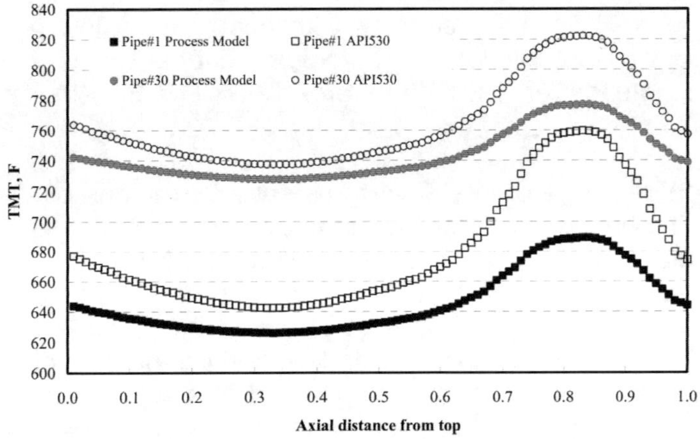

Figure 2. TMTs predicted by the process model and API-530.

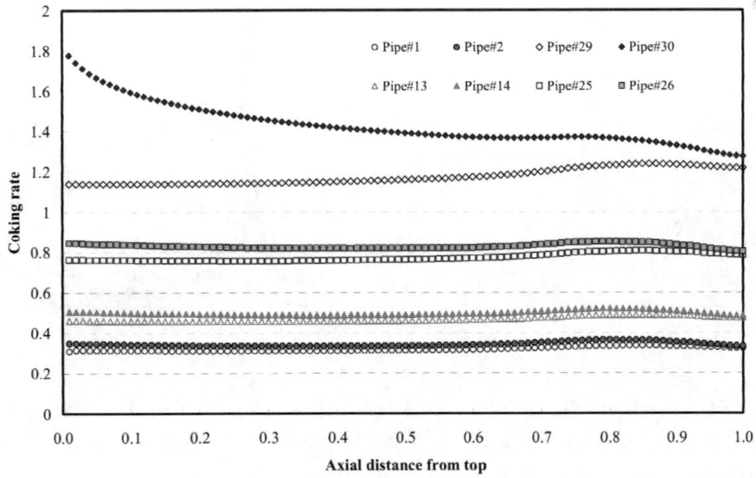

Figure 3. Coking rate (hr ft^2 F/Btu (10^{-4}) per day) distribution.

ANALYSIS OF U-BENDS

A considerable body of evidence exists in the literature that indicates that U-bends (that exist in process heaters) can have a major impact on the downstream fluid flow, heat transfer, and coking characteristics for distances up to 10-20 L/D or greater. This in turn could impact the location and magnitude of the TMT and thus the heater's safe operating range. An initial CFD study was therefore carried out to assess the major inertial effects on a two phase annular flow entering vertical U-bends for both up and down flow cases. While tube walls may be wet around the entire circumference in an annular flow entering a U-bend, the larger inertia of the liquid portion of the flow tends to force the liquid to the outer wall of the bend as the flow proceeds through the bend. A primary concern with this type of flow redistribution in fired heaters is that a region of wall dry out can be created on the inner wall of the exit portion of the bend. Dry wall regions (steady or oscillatory) are places where significant coke buildup generally occurs.

Test cases for the entire range of inlet flow rates showed regions of likely dry out on the inside wall at the exit of the bend in both up flow and down flow cases. The bulk of the mass of the liquid accumulates on the outside wall of the turn through the bend. Figure 4A shows the liquid-vapor distribution at a cross section of pipe two diameters up from the exit of a down flow U-bend. The mass flux in this case is 340 lb/s ft^2. The velocity vectors in the figure show the secondary flow pattern. The liquid occupies the dark region on the outside wall at the U-bend exit. Figure 4B shows the liquid distribution in the U-bend at a cut through its mid-plane. The liquid film on the wall at the entry is a little over one tenth of an inch thick and carries over 85 percent of the mass flow. A large accumulation of liquid in the bottom portion of the bend can be seen as well as an absence of liquid on the inner wall at the U-bend exit.

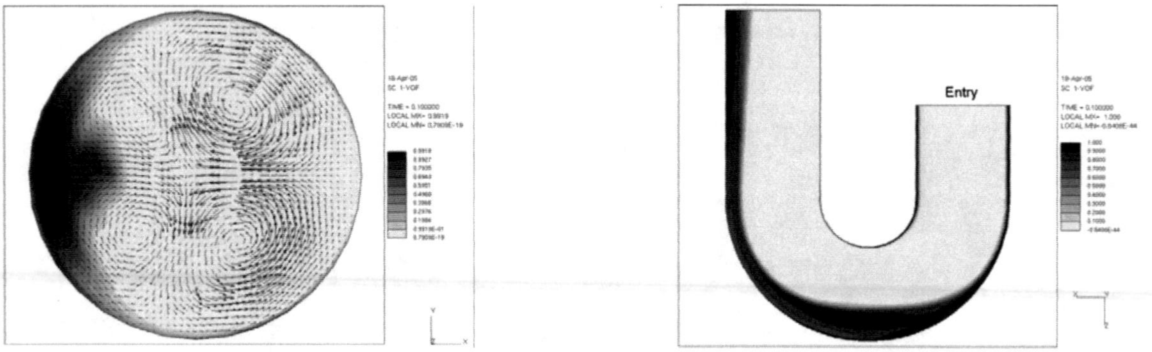

A B

Figure 4. A. Cross section at up flow exit of a down flow U-bend showing secondary flow patterns and accumulation of liquid (dark region) on the outer wall of the bend. B. Liquid distribution at mid-plane of a down flow U-bend.

CONCLUSIONS

The major conclusions from this analysis are as follows:

- The literature survey showed that there are new two-phase flow data and correlations that are applicable to pipe sizes typical of refinery heaters.
- Prediction correlations, especially for heat transfer coefficient, are flow-regime dependent. A single correlation cannot be applied to all flow regimes in horizontal and vertical (up flow and down flow) pipes.
- This analysis of a crude heater shows significant differences in predictions of heat transfer coefficients and TMTs by the API-503 and the pipe-side process model of this study.
- Further validation of this process model is needed with plant data from refineries over a wide range of operating conditions.
- Considerable heat flux variations exist in the high heat-flux zone; this signifies the need for a combined model to determine the interactive effects of localized coking and TMTs.
- A considerable body of evidence exists in the literature and operating experiences that indicates that U-Bend can have a major impact on the downstream fluid flow and heat transfer characteristics. Analytical methods, such as CFD model, are needed to predict circumferential variations of heat transfer coefficient.
- This process model focused on low exit vapor qualities typical of resid and crude heaters. It should be extended to high vapor qualities encountered in hydrotreating and reforming units, where regions of transition from annular to dispersed flow could have high propensity of localized coking.

ACKNOWLEDGEMENT

This study was funded by the DOE Office Energy Efficiency and Renewable Energy, the Office of Industrial Technologies Program. The authors acknowledge the support of Mr. R. Gemmer and Mr. J. Quinn of the Office of Industrial Technologies Program.

REFERENCES

Chen JC. 1966. A correlation for boiling heat transfer to saturated fluids in convective flow. I & EC Process Design and Development. 5, 322-329.

Chen TK et al., 1983. Flow pattern transition for gas-liquid two-phase flow in the bend of vertical hair-pin tubes. Proceedings of the third multi-phase flow and heat transfer symposium workshop. Miami, FL. Paper A3.6.

Chisholm D. 1973. Pressure gradients due to friction during the flow of evaporating two-phase mixtures in smooth tubes and channels. *Int J Heat Mass Transfer*. 16, 347.

Chisholm D. 1980. Two-phase flow in bends. *Int J Multiphase Flow*. 6, 361-367.

Farukhi MN. 1973. Two-phase flows in U-bends. *PhD Thesis*. Oklahoma State University, Stillwater, OK.

Friedel. 1979. New friction pressure drop correlations for upward, horizontal and downward two-phase pipe flow. *Presented at the HTFS Symposium*, Oxford, Hoechst AG Reference 372217/24 698.

Hewitt GF and Roberts DN. 1969. Studies of two-phase flow pattern by simultaneous X-ray and flash photography. *AERE-M 2159*.

Katan et al. 1998. Flow boiling in horizontal tubes: Part 1 – Development of diabatic two-phase flow pattern map; Part 2 – New heat transfer data for five refrigerants. Trans of the ASME. 120, 140-155.

Lawler D. 1979. Fouling of crude oil in refinery heat exchangers. *Proc. Conf on Fouling Science or Art*, The Institute of Chemical Engineers, UK. 155-165.

Moshfeghian M. 1978. Fluid flow and heat transfer in U-Bends. *MS Dissertation*. Oklahoma State University, Stillwater, OK.

Numrich R and Claus N. 1991. Druckabfall bel der kondensation in vertikalen rohren (in German). *Proceedings of Eurotherm Sem.* 18, 317-327.

Panchal CB and HuangFu EP. 2000. Effects of mitigating fouling on the energy efficiency of crude-oil distillation. *J Heat Transfer Engineering*. 21, 3-9.

Panchal CB, Kuru WC, Liao CF, Ebert WA, and Palen JW. 1997. Threshold conditions for crude-oil fouling. *Proceedings of the Engineering Foundation Conference: Understanding Heat Exchanger Fouling and Its Mitigation*, Begell House Publication, New York, NY. 273-281.

Kuru WC and Panchal CB. 1997. Application of single-phase fouling data to two-phase flows. *Proceedings of the Engineering Foundation Conference: Understanding Heat Exchanger Fouling and Its Mitigation*, Begell House Publication, New York, NY. 411-416.

Shah MM. 1976. A new correlation for heat transfer during boiling through pipes. *Trans. AHRAE 82*. 66-86.

Spedding PL, Woods GS, Raghunathan RS, and Watterson. 1998. Vertical two-phase flow - Part I: Flow regimes; Part II: Experimental semi-annular flow and hold-up; Part III: Pressure drop. *Trans Inst of Chem Eng*. 76, 620-634.

Spedding PL and Hand NP. 1995. Prediction of holdup and friction pressure loss in two-phase horizontal and downward co-current flow. *Proceedings of Int Symp of Two-Phase Flow Modeling* Experiment. Rome, Italy. 573-582.

Spedding PL and Nguyen VT. 1980. Regime maps for air water two-phase flow. *Chem Eng Sci*. 35, 779-793.

Steiner D and Taborek J. 1992. Flow boiling heat transfer in vertical tubes correlated by an asymptotic model. *Heat Transfer Eng*. 13, 43-69.

Taitel Y and Dukler AE. 1976. A model for predicting flow regime transitions in horizontal and near horizontal gas-liquid flow. *AIChE Journal*. 22, 47-54.

Wadekar VV and Kenning DBR. 1990. Flow boiling heat transfer in vertical slug and churn flow region. *Proceedings of Int. Heat Trans Conf.* Jerusalem. Israel. 449-454.

Nitration of Toluene in a Microreactor

Raghunath Halder, Woo Lee and Adeniyi Lawal
New Jersey Center for Microchemical Systems
Department of Chemical, Biomedical and Materials Engineering
Stevens Institute of Technology, Castle Point, Hoboken, NJ 07030

and

Reddy Damavarapu
Energetics and Warheads Division
Armament Research, Development and Engineering Center
U.S. Army - Picatinny Arsenal, NJ

Abstract:

Highly exothermic reactions such as aromatic nitration can be carried out in a highly efficient manner in microreactors due to the high heat and mass transfer rates. In this study, nitration of toluene was conducted in a microreactor immersed in a constant temperature bath using concentrated nitric acid as the nitrating agent. The effects of residence time, reaction temperature, and acid strength etc. on the reactor performance were studied. The results indicate that in the microreactor the nitration reaction proceeds fast, even without sulfuric acid, probably by self-protonation of nitric acid. Also, no by-products were detected. The isomeric distribution of the product nitrotoluenes was found to be identical to that generally found in the mixed-acid nitration. The results were compared with the data obtained from a batch reactor, and the microreactor reaction rates were much higher than the batch reactor rates.

Introduction:

Nitration of aromatics is an industrially important reaction for formation of intermediates of many compounds including pharmaceuticals, dyes, explosives, pesticides, etc. These nitration reactions are highly exothermic and can result in temperature runaway in a conventional reactor if not controlled properly. Microreactors have been shown to have very high heat transfer rates due to the high surface area to volume ratio, which enables the microreactors to control highly exothermic reactions very efficiently. Also, due to high mass transfer rates in the microreactors compared to the conventional reactors the reactor productivity or reaction rate can be increased greatly in a microreactor over the conventional reactors for a multiphase reaction in which the reaction is limited by mass transfer. Industrially, the nitration reaction of an aromatic compound is mainly carried out as a liquid-liquid multiphase reaction. In this study, the liquid-liquid nitration of toluene is conducted in a microreactor in order to demonstrate that reaction rate in a microreactor is superior to that in conventional reactors for this reaction. Originally, we planned to study the effect of different 'solid acid' catalysts, including ZSM 280, on the isomeric distribution in the nitration of toluene by concentrated nitric acid in a packed bed microreactor. We found that the isomeric distribution using different solid acids packed in the microreactor was very similar to the mixed acid nitration. It was also found that in the microreactor concentrated nitric acid reacted very fast even at ambient temperature without any sulfuric acid or solid acid catalysts.

Experimental Set-up and Procedure:

Figure 1 shows a sketch of the experimental set-up used to test the performance of different solid acid catalysts for the nitration of toluene. Toluene and concentrated nitric acid were pumped by two HPLC pumps and mixed in a 1/16" SS316L TEE. After mixing the two reactants in the TEE at room temperature, the reaction mixture was passed through a tubular microreactor packed with different solid acid catalysts. The reactor was made of SS316L stainless steel having an inner diameter of 775 micron. The total length of the packed reactor was 8.5 cm of which 6.0 cm was packed with the catalyst. Downstream of the catalyst bed, the remaining reactor length was filled with inert glass beads. The particulate catalyst size was in the range 75-150 micron. The glass beads were spherical and were in the size range 63-75 micron. The glass beads were used to prevent any fine catalyst particles from clogging the filter placed at the end of the reactor. In each of the experiments reported here, 13 mg of solid acid catalyst and about 20 mg of glass beads were used. The reactor was connected to the set-up with PTFE fittings. The reactor was then immersed in a constant temperature water bath to obtain uniform temperature. The tube connecting the reactor with the mixing tee had an inner diameter of 775 micron. From the constant temperature bath the mixture flowed to a product receiver where the product was collected in sodium carbonate solution to stop any further reaction. The tube used to connect the product receiver and the reactor had also an inner diameter of 775 micron. The total length of the flow path after the mixing tee was 220 cm and mainly consists of 775 micron inner diameter SS316L tube.

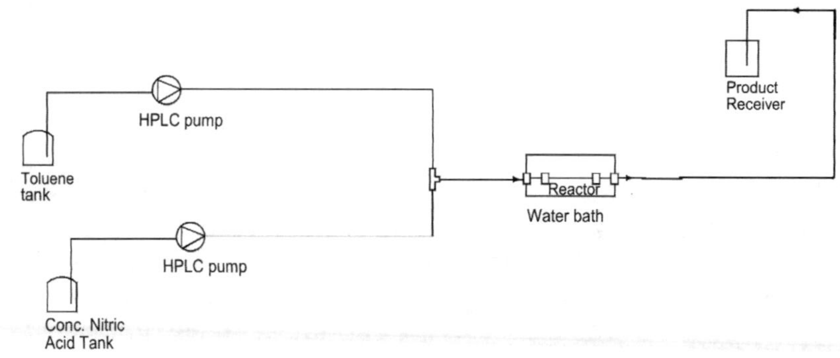

Fig. 1: Set-up for Toluene Nitration

Analytical Procedure

After attaining steady state during a run, the sample was directly collected in the product receiver containing sodium carbonate solution to neutralize the unreacted nitric acid and thus prevent further reaction. The organic phase was then separated from the aqueous phase. A fraction of the organic phase was diluted using excess toluene, and then the diluted sample was analyzed in a Shimadzu model GC-14B gas chromatograph using an Agilent DB5 column and a thermal conductivity detector with helium as the carrier gas. An internal standardization method was used following the procedure given by Grob [1]. The concentration of the unreacted toluene was calculated from the concentrations of the three nitrotoluenes since no other products or byproducts were found to form in the process. Nitric acid was used as the limiting reactant in all the runs, and since the mass flow rates of the aqueous and organic phases changed due to the reaction, the conversion of the limiting reactant nitric acid was calculated from a mass balance which is as follows:

$$w = MW_{NT} * C / \{MW_{NT} * C + (R - C) * MW_T\},$$

which when solved for the conversion gives,

$$C = MW_T * w * R / \{MW_{NT} - (MW_{NT} - MW_T) * w\} \quad (1)$$

Where, C is the conversion in fraction, MW_{NT} is the molecular weight of nitrotoluene (137), MW_T is molecular weight of toluene (92), w is the weight fraction of total nitrotoluenes (i.e., ortho-, meta- and para- nitrotoluenes) in the product organic phase and R (≥1) is the molar feed ratio of toluene to nitric acid. Determination of conversion by measurement of the nitric acid concentration in the aqueous phase was not done because we observed brown fumes, probably of nitrogen dioxide formed during the process, accumulating in the product receiver on top of the liquid phase.

Results and Discussions:

Figure 2 shows the results obtained with the reactor packed with different solid catalysts and Table 1 shows the isomeric distribution of the three nitrotoluenes obtained. Zeolite based Stevens catalysts A and B were prepared at Stevens Institute of Technology with different SiO_2 to Al_2O_3 ratio by sol-gel method. ZSM 280 in the hydrogen form was commercially obtained. It appears from the data that the conversion does not change much with residence time suggesting that the conversion is close to an equilibrium value. The conversion and the isomeric distribution appear to be very similar with all the three solid acids studied and the isomeric distribution is very similar to the isomeric distribution generally obtained in a mixed acid toluene nitration. The low conversion obtained in the experiments may be explained as follows:

Nitronium ion is responsible for the nitration of toluene with nitric acid. It has been found that the formation of nitronium ion is highly dependent on the nitric acid concentration [2]. As the reaction progresses the acid is diluted due to the formation of water in the reaction and nitronium ion concentration reduces. In fact, it has been observed that nitronium ion is nonexistent below certain concentration of nitric acid [2].

As discussed later in this work, it appeared that in the microreactor set-up, much of the conversion took place outside the solid catalyst bed and thus the effect of the solid catalysts on the isomeric distribution and conversion could not be established.

We conducted experiments in an empty tube of 775 micron inner diameter and 120 cm long (after the mixing tube) in which the whole length of the tube, except 5 cm at the exit portion for collecting sample, was immersed in the constant temperature bath. All the following data in the microreactor were taken from this reactor of length 120 cm, practically all of which was immersed in the constant temperature bath.

Effect of Feed nitric acid concentration on equilibrium conversion:

The effect of feed nitric acid concentration on the equilibrium conversion is shown in Fig 3. The data of Fig 3 show that nitric acid concentration has a strong effect on the conversion and this is due to the effect of water on nitronium ion concentration as mentioned above. This limits the use of commercial nitric acid (~70 wt%) alone in the nitration of aromatics.

Fig 2: Effect of residence time on conversion (reactor at 90°C, HNO_3 conc = 90%(w/w), toluene/HNO_3 molar ratio = 1.3, 13 mg of catalyst)

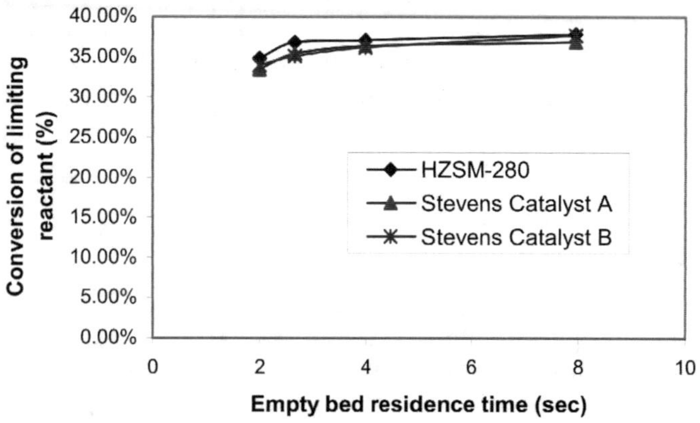

Table 1: Isomeric distribution with different 'solid acid' catalysts

Catalyst	HZSM-280	Stevens catalyst A	Stevens catalyst B
o-Nitrotoluene	55.82%	55.57%	54.94%
p-Nitrotoluene	39.58%	40.10%	40.23%
m-Nitrotoluene	4.60%	4.33%	4.82%
o-/p- Ratio	1.41	1.39	1.37

Effect of reaction temperature on equilibrium conversion

The effect of temperature on conversion is shown fig. 4. The flow rates were sufficiently low that the conversion was found not to increase with decrease in flow rates. It appears from these results that with decrease in temperature the equilibrium conversion somewhat increases. This may be due to increase in nitronium ion concentration with decrease in temperature as reported by Edwards and Fawcett [3].

Fig. 3: Effect of nitric acid concentration on conversion (toluene/HNO_3 molar ratio = 1.3, reactor temp. = 60°C)

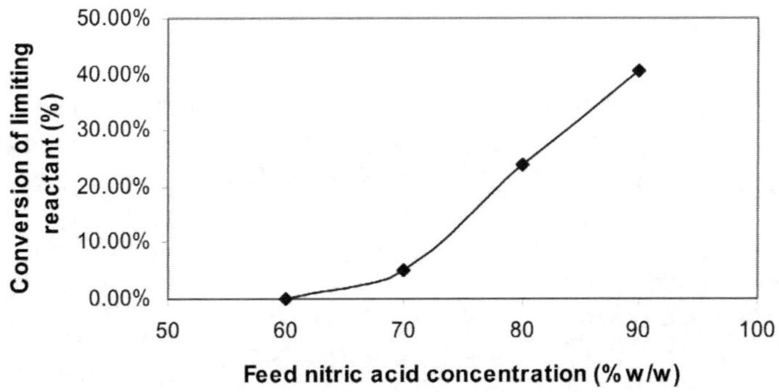

Fig 4: Effect of reaction temperature on equilibrium conversion (toluene to nitric acid molar flow ratio = 1.3, 0.15 ml/min of toluene, 0.05 ml/min of 90 wt% HNO_3)

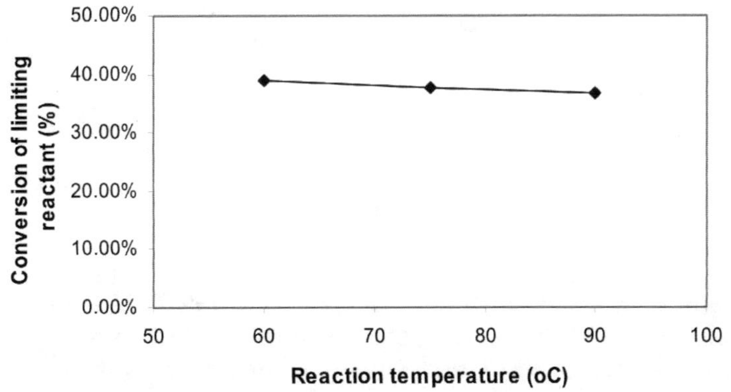

We also conducted experiments by immersing the reactor in the water bath at 22°C to find out the extent of the conversion taking place at ambient temperature without any solid acid. As mentioned above, 115 cm length of the reactor was immersed in the water bath keeping 5 cm out of the bath to collect samples. Figure 5 shows the conversion near ambient temperature (room temperature was 20°C during the runs). In each of the run the bath temperature did not deviate by more than 0.5°C from 22°C . It appears from these data that the nitration of toluene with 90 wt% nitric acid proceeds fast even at ambient temperature and without any other catalyst. This is due to nitric acid acting as a self-catalyst in which one molecule of nitric acid protonates another leading to the formation of the nitronium ion according to the following reaction [4]:

$$2\,HNO_3 \leftrightarrow NO_2^{+} + NO_3^{-} + H_2O$$

For a comparison of this conversion with the conversion at the higher temperature of 60°C, the 60°C data are also included in fig. 5. Although the data at 60°C suggest that equilibrium conversion was attained at 160 sec residence time, data for the 22°C run show that conversion was still increasing with residence time and if the trend of fig. 4 continues, the equilibrium conversion would probably have been more than that of 60°C.

These data might explain the reason why we were not obtaining our desired increase in the selectivity of para-nitrotoluene with solid acids, particularly with ZSM 280 catalyst. The data suggest that much of the conversion was taking place outside the solid acid catalyst bed.

Fig. 5: Conversion in empty tube with residence time (HNO_3 conc = 90%(w/w), toluene/HNO_3 molar ratio = 1.3, 120 cm reactor length))

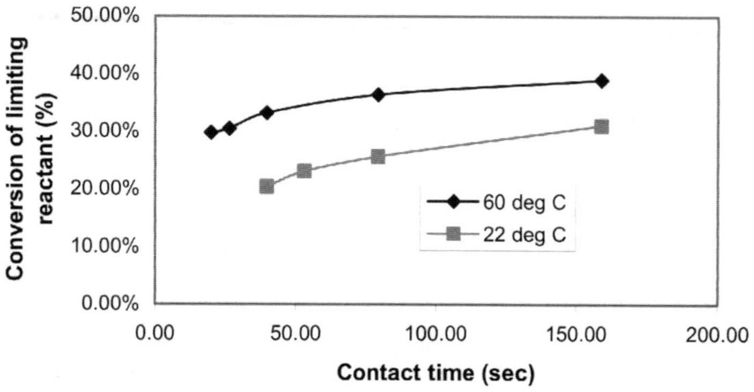

Experimentation in a Batch Reactor:

To compare the reaction rates of the microreactor and the batch reactor, we conducted the nitration reaction in a batch reactor made by PARR Instrument Company, model 5511 having a volume of 25 ml. It is made of stainless steel 316 and equipped with a magnetic drive stirrer. It has a heating system and also a jacketed cooling system. The reactor comes with, among other devices, a controller which can control the stirrer speed and reactor temperature, a pressure transducer and a sampling valve. For a batch run, initially 5 ml of 90 wt% nitric acid was charged to the reactor. From a constant temperature water bath, water was circulated through the jacket at a temperature which would give the desired reactor temperature. After the reactor attained the desired temperature, 15 ml of toluene was introduced into the reactor by a syringe through the sampling port. The total volume of the reactants was thus 20 ml. After adding the toluene, the temperature of the reactor was found to rise immediately, however very quickly it came back to the desired reactor temperature due to the cooling by the jacket. The reactor temperature was maintained constant by circulating water from constant temperature bath through the jacket. The stirrer speed was kept constant at 400 rpm. Although this is a liquid-liquid reaction, during a run the pressure of the reactor was found to increase with time even in a run made at 22°C. We found some brown fumes to be present in the reactor after the completion of the run. This is possibly due to the formation of some nitrogen dioxide which increased the reactor pressure. Under this gas pressure, samples were collected at different times through the sampling port. The conversion and isomeric distribution were determined in a gas chromatograph as mentioned before.

Comparison of the Batch and Microreactor Reactors Data:

Figures 6-9 show the comparison of the conversion and the average reaction rates for the batch and the microreactor at two temperatures of 60°C and 22°C. The microreactor data were obtained in a 120 cm long 775 micron inner diameter empty microreactor as mentioned before.

The ratios of the reaction rates at 60°C and 22°C as shown in fig. 7 and fig. 9 are approximately 30 and 15 respectively. This suggests that mass transfer rates influence the rate in the batch reactor. The higher ratio of the reaction rates at higher temperature could be due to the fact that at higher temperature, intrinsic kinetics reaction rate becomes faster compared to the mass transfer rate and thus the influence of mass transfer rate becomes more prominent. Thus, in the microreactor both mass transfer and intrinsic kinetics probably influence the reaction rate at lower temperature levels, but the reaction rate is more influenced by mass transfer at higher temperature levels.

Fig 6: Conversion with time in microreactor and batch reactor at 60°C (Toluene: HNO_3 molar feed conc ratio = 1.3)

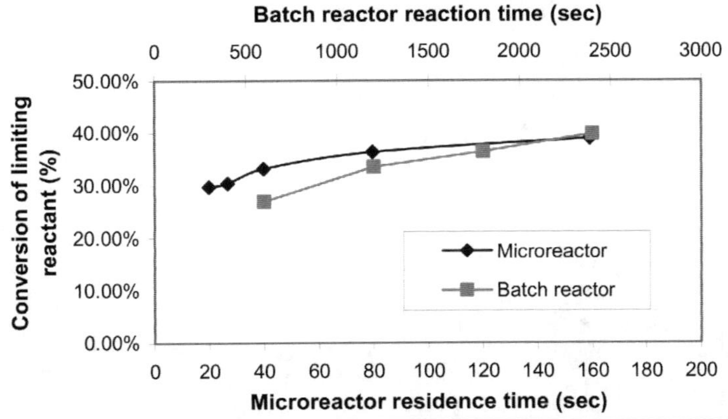

Fig. 7: Average reaction rate in microreactor and batch reactor at 60°C

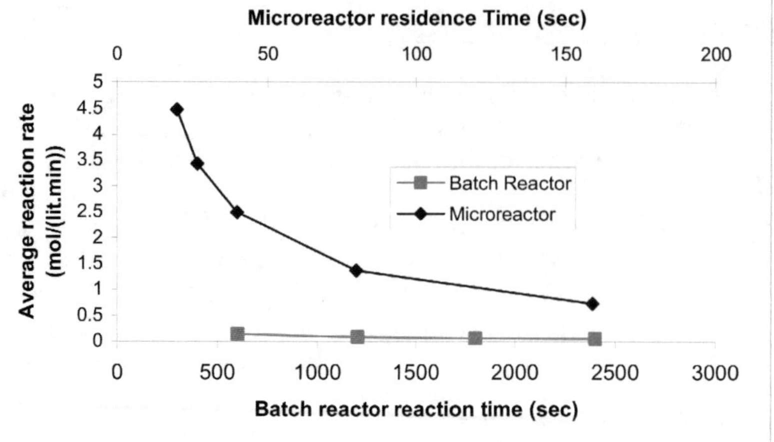

Fig 8: Conversion with time in microreactor and batch reactor at 22°C (Toluene:HNO3 molar feed conc. Ratio = 1.3)

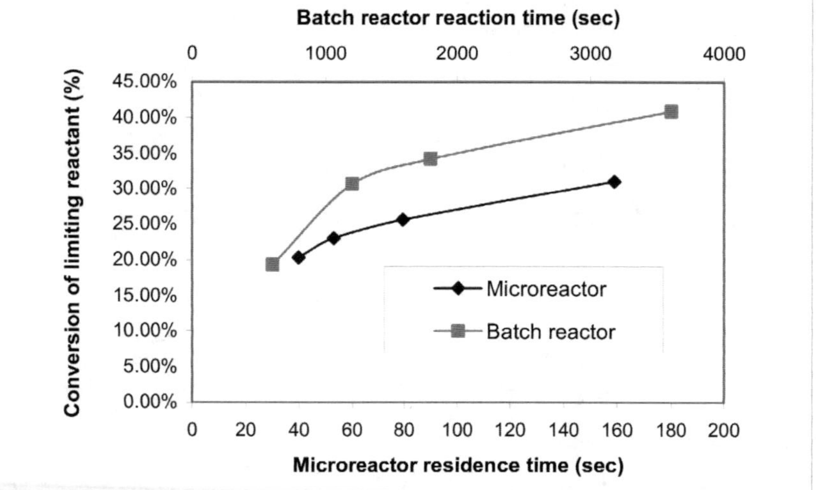

Fig. 9: Average reaction rate in microreactor and batch reactor at 22°C

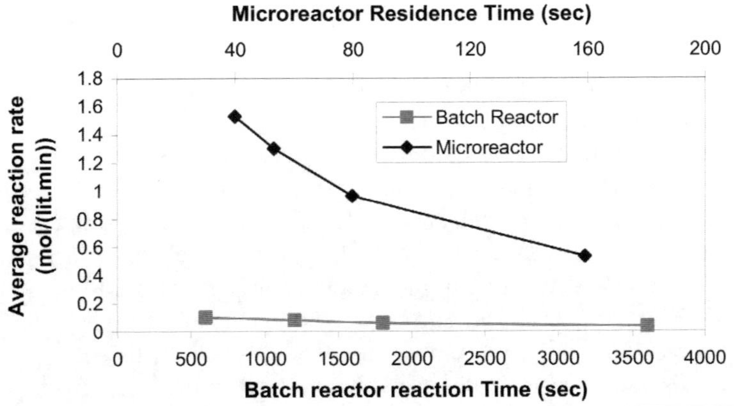

Conclusions:

The results obtained in this work suggest that in the multiphase liquid-liquid nitration of toluene with 90 wt % nitric acid, the mass transfer rate influences the rate of reaction in a conventional reactor. In a microreactor both mass transfer and intrinsic kinetics probably influence the reaction rate at lower temperature levels and the reaction rate is more influenced by the mass transfer at higher temperature levels. Thus for this reaction much higher reaction rate, in orders of magnitude, could be obtained in a microreactor than in the conventional reactors while at the same time preventing any temperature runaway due to its superior heat and mass transfer coefficients. However, with nitric acid alone as the nitrating agent, the yield of the nitrotoluenes is low due to the adverse dependency of water formed in the reaction on the nitronium ion concentration.

Acknowledgement

This work is funded by US Army ARDEC under contract W15QKN-05-D-0011-0009. The authors gratefully acknowledge the support.

References

1. Grob, R. L. (Editor), Modern Practice in Gas Chromatography, John Wiley and Sons, New York (1977) pp 181-184.
2. Marziano, N. C., Ronchin, L., Ronchin, S. and Ferrari, M., Raman and Kinetic Evidence of NO_2^+ in solid acid catalysts, Catalysis Comm. 1(2000) pp 25-31.
3. Edwards, H. G. M. and Fawcett, 'Quantitative Raman spectroscopic studies of nitronium ion concentrations in mixture of nitric acid and sulfuric acid' J. of Molecular Structure, 326, (1994) pp131-143.
4. De La Mare, P. B. D. and Ridd, J. H., 'Aromatic Substitution", Academic Press, New York (1959), p 58.

27d Catalytic Selective Oxidations in Microchannel Reactors

Terry Mazanec, Project Management, Velocys, Inc., 7950 Corporate Boulevard, Plain City, OH 43064, Frank Daly, Catalyst Development, Velocys, Inc., 7950 Corporate Boulevard, Plain City, OH 43064, and Micheal Lamont, Experimental Operations, Velocys, Inc., 7950 Corporate Boulevard, Plain City, OH 43064.
Catalytic selective oxidations pose a number of processing challenges, such as uniform heat removal and adequate oxygen mixing. The salient consequences of these challenges are shorter catalyst life, and poor conversion and selectivity. The decreased catalyst life is due to hot spots that form in the reactor where the heat generated from the selective oxidation reaction can not be sufficiently dissipated to the heat exchange medium. Low conversion and selectivity arise from a combination of unreacted reagents and fully combusted products. Dilution is the tactic commonly used to overcome both of these challenges, but this solution comes at the price of higher capital and operating costs from increasing the size of the reactor and other equipment, and the large volume of gas that must be recycled in the process.

Microchannel reactors offer a number of key features for improving catalytic selective oxidation processes and overcoming the challenges described above. These include: 1) Tailored mixing of reactants to prevent inappropriate oxygen concentrations that limit both conversion and selectivity to the desired product, 2) Enhanced heat transfer to eliminate hot spots in the reactor; thereby prolonging catalyst life, 3) Rapid quenching after the reaction takes place to limit side reactions and improve overall selectivity, 4) High surface area to volume ratio that intensifies the reaction, decreasing the size and cost of the reactor, 5) Capability to work with otherwise explosive or flammable mixtures by mixing them internally, and 6) Potential for higher pressure partial oxidation, which would further intensify the process.

This presentation will describe the development efforts by Velocys, and its partnering organizations (Dow Chemical, Pacific Northwest National Laboratory and others) regarding the viability of the microchannel approach to catalytic selective oxidation. Early laboratory experiments with microchannel reactors have shown very encouraging results. Per pass conversions of for one selective oxidation process could be increased to more than double that of state-of-the-art plant, while selectivity was comparable to that seen in the commercial process. It should also be noted, in microchannel devices the chemistry and flow dynamics inside the reactor do not change during scale-up, so results similar to those observed in the laboratory are also expected for industrial scale selective oxidation processes.

27e Kinetic Investigation of H_2O_2 Synthesis by Controlled H_2/O_2 Reaction in a Microchannel Reactor

Yury Voloshin[1], Raghu Halder[1], E. Dada[2], Woo Lee[1], and Adeniyi Lawal[1]. (1) Department of Chemical, Biomedical and Materials Engineering, Stevens Institute of Technology, New Jersey Center for Microchemical Systems, 1 Castle Point, Hoboken, NJ 07030, (2) FMC Corporation, US 1 & Plainsboro Road, Princeton, NJ 08543
Hydrogen peroxide was formed in a packed-bed microreactor in a three-phase process involving the direct combination of hydrogen and oxygen over a solid catalyst and with the hydrogen peroxide dissolved in a liquid solvent. Hydrogen and air were combined in all proportions, including explosive regime, over platinum group catalyst supported on various oxides. The ultimate goal of the work presented here is to determine the intrinsic kinetics of the direct combination (DC) reaction in the packed-bed microchannel reactor. The DC process comprises multiple reactions involving formation of hydrogen peroxide, formation of water, and consumption of peroxide either by reduction or decomposition. The strategy for determining the overall kinetics of this reaction is to determine the rate expressions for each constituent reaction individually and then to combine the expressions into an overall rate law. Experimental results and calculations showed that the reaction rate is controlled by intrinsic kinetics. Rate expressions for the reduction of hydrogen peroxide by hydrogen, and for the formation of hydrogen peroxide are determined as two of the steps toward determining the kinetics of the overall DC reaction.

61a Structured Microreactors for Intensification of Catalytic Process

Lioubov Kiwi-Minsker and Albert Renken. Chemistry, École Polytechnique Fédérale de Lausanne, Lausanne, CH-1015, Switzerland

The manufacture of chemicals in catalytic microreactors has become recently a new branch of chemical reaction engineering focusing on process intensification and safety. Chemical microstructured reactors (MSR) are devices containing open paths for fluids with dimensions in the sub-millimeter range. Mostly MSR have multiple parallel channels with diameters between ten and several hundred micrometers where the chemical transformation occur. This gives a high specific surface area in the range of 10000 to 50000 m2/m3 and allows an effective mass and heat transfer if compared to more traditional chemical reactors having usually ~100 m2/m3. Another important feature of MSR is that the heat exchange and the reaction are often performed in the same devise. MSR are operated under laminar flow with the heat transfer coefficient for liquids about 10 kW/(m2×K). This is one order of magnitude higher than in the traditional heat exchangers allowing to avoid hot-spots formation, to attain higher reaction temperatures and to reduce reaction volumes. This in turn improves the energy efficiency and reduces the operational cost. An integrated heat exchange makes the key difference between MSR and other structured reactors like honeycombs. Isothermal conditions combined with short residence times and narrow residence time distributions are the main characteristics of MSR. Small reactor dimensions facilitate the use of distributed production units at the place of consumption. Moreover, the small inventories of reactants and products lead to an inherent safety during the reactor operation even under conditions within the explosion regime. This avoids the high cost of transport and storage of dangerous materials.

In the present paper novel designs of MSR and the incorporation of catalytically active phase will be presented. The beneficial effect of improved heat transfer due to reactor miniaturization will be demonstrated through some case studies.

61b Development of Integrated Ceramic Microreactors for Production of Hydrogen

Paul J.A. Kenis, nfn Christian, and Michael Mitchell. Chemical & Biomolecular Engineering, University of Illinois at Urbana-Champaign, 600 S. Mathews Ave, Urbana, IL 61801
The development of microreactors for the catalytic reforming of fuels, such as methanol and hydrocarbons, for on-site H2 production has grown rapidly in the past few years. The challenges encountered in the reforming of fuels are: (i) to avoid coking of the catalyst by operating at temperatures greater than 800 °C; (ii) to achieve high conversion in a small reactor volume; and (iii) to minimize the pressure drop across the reactor. A suitable microreactor, to meet these challenges, must be compatible with high operating temperatures, have a high surface area-to-volume ratio, and have a high porosity.

To meet these requirements, we have fabricated highly porous inverted beaded catalyst support structures made from silicon carbide (SiC) and silicon carbonitride (SiCN) [1], and have integrated these structures within high-density, non-porous alumina housings. The integrated ceramic microreactors show excellent thermal and chemical stability up to 1200 °C in air, and the SiC or SiCN support structures have geometric surface areas between 105 and 108 m2/m3. These catalyst support structures have a void fraction of 0.74, significantly reducing the pressure drop compared to packed catalytic beds. Additionally, the channeling of reactants occurring with packed particles is avoided by the use of these inverted beaded porous monoliths. We will show our efforts in the fabrication and characterization of the SiC and SiCN porous structures, and their integration into alumina reactor housings. We will also show the conversion data for these integrated ceramic microreactors at temperatures up to 1000 °C for the decomposition of ammonia and the steam reforming of propane at temperatures above 800 ºC.

[1] Fabrication of macroporous SiC and SiCN materials for high temperature microreactors, I.-K. Sung, Nfn. Christian, M. Mitchell, D.-P. Kim, P.J.A. Kenis, Advanced Functional Materials, 2005, 15, 1336-1342.

Methanol Steam Reforming on Cr_2O_3-ZnO Catalyst in a Microreactor

Weiqiang Cao, Guangwen Chen*, Shulian Li, Jiansheng Chu, Quan Yuan

Dalian Institute of Chemical Physics, Chinese Academy of Sciences, 457 Zhongshan Road, Dalian 116023, China

Abstract: The Cr_2O_3-ZnO catalysts for methanol-steam reforming were prepared by coprecipitation method, and then were characterized by means of X-ray diffraction (XRD) and N_2 adsorption. Addition of Cr to ZnO resulted in the formation of $ZnCr_2O_4$ spinel and promoted the activity and selectivity of the catalyst, although the BET surface area decreased. However, increasing Cr content resulted in increasing CO selectivity. The Cr_2O_3-ZnO catalysts calcined above 500 ℃ exhibited better catalytic performance than those calcined at 400 ℃.

Key words: methanol, zinc oxide, Cr oxide, hydrogen, steam reforming

1 Induction

Fuel cells are expected to play a major role in the future for their high efficiency and low emission of pollution. However, the storage and handling of on-board hydrogen in the fuel cell vehicles is still an unsolved issue. One of the solutions is to use hydrocarbon fuels as hydrogen carrier. Among all possible choices of fuels, methanol is considered to be the most favorable candidate due to its high ratio of hydrogen to carbon and low reaction temperatures [1].

The current investigation of hydrogen production from methanol mainly focuses on Cu-based [2-10] and Pd-Zn catalysts [11-17]. However, Cu-based catalysts deactivate quickly [7,10,18] and Pd-Zn catalysts are extremely expensive, though they have a high activity and selectivity. Therefore, they are not suitable for on-board production of hydrogen. It is well known that Cr_2O_3-ZnO catalysts have been used for alcohol synthesis from synthesis gas [19-21]. Our studies show that Cr_2O_3-ZnO catalysts are promising catalysts for the hydrogen production from methanol.

Generally, steam reforming was conducted in a conventional fixed bed reactor heated in a furnace. However, one of the most significant drawbacks in such experiments is the nonisothermality of the catalytic bed caused by the endothermic SRM reaction. For strongly endothermic chemical reactions, the cold spots

* Corresponding author. Tel: +86-411-437-9031, Fax: +86-411-469-1570

E-mail address: gwchen@dicp.ac.cn (G. Chen).

in the packed catalyst bed usually result in underestimation of the catalyst activity. Microchannel reactors are able to offer an isothermal environment because of the reduction of the channel sizes [22].

2 Experimental

2.1 Catalyst preparation

Several Cr_2O_3-ZnO catalysts were prepared via the coprecipitation method. After defining the desired catalyst composition, a solution was prepared via addition of appropriate quantities of $Cr(NO_3)_2{\cdot}9H_2O$ and $Zn(NO_3)_2{\cdot}6H_2O$ to distilled water. 1 M NH_3 solution was added dropwise to the stirred beaker of the Cr-Zn solution maintained at PH=8 and 50 ℃. The co-precipitated materials were filtered and washed repeatedly with distilled water. Subsequently the precipitations were dried at 110 ℃ and finally calcined at a defined temperature (400~500 ℃) for 3 h.

2.2 Catalyst characterization

Powder X-ray diffraction (XRD) spectra were obtained by an X-ray diffractometer (Rigaku, D/max-rB) using CuKα radiation (30 kV, 15 mA). Volume-weighted crystallite sizes were estimated from XRD spectra using the Scherrer equation. N_2-BET analysis (ASAP2010, Micromeritics) was used to determine the surface areas of the catalyst samples after evacuation at 350 ℃ for 4~5 h.

2.3 Apparatus for methanol steam reforming

The balance gas N_2 (99.99%) was precisely controlled and delivered into the vaporizer by mass flow controllers. The liquid methanol (99.5%) and water mixture was pumped into the vaporizer with a precise P230 constant flow pump. The reactants were vaporized and preheated to a defined temperature in the vaporizer, and subsequently reacted in the microreactor. The reactor contained 30 parallel channels (1 mm wide, 1.2 mm deep and 30 mm long) which contained 1.6 g catalyst. The reactor effluent passed through a cold trap and the dry gaseous products were analyzed on-line by a gas chromatograph (GC4000A, Beijing East & West Analytical Instruments Inc.) equipped with a thermal conductivity detector (TCD). A carbon molecular sieve column was used to analyze the dry gaseous components. The flow rate of the dry gaseous products was measured by a soap bubble flow meter.

3 Results and discuss

3.1 *Catalyst characterization*

Power XRD spectra of the Cr_2O_3-ZnO catalysts after calcinations for 3 h at 500 ℃ are shown in Fig. 1.

$ZnCr_2O_4$ spinel was obviously observed in the Cr_2O_3-ZnO catalysts calcined at 500 ℃. Increase in catalyst Cr content from 0 to 17.6 wt.% resulted in a decrease in the intensities of the diffractions from ZnO and concomitant increase in the intensities of the diffractions from $ZnCr_2O_4$. In addition, Cr_2O_3 was not observed in the Cr_2O_3-ZnO catalysts even though Cr content reached 17.6%, which showed that Cr_2O_3 was efficiently dispersed into the bulk ZnO.

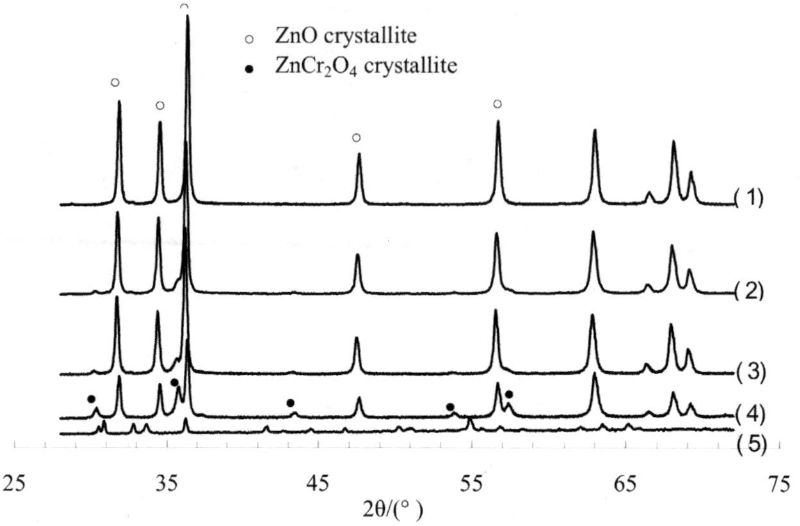

Fig 1 XRD patterns of different Cr_2O_3-ZnO catalyst samples calcined at 500 ℃

Cr content (wt.%): (1) 0, (2) 2.6, (3) 5.3, (4) 17.6, (5) 100

Surface areas and the crystallite sizes of the catalysts are shown in Table 1. The data demonstrated that addition of Cr to ZnO resulted in a decrease in catalyst surface area, although ZnO crystallite size kept almost constant. One of the possible reasons was that Cr with a low surface area was dispersed onto the surface of ZnO. In addition, the formation of $ZnCr_2O_4$ spinel also probably resulted in a decrease in catalyst surface area due to its large crystallite size.

Table 1 Surface areas and particle size of different Cr_2O_3-ZnO catalysts

Cr content /(%)	A /(m^2/g)	ZnO(101) crystallite size /(nm)	$ZnCr_2O_4$ (311) crystallite size /(nm)

0	20.20	6.90	—
2.6	13.96	6.26	17.22
5.3	13.25	6.21	9.81
17.6	15.65	7.65	17.22

Cr_2O_3-ZnO catalysts were calcined at 500 ℃.

Fig. 2 shows the XRD profiles of Cr_2O_3-ZnO catalysts calcined at different temperature. $ZnCr_2O_4$ spinel was not observed in the catalyst calcined at 400 ℃, while $ZnCr_2O_4$ spinel was obviously observed in the catalysts calcined above 500 ℃. Moreover, the intensities of the diffractions from $ZnCr_2O_4$ increased with increasing calcined temperature.

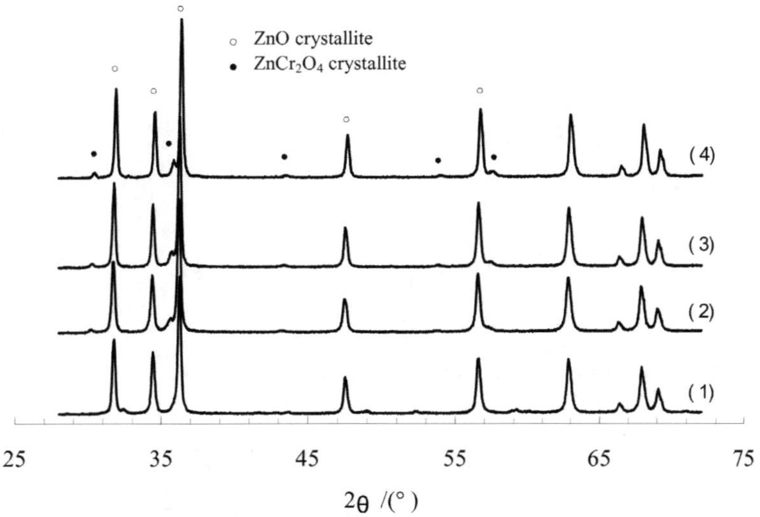

3.2 Effects of Cr content on the catalytic performance

Fig. 3 summarizes the effect of Cr content in the Cr_2O_3-ZnO catalyst on the catalytic performance. ZnO catalyst indicated a high activity while Cr_2O_3 showed a low activity showing that ZnO was the active species in the catalyst. Increasing Cr in the Cr_2O_3-ZnO catalyst resulted in an increase in catalyst activity. However, the BET results indicated that addition of Cr to ZnO resulted in a decrease in catalyst surface area. It could be inferred that the addition of catalyst activity could be attributed to the formation of $ZnCr_2O_4$ spinel. Moreover, the Cr_2O_3-ZnO catalyst gave a much more stable methanol conversion than either ZnO or Cr_2O_3, which was also probably ascribable to the formation of $ZnCr_2O_4$ spinel. In addition, the CO selectivity increased noticeably with Cr content and reached the highest point on the pure Cr_2O_3 catalyst, which demonstrated that methanol was more easily decomposed into CO on the Cr_2O_3 catalyst than on the

the ZnO catalyst.

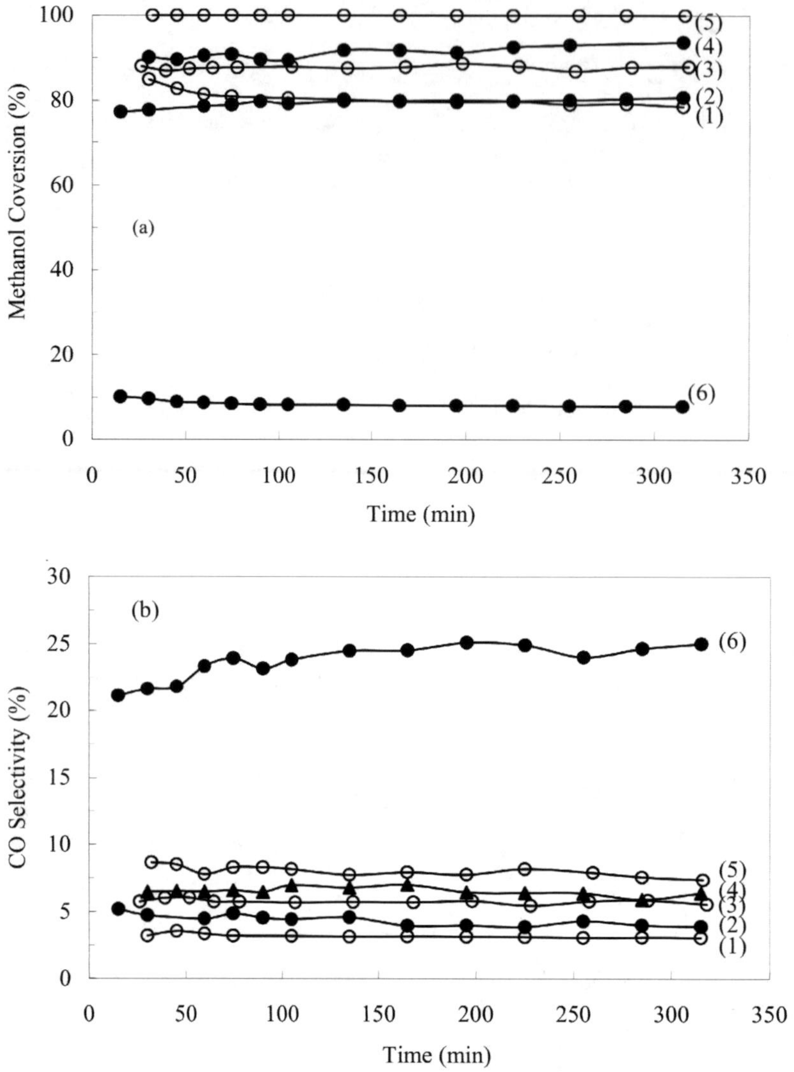

Fig 3 Methanol conversion (a) and CO selectivity (b) with time on stream on Cr_2O_3-ZnO with different Cr contents. Cr (%): (1) 0, (2) 2.6, (3) 5.3, (4) 8.1, (5) 17.6, (6) 100

(Reaction condition: Reaction temperature θ= 450 ℃, n (H_2O)/n (CH_3OH) = 1.4, n (N_2)/n (CH_3OH) = 1.05, GHSV based on methanol = 9400 h^{-1}.)

3.3 Effects of calcined temperature on the performance of the catalyst

Fig. 3 summarizes the effect of the calcined temperature on the catalytic performance. The results showed that the catalyst activity and selectivity decreased with calcined temperature. This is probably ascribable to the reason that increasing calcined temperature resulted in a decrease in catalyst surface area. In addition, the Cr_2O_3-ZnO catalysts calcined above 500 ℃ gave a much more stable methanol conversion than those calcined at 400 ℃. Fig. 2 indicated that $ZnCr_2O_4$ spinel was formed until the calcined temperature reached 500 ℃. These results furth showed that the improvement of catalyst stability could be attributed to the formation of $ZnCr_2O_4$ spinel.

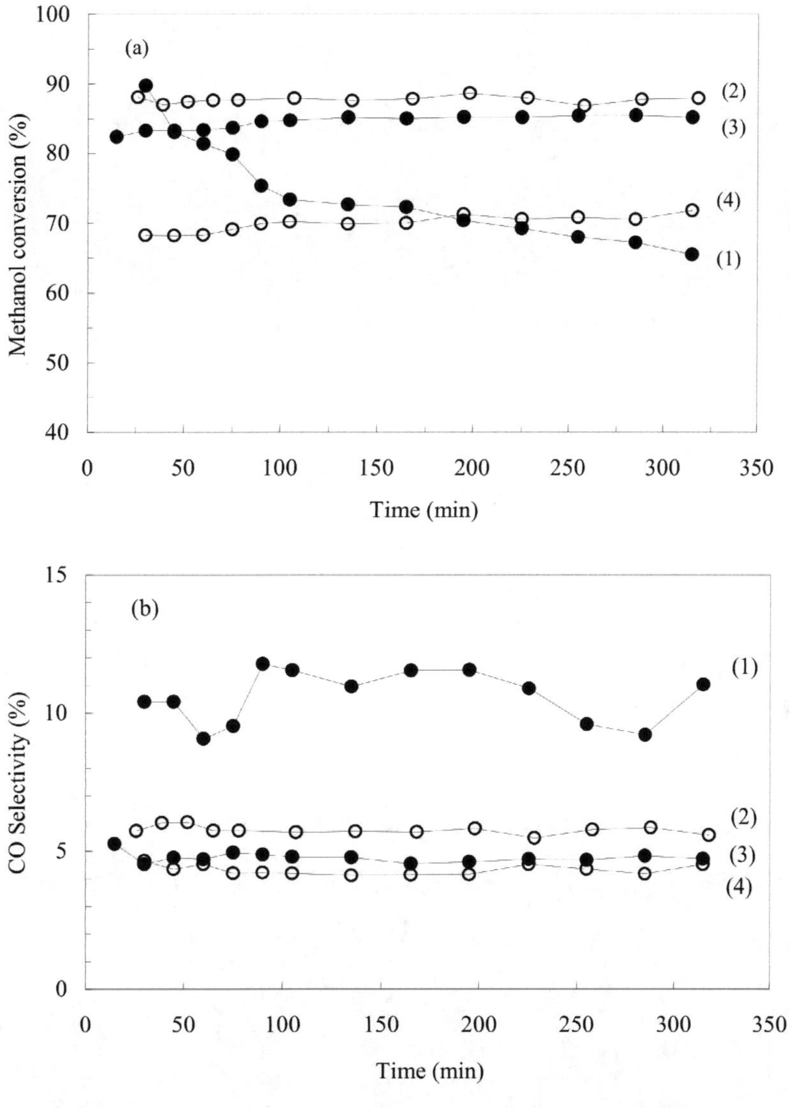

Fig 4 Methanol conversion (a) and CO selectivity (b) with time over Cr_2O_3-ZnO catalysts calcined at different temperatures (Cr% = 5.3 %, reaction conditions are the same as in Fig 3.)

(1) 400 ℃, (2) 500 ℃, (3) 700 ℃, (4) 800 ℃

4 Conclusion

Several Cr_2O_3-ZnO catalysts prepared via co-precipitation were characterized by XRD, BET, and activity for methanol steam reforming in a microreactor at 450 ℃. Data revealed that ZnO was the active species and addition of Cr to ZnO promoted the activity and stability of the catalyst, although catalyst surface area decreased. Theses promotions could be probably attributed to the formation of $ZnCr_2O_4$ spinel. However, increasing Cr content resulted in an increase in CO selectivity. In addition, the Cr_2O_3-ZnO catalysts calcined above 500 ℃ exhibited better catalytic performance than those calcined at 400 ℃.

References

[1] Lindström B, Pettersson LJ. Inter J of Hydrogen Energy, 2001 (9), 26: 923

[2] Idem RO, Bakhshi NN. Ind Eng Chem Res, 1994 (9), 33: 2047

[3] Idem RO, Bakhshi NN. Ind Eng Chem Res, 1995 (5), 34: 1548

[4] Purnama H, Ressler T, Jentoft RE, Soerijanto H, Schlögl R, Schomäcker R. Appl Catal A, 2004, 259 (1): 83

[5] Breen J P, Ross JRH. Catal Today, 1999 (3-4), 51: 521

[6] Agrell J, Birgersson H, Boutonnet M, Melián-Cabrera I, Navarro RM, Fierro JLG. J of Catal, 2003, 219 (2): 389

[7] Choi Y, Stenger HG. Appl Catal B, 2002, 38 (4): 259

[8] Li ShL, Chen GW, Jiao FJ, Li HQ. Chin J Catal, 2004, 25(12): 979

[9] Peppley BA, Amphlett JC, Kearns LM, Mann RF. Appl Catal A, 1999, 179 (1-2): 21

[10] Matter PH, Braden DJ, Ozkan US. J of Catal, 2004, 223 (2): 340

[11] Liu S, Takahashi K, Uematsu K, Ayabe M. Appl Catal A, 2005, 283 (1-2): 125

[12] Cubeiro ML, Fierro JL G. J of Catal, 1998, 179 (1): 150

[13] Iwasa N, Mayanagi T, Nomura W, Arai M, Takezawa N. Appl Catal A, 2003, 248 (1-2): 153

[14] Kapoor MP, Raj A, Matsumura Y. Microporous and Mesoporous Mater, 2001, 44-45: 565

[15] Chin YH, Wang Y, Dagle RA, Li XS. Fuel Process Technol, 2003, 83 (1-3): 193

[16] Ranganathan ES, Bej SK, Thompson LT. Appl Catal A, 2005, 289 (2): 153

[17] Iwasa N, Takezawa N. Top Catal, 2003, 22 (3-4): 215

[18] Farrauto R, Hwang S, Shore L, Ruettinger W, Lampert J, Giroux T, Liu Y, Ilinich O. Annu Rev Mater Res, 2003, 33: 1

[19] Bradford MCJ, Konduru MV, Fuentes DX. Fuel Proc Technol, 2003, 83 (1-3): 11

[20] Epling WS, Hoflund GB, Hart WM, Minahan DM. J of Catal, 1997, 172 (1): 13

[21] Minahan DM, Epling WS, Hoflund GB. Appl Catal A, 1998, 166 (2): 375

[22] Cao C, Xia G, Holladay J, Jones E, Wang Y, Appl. Catal. A 262 (2004) 19-29.

61d Catalyst Coatings in Ceramic Microreactors for Steam Reforming of Methanol

Travis R. Conant, Ayman Karim, and Abhaya Datye. Chemical and Nuclear Engineering, University of New Mexico, 1 University of New Mexico MSC 01 1, 203 Farris Engineering Center, Albuquerque, NM 87131
Compact fuel cell systems operating with hydrocarbon fuels possess high volumetric and gravimetric energy density and could outperform batteries in low-power, portable electronics [1]. Among the fuels considered, methanol is an attractive candidate since it is sulfur free, and most importantly, can be reformed at low temperatures (200-250°C). For a 20W fuel cell system operating at a 25% overall efficiency, about 3g of catalyst would be needed for the reformer operating isothermally at 230°C. In order to overcome the pressure drop, non-isothermality and repacking problems associated with micro-packed bed reactors, a wall-coated reactor offers a superior geometry [2, 3]. We have previously shown the feasibility of coating non-porous surfaces with a commercial BASF catalyst and reported coatings of up to 10μm thick in a 530μm fused silica capillary [4].

The challenge is to minimize the size of the wall-coated reactor, i.e. increase the catalyst loading per unit volume, which requires coatings thicker than 10μm. G.I. Taylor reported that the thickest coating possible occupies 57% of the total cross sectional area, independent of channel diameter. However, Taylor performed his study with a Newtonian fluid in channel diameters ranging from 1.5-4mm [5]. In this work we performed an experimental study to determine the maximum coating thickness feasible for non-Newtonian fluids in channels ranging from 250 to 530μm in diameter. In contrast to Taylor's findings, our data shows the fraction coated to be dependent on the channel diameter due to inertial effects within the coating fluid. The maximum fraction coated was drastically lowered, e.g. 12% in 250μm diameter channel.

It is imperative to minimize the effects of surface tension while concurrently minimizing the inertial effects within a given coating structure. We were able to increase the fraction coated by minimizing these effects through careful manipulation of the viscosity of the coating fluid. This was achieved using shear-thickening additives.

The presentation will show data on coating multi-channel ceramic microreactors. The microreactors were coated using a commercial $Cu/ZnO/Al_2O_3$ catalyst from BASF and the catalyst reactivity was measured and compared to that in a packed-bed.

This work has been funded by the U. S. Army Research Laboratory under the Collaborative Technology Alliance Program, Cooperative Agreement DAAD19-01-2-0010

References

1. Hu, J., Wang, Y., VanderWiel, D, Chin, C., Palo, D., Rozmiarek, R., Dagle, R., Cao, J., Holladay, J., Baker, E. Chem. Eng. J. 93, 5560 (2003)

2. Karim, A., Bravo J., Datye, A. Appl. Catal. A. 282, 101 (2005)

3. Karim, A., Bravo J., Gorm, D., Conant, T., Datye, A Catal. Today (in press).

4. Bravo, J., Karim, A., Conant, T., Lopez, G.P., Datye, A. Chem. Eng. J. 101, 113 (2004)

5. G.I. Taylor, Fluid. Mech., 10, 161 (1961)

61e Integrated Micro System for Controlled Delivery of Hydrogen from Metal Ammines

Rasmus Zink Sørensen[1], Asbjørn Klerke[1], Ulrich Quaade[2], Søren Jensen[3], Ole Hansen[3], Tue Johannessen[4], Jens K. Nørskov[2], and Claus Hviid Christensen[1]. (1) Center for Sustainable and Green Chemistry, Department of Chemistry, Technical University of Denmark, Kemitorvet 207, Lyngby, DK-2800, Denmark, (2) Center for Individual Nanoparticle Functionality, Department of Physics, Technical University of Denmark, Lyngby, DK-2800, Denmark, (3) Department of Micro and Nanotechnology, Technical University of Denmark, Lyngby, DK-2800, Denmark, (4) Amminex A/S, Kemitorvet 206, Lyngby, DK-2800, Denmark

Efficient and safe storage of hydrogen for energy production has received significant attention over the last decades. However, it still remains a substantial challenge for the prospects of a hydrogen economy to provide a safe, reversible and dense storage method for hydrogen. Most research is focused on the so-called complex hydrides based on alanates or borates. Some hydrides have promising properties, but most suffer from a too low hydrogen density, too slow kinetics or lack of reversibility.

Ammonia is an efficient hydrogen carrier. The "reforming" of ammonia into hydrogen and nitrogen, the reverse ammonia synthesis, is well understood and can be done at a temperature level close to 650K [Sørensen et. al., 2005]. Pressurized ammonia condenses above 8 bars to a liquid with as much as 110 kgH2/m3, and as such it is a very interesting hydrogen carrier. However, the common method of storing ammonia in a pressure vessel is impractical to incorporate into a micro system, and for many applications the toxicity of ammonia can be a problem. Consequently, ammonia has great potential as hydrogen rich fuel, when combined with a safe and practical, high density storage method.

In earlier work [Christensen et. al. 2005] we have demonstrated the use of solid metal ammine complexes as hydrogen storage materials. Ammonia can be bound into a metal salt and with focus on $MgCl_2$ as one of the most promising materials we will outline the potential of such a storage method. $Mg(NH_3)_6Cl_2$ contains 9.1 wt % hydrogen at 109 kgH_2/m^3 in the form of ammonia. The ammonia storage is completely reversible and by combining it with an ammonia decomposition catalyst hydrogen can be delivered at temperatures below 650 K. The reversibility has been demonstrated by successively absorbing and desorbing ammonia from $MgCl_2$. Temperature Programmed Desorption of successive runs are almost identical and the saturation reaches 99-100% within the experimental uncertainty. Thermal decomposition of the complex starts around 350 K. At 500 K, two thirds of the ammonia has been released. The remaining third is released below 620 K.

The present work presents a micro structured system, which is able to desorb ammonia from the compact metal ammine, decompose it catalytically into H_2 and N_2 and at the same time supply heat for both reactions by catalytic combustion of part of the decomposed ammonia or a simulated purge gas from a fuel cell running on decomposed ammonia.

Hydrogen production by decomposition of ammonia in micro reactors is a known process [Sørensen et. al., 2005], as is ammonia release from metal ammines [Christensen et. al. 2005] and catalytic combustion of hydrogen in air [Ryi et. al. 2005]. This work, however, is the first to combine these processes in an integrated catalytic unit. It is shown that heat for driving the reactions can be supplied either by part of the fuel before it enters the fuel cell, purge gas from said fuel cell or a mixture thereof.

A basic flow sheet of the process is given in fig 1, while fig 2 is a schematic representation of the setup. Initially the decomposition reactor is heated to its operating temperature. The heat dissipating from the reactor will trigger desorption of ammonia from the storage material, and the ammonia will be delivered directly to the decomposition reactor. The hydrogen rich stream leaving the reactor is then split into a part bypassing the fuel cell to supply heat for more decomposition and ammonia release, and the desired part, which fuels the fuel cell. As the hydrogen is diluted by nitrogen, some must be purged from the fuel cell at all times. This purge gas will still contain some hydrogen, and is passed through the catalytic combustion reactor to provide heat for the other reactions and to avoid release of hydrogen gas to the environment.

Figure 1: Flow sheet of a system for delivery of hydrogen to a fuel cell, employing $Mg(NH_3)_6Cl_2$ for hydrogen storage.

Figure 2: Schematic representation of a catalytic micro system for delivery of hydrogen to a fuel cell, employing $Mg(NH_3)_6Cl_2$ for hydrogen storage. Two catalytic micro reactors are combined with a small chamber holding the solid hydrogen storage material.

References

Claus Hviid Christensen, Rasmus Zink Sørensen, Tue Johannessen, Ulrich J. Quaade, Karoliina Honkala, Tobias D. Elmøe, Rikke Køhler, Jens K. Nørskov, Metal ammine complexes for hydrogen storage, Journal of Materials Chemistry, 15, 4106-4108, 2005.

Rasmus Zink Sørensen, Lærke J.E. Nielsen, Søren Jensen, Ole Hansen, Tue Johannessen, Ulrich Quaade and Claus Hviid Christensen: Catalytic ammonia decomposition: miniaturized production of COX-free hydrogen for fuel cells, Catalysis Communications, 6 (3), 229-232, 2005.

Shin-Kun Ryi, Jong-Soo Park, Seung-Hoon Choi, Sung-Ho Cho, Sung-Hyun Kim: Novel micro fuel processor for PEMFCs with heat generation by catalytic combustion, Chemical Engineering Journal, 113, 47–53, 2005.

Figure 1

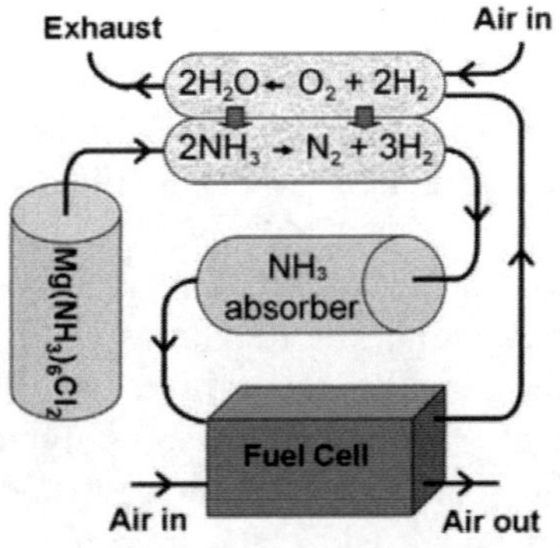

Figure 2

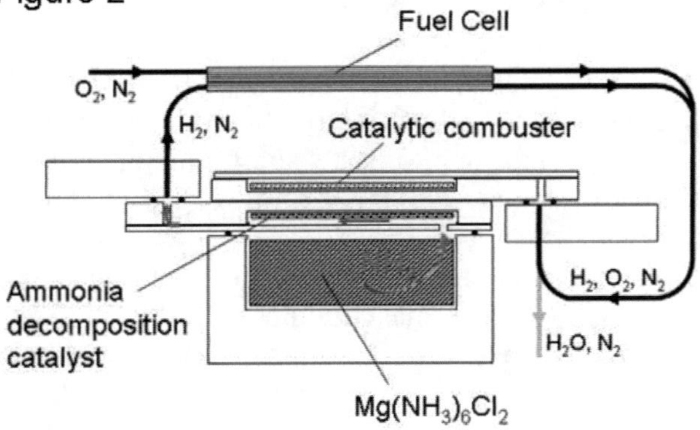

61f Robust Ceramic Sic-Microreactors with Catalytic Coatings

Frank A. Meschke[1], Volker Hessel[2], Jochen Schürer[2], and David W. Agar[3]. (1) ESK Ceramics GmbH & Co. KG, Max-Schaidhauf-Strasse 25, D-87474 Kempten, Germany, (2) Institut für Mikrotechnik Mainz GmbH, Carl-Zeiss-Straße 18-20, Mainz, D-55129, Germany, (3) Chemical Engineering, Dortmund University, Emil-Figge-Strasse 66, D-44227 Dortmund, Germany

Microreaction technology has become very popular with chemical engineers over the past decade due to the enhanced safety features, modular architecture, flexibility, improved performance and other benefits it offers. Hermetically gas-tight microreactors made from sintered silicon carbide ceramics complement the metal, glass and polymer systems currently available. They are characterised by excellent high temperature stability and almost complete resistance to both corrosion and erosion. SiC-microreactors can be manufactured as a ceramic monolithic structure for almost any internal geometry desired. High heat transfer rates, the absence of leakage and stability at high pressures and at temperatures well above 800°C are key features.

The aim of this presentation is to demonstrate the potential of monolithic SiC microreactors and to outline the material properties and the heat exchange characteristics. The gas/gas heat exchange performance in pilot-scale studies on three different hermetically gas-tight SiC heat-exchangers at various throughputs are presented.

The suitability of SiC and its modifications as a substrate for catalytic coating, combining the superior adhesive properties of conventional ceramics with the mechanical robustness of metallic materials, will also be discussed. Furthermore the application of novel numerical and experimental tools for determining flow fields and residence time distributions in order to optimise the monolithic structures will be described.

61g Scale-up of microtechnology for fuel processing applications: comparison between fixed-bed and microchannel reactor systems

Jaap C. Schouten, Erik R. Delsman, and Mart H.J.M. de Croon. Chemical Engineering and Chemistry, Eindhoven University of Technology, Helix Building, Eindhoven, 5600 MB, Netherlands

Fuel cells have great application potential as stationary power plants, as power sources in transportation, and as portable power generators for electronic devices. Most fuel cells currently being developed for use in vehicles and as portable power generators require hydrogen as fuel. Chemical storage of hydrogen in liquid fuels is considered to be one of the options for supplying hydrogen to the fuel cell. In this case a so-called fuel processor is needed to convert the liquid fuel into a hydrogen-rich stream. Microreactors show large promise in the development of miniature fuel processors for small scale electricity production, where several unit operations can be integrated to form a micro hydrogen producing plant. The question now rises whether microreactors can compete with or even replace current conventional reactors for larger scale fuel processing applications. In order to answer this question a comparative study is presented between between microreactor (in particular microchannel) technology and conventional (in particular fixed-bed) reactor technology for the case of methanol fuel processing. The study is limited to the chemical reactor devices of the methanol fuel processor: a reformer-burner (RefBurn) reactor, with coupling of endo- and exothermal reactions, and a preferential oxidation reactor with integrated heat exchangers (ProxHeatex), with integration of reaction and heat exchange. Detailed system designs are presented for 100 We and 5 kWe power output. The 100 We case is at the scale typical for portable electronics applications, while the 5 kWe case is typical of auxiliary power units, which are used for automotive and domestic power applications. Two RefBurn designs and two ProxHeatex designs are evaluated on four comparison criteria: system volume, insulation volume, system weight, and required catalyst mass. Fixed design critera are used for reactor conversion, maximum reactor temperature, pressure drop, and heat loss. In this way, differences in heat and mass transfer characteristics between both reactor technologies show up as differences in the required catalyst mass and ultimately in the system's size and weight. The conventional fixed-bed ProxHeatex devices at both power outputs consist of three separate units: a fixed-bed reactor and two compact plate-fin heat exchangers, which are the first choice for gas-to-gas heat exchange. The fixed-bed reactors are designed using a two-dimensional pseudo-homogeneous model including heat transfer resistances in the catalyst bed and at the tube walls and including mass transfer resistances in the catalyst pellets. The microreactors at both power outputs are considered as being constructed as a stack of microstructured metal plates. The microreactors are designed using dedicated microreactor models including heat conduction in the microstructured plates and mass transfer limitations in the coated catalyst layer on the microchannel walls. On both power output scales, the microreactor designs outperform the conventional fixed-bed designs, leading to significantly lower reactor volumes and weights. The scaling factors of reactor volume and reactor weight are larger for the microreactor systems as for the conventional fixed-bed systems, indicating that at larger scales the fixed-bed reactors will ultimately outperform the microreactor designs. The study shows that microreactor technology may be a viable alternative for conventional reactors for the design of relatively small-scale reactors in which heat exchange is important. Furthermore, the use of a fuel processor and fuel cell system shows a clear weight benefit over other fuel cell systems in appllications where a large energy storage capacity is required.

62a Novel Tubing Microreactor for Monitoring Chemical Reactions

Charles A. Nielsen and Dayroosh D. Beigzadeh. Research Engineering Science, The Dow Chemical Co., Building 1897, Door 10, Midland, MI 48667
There is an expanding interest in small-scale methods to evaluate catalysts and chemical reactions at a variety of conditions, ranging up to 6.9 MPa (1000 psig) and 300 °C. Multiwell parallel batch techniques are most commonly applied in high-throughput screening systems. In contrast, we describe here a rapid, serial, highly controllable method based on LC-type steel tubing rated for high pressures. The tube, containing a variety of flowing ingredients, such as carrier solvents, catalyst formulations, and reactants can be heated by either immersing in an oil bath or self-heated ohmically using electrical current from a power supply monitored and regulated with a precision of 0.01%. An array of voltage taps arranged along its length serves to sense the real-time temperature profile of the tube. Reactions are seen as temperature pulses progressing through the reactor, in zones of 200 µL each, and tracked with a temperature precision of 0.1 °C. A unique pressure controller was devised to maintain constant reactor pressures despite effluent viscosity fluctuations due to polymerization.

Several chemical reaction systems have been characterized to date, including decomposition reactions of di-tert-butyl peroxide, polymerizations of styrene, formation of polyethylene from ethylene, and copolymerization of ethylene with 1-octene. For ethylene polymerization, the amount of the mass of polymer formed is proportional to the responses observed.

Also explored were alternative modes of operation to simulate other types of reactors, for example, a semi-batch reactor. Unlike a semi-batch reactor, tubular reactors are in nature similar to batch systems where the concentration of monomer declines along the length of the tube. Operating conditions of the tubular reactor were modified from single-liquid to gas-liquid phase. This simulates semi-batch operation where the ethylene in the gas phase replenishes the ethylene consumed in the liquid. As expected, significantly higher molecular weights were obtained in the two-phase versus single-phase experiments.

AMINO ACID POLYMERIZATION IN MICRO-REACTOR

Masaya Miyazaki,[1,2] Takeshi Honda,[1] Hiroyuki Nakamura,[1] and Hideaki Maeda[1,2]**
[1]Nanotechnology Research Institute, National Institute of Advanced Industrial Science and Technology (AIST), Tosu, Saga, Japan
[2]Interdisciplinary Graduate School of Engineering Sciences, Kyushu University, Kasuga, Fukuoka, Japan

Biography
[1]Micro- & Nano-space Chemistry Group, Nanotechnology Research Institute, National Institute of Advanced Industrial Science and Technology (AIST), 807-1 Shuku, Tosu, Saga 841-0052, Japan; Tel: +81(942)81-3676; Fax: +81(942)81-3657
[2]Department of Molecular and Material Sciences, Interdisciplinary Graduate School of Engineering Sciences, Kyushu University, 6-1 Kasuga Koen, Kasuga, Fukuoka 816-8580, Japan; Tel & Fax: +81(92)583-8891
**Corresponding authors: E-mail: m.miyazaki@aist.go.jp or maeda-h@aist.go.jp*

A microreaction system consisting from micromixer and capillary reaction part was developed for amino acid polymerization. A chip-type micromixer which tethering three dimensional microchannel network was fabricated by combination of maskless technique, dry-etching, and sequential anode bonding. Resulting microreactor showed excellent controllability of molecular weight distribution of poly(L-Lysine) synthesis. Microreactor without micromixer yielded worth controllability, demonstrated that rapid micromixing is important for reaction control. Also, the microreactor produced amino acid copolymers with highly controlled molecular weight distribution than that of batchwise reaction. These features are suitable for high-throughput combinatorial preparation of amino acid polymers.

Introduction

Recent intense research in biomaterial science has evolved to meet the scientific and industrial needs. In particular, synthetic biopolymers including polypeptides and related artificial poly(amino acid)s are very important in fields such as biomedical and tissue engineering [1]. Such functional polymers have been significantly developed owing to their specific properties such as various biological activities, biodegradability, and biocompatibility. Several useful polymers were developed; however, it is usually effort and time consuming to obtain useful polymers which can be commercialized. One of the underlying problems deals with design and screening of polymer molecules. Although there have been demands for high-throughput and combinatorial synthesis system to rapidly and effectively obtain useful polymers, it is generally difficult to control polymer properties such as average molecular weight and molecular weight distribution, thus, a method which would enable control of polymerization reactions is desirable.

Microreaction systems are expected as an efficient reaction apparatus that enables high controllability of chemical reactions [2]. Such controllability results from efficient heat transfer, mass transport, and/or larger surface/interface area. Recent studies have shown the potential benefits of applying the various chemical reactions to microfluidic reactors. Several polymerization reactions have been performed using microreaction systems [3]. Controlling of rapid reactions such as living polymerization was successfully performed.

Our group has been interested in chemical reactions and analysis methods utilizing hydrodynamic behaviors generated characteristically in microfluids, including laminar flow, secondary flow, and shearing force [4-6]. We have shown that biopolymer chain can be expanded and oriented in microfluidic system using DNA as a model [6]. Such behavior demonstrates the utility of microreactor as novel reaction apparatus for biopolymers. Taking into account the properties of microfluidic systems, we have decided to apply microreaction technology in biopolymer synthesis. In the present study, we performed polymerization reactions of amino acid N-carboxyanhydride (NCA) as a model. Amino acid polymerization by NCA is an anion polymerization which has two steps: initiation and propagation (Scheme 1). We examined effect of microchannel system on this reaction.

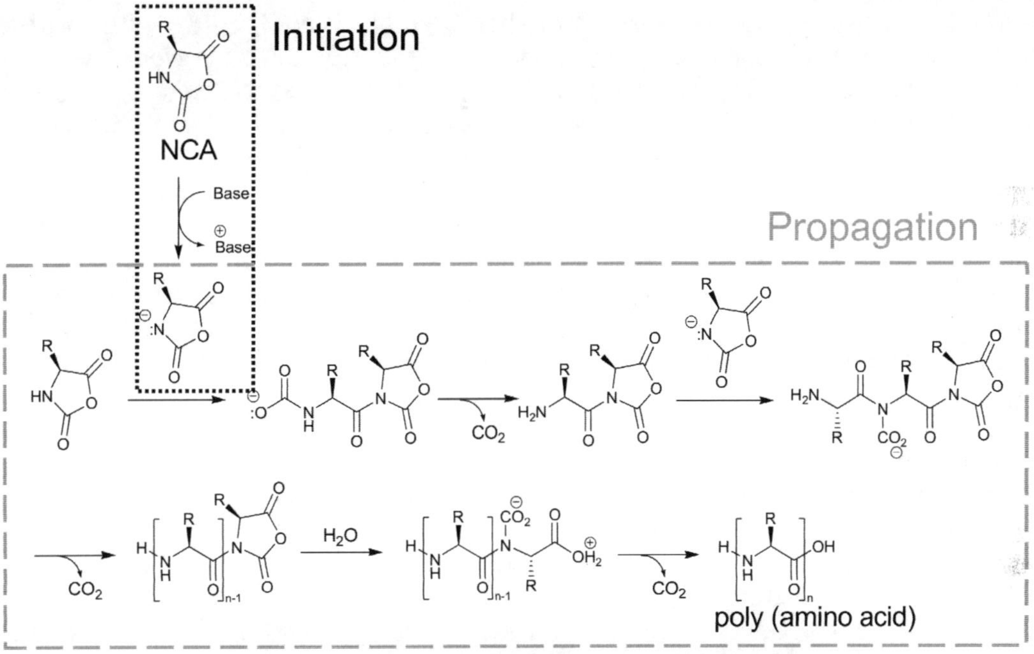

Scheme 1 NCA-polymerization reaction process.

Experimental

Materials

*N*ε-Benzyloxycarbonyl-L-lysine (Lys(Z)), alanine (Ala), leucine (Leu), 30% HBr solution in glacial acetic acid, *m*-cresol, thioanisole, trifluoroacetic acid (TFA), and trimethylsilyl bromide (TMSBr) were purchased from Watanabe Chemical Industries Ltd. (Hiroshima, Japan). Trichloromethylchloroformate (TCF) was obtained from Merck-Schuehardt (Hohenbrunn, Germany). Sigma-Aldrich Co. (St. Louis, MO) supplied dichlorodimethylsiloxane, tetraethoxysilane, and anhydrous organic solvents including triethylamine (TEA), tetrahydrofuran, diethyl ether, ethyl acetate, *n*-hexane, ethanol and *N,N*-dimethylformamide (DMF). Powder charcoal used as catalyst was from Wako Pure Chemical Industries Ltd. (Osaka, Japan). Poly(tetrafluoroethylene) (PTFE) microtube (250 μm of inner diameter) and PTFE

T-shaped connector were obtained from Flon Chemical Inc. (Osaka, Japan). Cellulose acetate dialysis tube was purchased from Spectrum (Houston, TX, USA).

Fabrication of Microreactor

Silicon plate (70 mm x 30 mm x 1 mm) was prepared by cutting 12 inch wafer using diamond cutter. Microchannel pattern was projected using a Mask-less lithography system (Nanosystem Solutions, Inc., Tokyo, Japan) [4], followed by Cr-sputtering. The plate was subjected to dry-etching, and then the size of fabricated microchannel was analyzed by Laser microscopy (Keyence Co., Tokyo, Japan). The plate was covered by glass plates to give micromixer as shown in Figure 1.

Microchannel Reaction

Amino acid NCA was prepared by well established method using TCF [5]. After a series of recrystallization, NCAs were subjected to reaction. NCA and TEA solutions were each charged into a 1 ml Hamilton Gastight Syringe (Hamilton Co., Reno, NV). The solutions were supplied to the microreactor using a KDS 230 syringe pump with a parallel syringe holder (KD Scientific Inc., New Hope, PA). The two syringes were set at identical pumping rates so that the final mixed solution would have a constant reagent concentration. The flow rate affects the reaction time, that is, the residence time in which the reaction mixture flowed through the microreactor. During the reaction, both NCA and TEA solutions stored in syringes were cooled at 0 ^{o}C. The reaction mixture was kept at 30 ^{o}C in a water bath, and a 100 μl product was collected. The microreactor setup operated under a dry argon atmosphere. The obtained product was immediately quenched with aqueous 0.1M hydrochloric acid and the resulting suspension was evaporated, dried, and a final product of white pellet was obtained.

Gel Filtration Chromatography (GFC) Analysis

After removal of protecting group by 30% HBr in AcOH or 1M TMSBr in TFA solutions followed by dialysis against water using dialysis tube (MWCO=500da), the synthesized polymers were subjected to chromatographic analyses. Gel filtration chromatography analysis was performed by Alliance 2965 system (Waters Corporation, Milford, MA) with UV detector under the following conditions: YMC-Pack Diol-120 and -300 columns (8.0×300 mm) and 50 mM phosphate buffer (pH 6.0) eluent at a flow rate of 1 ml/min at 30 ^{o}C. The calibration curves were obtained by using commercially available poly(L-Lys) as standard. Weight-average molecular weight (Mw), number-average molecular weight (Mn), and polydispersity (Mw/Mn) for each polymer product were determined from GFC data.

Results

Microreactor Fabrication

First, we prepared micromixer chip. Basic concept was the same as our previous report using PMMA as the material [6]. Such three dimensional structureis difficult to prepare in chip-type microreaction devices, especially positional alliance is quite difficult when assembling microchannel. We developed rapid preparation technique for such complicated three dimensional microchannel network. First, the channel pattern was projected using a Mask-less lithography technique. This method gave strict positioning of channels. We took this advantage and designed microchannel on both side of Si plate (Figure 1a). Obtained plate was covered with top and bottom glass plates that have positioned bier hall prepared using microdrill, by step-by-step anode bonding to produce the chip-type micromixer. Using this micromixer, microreaction system was assembled as shown in Figure 1b. Mixing efficiency was almost the same as our previous study (data not shown) [6].

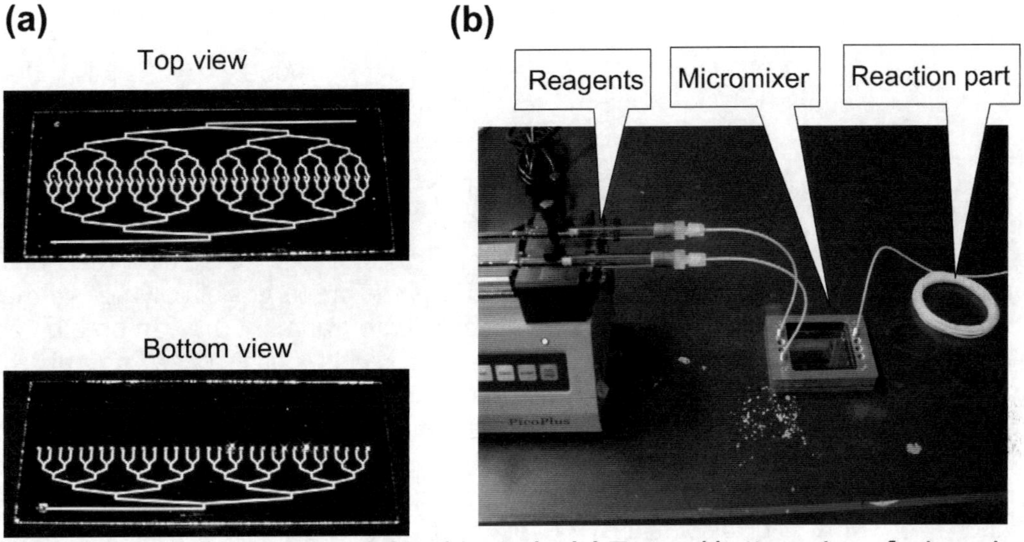

Figure 1. Microreaction system used in this study. (a) Top and bottom view of micromixer chip and (b) assembled microreaction system.

Polymerization of N^{ε}*-benzyloxycarbonyl-L-lysine*

Polymerization reactions were performed in DMF and initiated by TEA. Each DMF solutions of NCA and initiator were loaded in syringe and separately injected into the reaction system by using syringe pumping, as shown in Figure 1c. The GFC analysis of resulting polymer molecules showed higher molecular weight species can be obtained by decreasing flow rates, that is, as a consequence of the increase in time required for polymerization to occur. It was found that the molecular mass of the polymer produced using the microreactor is simply governed by the flow (pumping) rate.

Next, properties of polymer products synthesized using the microreactor was compared with those obtained by the batchwise system. The reactions were conducted at 0.3 M NCA concentration at various reaction times, and the mole ratio of monomer and initiator was 10:1. Results were summarized on Table 1. There were no significant differences in yields between the two systems at each reaction time. The values for number-average molecular weight of polymer in microfluidic system (Mn = ca. 20,000) were slightly larger in relation to those in batchwise system (Mn = ca. 18,000) at 200 min reaction time. It was noted that the obvious differences between the two systems were observed with respect to polydispersity values. The molecular weight distributions in the microfluidic system were narrower than those in the batchwise system. The NCA-polymerization using the microreactor resulted in a better molecular weight distribution control as compared to the batchwise system.

Table 1 Comparison of polymer properties in poly(L-Lysine) synthesis.

	batchwise reaction		microreactor	
Time/min	Mn	Mw/Mn	Mn	Mw/Mn
1	7833	1.42	6960	1.33
5	9818	1.43	9426	1.27
10	13755	1.40	13755	1.22
50	16824	1.42	18530	1.19
100	17333	1.47	19069	1.17
200	18199	1.47	20073	1.17

Controlling the molecular weight distributions would become increasingly harder when using NCA solutions of higher concentrations, because gradients of reagent concentration tend to generate locally. Thus, controllability of the microreactor for polymerization reactions was investigated at various NCA concentrations. Table 2 shows the properties of polymer synthesized at 0.1 M – 0.5 M of Lys(Z)-NCA for 200 min (monomer:initiator ratio of 10:1). In both batchwise and microfluidic systems, the slight increase in yields was observed at higher NCA-concentrations. There was no significant difference in yields between the two systems at all concentrations (data not shown). The number-average molecular weights also had a tendency to increase with a rise in NCA concentration. The polymers produced using the microreactor had slightly larger number-average molecular weights than those using batchwise system at all concentrations of NCA. Polydispersities of polymers increased linearly with increasing concentration of NCA (R2 = 0.9902) in batchwise system. However, there was only a very slight increase in the molecular weight distributions of polymers in the microfluidic system. This result emphasizes one of the advantages of the microreactor. In a situation wherein control of polymerization is unfavourable, the micromixer may offer a significant role in controlling the molecular weight distribution.

Table 2 Effect of NCA concentration on polymer properties in poly(L-Lysine) synthesis.

	batchwise reaction		microreactor	
Conc. (mM)	Mn	Mw/Mn	Mn	Mw/Mn
0.1	17424	1.15	17734	1.11
0.2	17292	1.32	18211	1.17
0.3	18199	1.47	20073	1.17
0.4	18326	1.66	20548	1.21
0.5	20795	1.90	23209	1.18

We also examined effect of the micromixer on reaction. Polymerization using a microreactor attached to a simple T-shaped connector in place of the micromixer was carried out. The reaction was carried out using 0.5 M of NCA solution at monomer:initiator ratio of 10:1. The polydispersity value of microreactor attached to a T-shaped connector demonstrated a higher polydispersity value (Mw/ Mn = 1.75) than the micromixer system (Mw/ Mn = 1.18), indicating poor control of molecular weight distribution by the microreactor attached to a T-shaped connector. Slight improvement of T-shaped connector system than that of batchwise reaction was obserOn the other hand, lower polydispersity values of the micromixer reflected a better

Table 3 Comparison of polymer properties in amino acid copolymer synthesis.

		Mn	Mw/Mn
poly(Lys, Ala)	batchwise reaction	17,700	2.56
	microreactor	18,800	1.64
poly(Lys, Leu)	batchwise reaction	18,100	2.13
	microreactor	19,200	1.54

control of molecular weight distribution. Thus, strict reaction control by rapid micromixing in our microreactor is suitable for NCA polymerization.

Synthesis of amino acid copolymer in microreactor

Next, copolymerizations of NCAs were applied to the microfluidic reaction. Copolymerization mechanism is known to be more complex due to differences in the reactivities of individual NCAs. The properties of copolymers obtained by the microfluidic and batchwise systems were indicated in Table 3. The reactions were performed at 0.5 M of total NCA concentrations for 200 min (monomer:initiator ratio of 10:1). The synthesized copolymers were Lys-based polymer mixed with Ala (or Leu) with the mole ratio of Lys(Z)-NCA to Ala-NCA (or Leu-NCA) at 7:3. In a batchwise system, the product yield obtained from Lys/Leu-copolymerization (66%) was higher than that of Lys/Ala-copolymerization (59%). In the two copolymers, there were no significant differences in the yields for batch and microreactor systems [poly(Lys/Ala) 59% and 58%; poly(Lys/Leu) 66% and 67%, respectively]. Lys/Ala- and Lys/Leu-copolymers as well as homo-polymerization of Lys(Z)-NCA synthesized using the microreactor had slightly larger average molecular weights as compared to that of batchwise system. Batchwise copolymerizations showed much poorer molecular weight distribution control than batch-homopolymerization. In terms of molecular weight distribution, copolymerization often has a negative effect. The microreactor provided optimized control of molecular weight distributions as indicated by the low value obtained from each copolymer (1.64 Mw/ Mn value for Lys/Ala, 1.54 Mw/Mn value for Lys/Leu). Compared to the molecular weight distributions of homopolymers, copolymers produced by microreactor showed lower values than those in batchwise system.

Discussion

Our study demonstrated the usefulness of microreactor in controlling NCA-polymerization reaction. Furthermore, this microreaction technology offers several advantages in the production of amino acid polymers. First, polymer with low Mw/Mn values can be obtained at relatively fast reaction rate in base-initiated polymerization. In NCA-polymerization by base initiation such as tertiary amine, the NCA anions rapidly polymerize in the same way as step polymerization. On the other hand, polymerization using nucleophile such as primary amine occurs at slower reaction rate through nucleophilic attack on NCA. Molecular weight distributions of the obtained polymers tend to be broad in the former reaction mechanism, but narrow in the later. NCA-polymerization using tertiary amine in combination with microfluidic system enabled rapid production of polymer with narrower distribution. Second, the microreactor allowed narrow molecular weight distribution even with high concentration of NCA. This leads to increase in product yield per unit time, and could be of great advantage in high-throuput synthesis. In capacity of production, our microreactor could synthesize polylysine (Mw = 20,000, Mw/Mn = 1.17) at the yield of 100 mg/min; a single reactor could produce approximately 6 g of the polylysine for 10 hours. Also, our method is applicable for efficient copolymer synthesis. Thus, our microreaction system can be applicable high-throughput screening of amino acid polymers. Third, this microreactor can be produced rapidly, by using mask-less technique. This is advantageous for optimization of microreactor structure. Any structure larger than the resolution of the mask-less system (> 1μm) can be designed simply by CAD system. Also, sequential anion bonding enables three dimensional construction of microchannel even in chip system. Such feature might be advantageous for developing microreactor with microfluidic operation at higher degree.

We have shown that polymer chain can be expanded and oriented in microfluidic system using DNA as a model. In this study, we could not demonstrate advantages of such conformational change, because only molecular weight distribution was dramatically improved than that of batchwise system. Also, microreactor without micromixing part showed broad molecular weight distribution, even sharing force might affect on conformational change. Such conformational change induced by laminar flow system might not play important role in stepwise condensation. Terminal part of polymers usually locates at the surface of the molecule and easy to access. Therefore, reactivity was not much improved in microreactor. Expansion and orientation of polymers might be advantageous for substitution of polymers or graft polymerizations, which target of reaction locates middle of the polymer molecule.

Conclusion

In this study, we developed microreaction system for amino acid polymerization. Combination of maskless technique, dry-etching, and sequential anode bonding enables rapid production of micromixer chip which has three dimensional microchannel networks. Resulting microreactor showed excellent controllability of molecular weight distribution of amino acid polymers, homopolymer and some copolymers. Although conformational change of macromolecules does not show influence, strict reaction control yielded much narrow molecular weight distribution than that of batchwise reaction or microreaction without micromixer. These features are suitable for high-throughput combinatorial preparation of amino acid polymers.

Acknowledgment

We thank Prof. R. Katakai for advices in NCA preparation. Part of this work was supported by Grant from Production, Analysis and Measurement System for Microchemical Technology Project of NEDO.

Literature Cited

[1] B. D. Ratner and S. J. Bryant, *Annu. Rev. Biomed. Eng.*, **6**, 41-75 (2004).

[2] W. Ehrfeld, V. Hessel, and H. Lowe, *Microreactors, New Technology for Modern Chemistry*, Wiley-VCH, Weinheim 2000.

[3] V. Hessel, C. Serra, H. Lowe, and G. Hadziioannou, *Chemie Ingenieur Technik*, **77**, 1693-1714 (2005).

[4] M. Miyazaki, K. Yamashita, Y. Yamaguchi, T. Honda, H. Nakamura, M. Fujii, and H. Maeda,

New J. Chem., **28**, 1622-1627 (2004).
[5] K. Yamashita, Y. Yamaguchi, M. Miyazaki, H. Nakamura, H. Shimizu, and H Maeda, *Lab Chip*, **4**, 1-3 (2004).
[6] K. Yamashita,Y. Yamaguchi, M. Miyazaki, H. Nakamura, H. Shimizu and H. Maeda, *Anal. Biochem.*, **332**, 274-279 (2004).
[7] http://www.nanosystem-solutions.com/
[8] R. Katakai and Y. Iizuka, *J. Org. Chem.*, **50**, 715-716 (1985).
[9] Y. Yamaguchi, K. Ogino, K. Yamashita, and H. Maeda, *J. Chem. Eng. Jpn.*, **37**, 1265-1270 (2004).

Convergent Synthesis of Dendrimers Using a Continuous Flow Microreactor

Shuhong Liu[1], C.-H. Chang[1], Brian K. Paul[2], Vincent T. Remcho[3], Bindiya Abhinkar[2]
1. Chemical Engineering, Oregon State University, 102 Gleeson Hall, Corvallis, OR 97331
2. Industrial and Manufacturing Engineering, Oregon State University, 118 Covell Hall, Corvallis, OR 97331
3. Chemistry, Oregon State University, 153 Gilbert Hall, Corvallis, OR 97331

Dendrimers are nanoscale macromolecules that have highly branched, core-shell structures. Higher generation dendrimers have close-packed peripheral functional groups and a hollow interior. The chemistry of the core and the terminal functionalities can be tailored according to the specific application. These unique features provide dendrimers with great potential for a wide variety of applications including catalysis, sensors, drug delivery and light harvesting. They are synthesized in a stepwise manner to higher generations using a number of chemical reactions, so carrying out the synthesis in conventional reactor is time-consuming. Therefore the dendrimer products are extremely expensive. In this work, convergent synthesis of dendrimer was carried out in a continuous flow microreactor. The short pathway and high surface to volume ratio features of microreactors improve reactions tremendously due to rapid diffusion and heat transfer. In comparison to conventional batch processes, microreactors demonstrate notable advantages such as a much higher reaction rate: 3 seconds of residence time in a microreactor versus several hours of residence time in a conventional flask reactor for similar yield. Microreactors have a promising future for dendrimer synthesis.

Introduction:

Dendrimers[1] [2] are highly-branched molecules with fractal morphologies. Dendrimers consist of a core-unit, branching units, and peripheral end groups. Higher generation dendrimers have close-packed peripheral functional groups and a hollow interior. This unique feature provides dendrimers with the capacity to serve as hosts to encapsulate guests in the interior and to conjugate molecules on the surface. There are two major strategies to synthesize dendrimers: the *divergent* approach and the *convergent* approach. The convergent approach starts from the peripheral functional groups and synthesizes inward to form higher and higher generations of dendrons. Finally, the dendrons react with a core molecule to generate dendrimer. Dendrimers can be synthesized with great precision, thus, ideally, a certain generation of dendrimer has a single size and molecular weight rather than the broad molecular weight distribution characteristic of linear polymers [3]. Their dendritic architecture has shown great potential for a wide variety of applications including catalysis, sensors, drug delivery, light harvesting, MRI imaging and gene transfer techniques [4] [5] [6]. However, the synthesis of dendrimers is a tedious and time-consuming process, for example, some reactions require days to complete. Therefore the limiting factor on the application of dendrimers is often the cost of their production. For dendrimers to realize their full potential, methods must be developed by which the uniformity and efficiency can be closely approximated in the production of macromolecules in nature.

Microreactors are well recognized for their ability to enhance mixing and heat transfer due to their short diffusion pathways and large interfacial areas per unit volume (10,000~50,000m^2/m^3). In contrast, conventional reactors have surface area to volume ratios of 100 m^2/m^3[7]. These two features of microreactors are key to improving yield and selectivity, specifically for mass-transport controlled reactions, highly exothermic or endothermic reactions, and reactions with inherently unstable intermediates. In addition to the benefits mentioned above, another attractive advantage is the ability to "number-up" laboratory-scale reactors by simply arraying the identical microreactors without a need for further process development and parameterization. This numbering-up process has been demonstrated by Clariant, where a chemical was created in the quantity of 80 tons per year within a microchannel format [8].

We have demonstrated the advantages of using a continuous microreactor to synthesize EDA–cored PAMAM dendrimers [9]. We were able to achieve high-yield synthesis of generation G-0.5 PAMAM and generation G0.0 PAMAM. Most importantly, the mean residence time for the synthesis in the microreactor is seconds in comparison with days in a conventional batch reactor. We believe dendrimer synthesis can benefit greatly from implementation in a highly-parallel, process-intensified microsystem format. In this paper, we report our progress of extending our success to convergent synthesis of dendrimers.

Experimental:

To explore the potential benefits of microreactors for dendrimer production, we have investigated the convergent approach to dendrimer synthesis using a continuous flow micromixer. The reactions are illustrated in Scheme 1 [10]. The convergent approach used 3, 5-bis(4-aminophenoxy)-benzoic acid (coumpound 1) as building blocks and N-methyl-2-pyrrolidinone as solvent. Thionyl chloride was used as an activating agent, while 4,4'-oxydianiline served as a core molecule for the synthesis of dendrimers. The synthesis of each individual generation dendron – except for dendron G1 – includes two steps: activation by thionyl chloride and coupling with building block 1. The synthesis of each dendrimer consists two similar steps: activation by thionyl chloride and coupling with core 2. This synthesis strategy has been fulfilled in a conventional in vitro approach by Washio [10], with exception of synthesis of dendrimer G1 and dendrimer G2. However the reactions of each generation of dendron and dendrimer required cooling and inert gas protection, and took about 4-6 hours to complete.

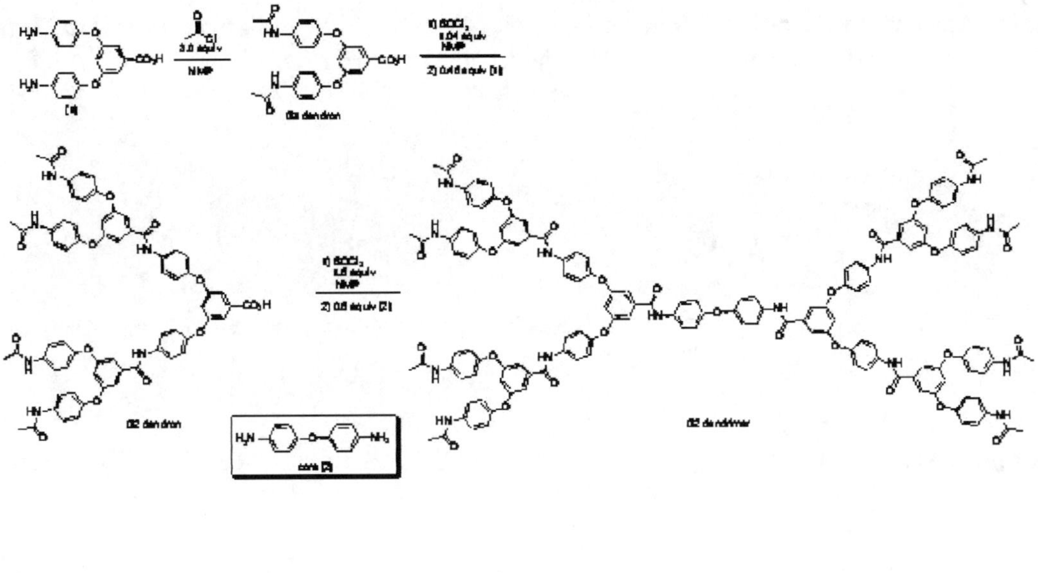

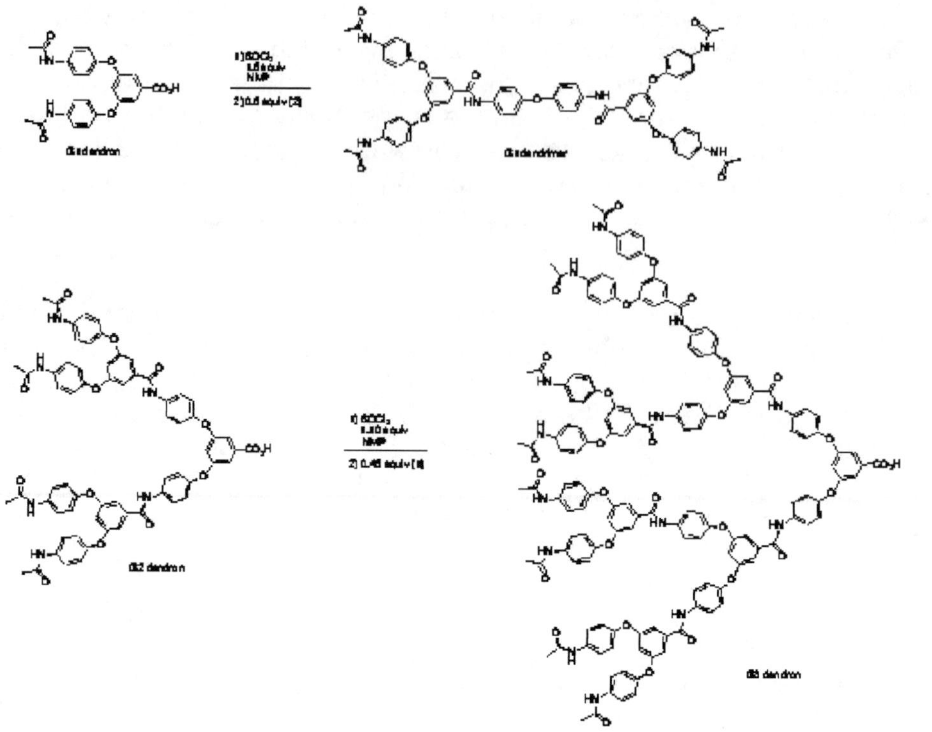

Scheme 1. Reactions of the convergent approach of dendrons and dendrimers

A schematic diagram of the micromixer is given in Figure 1A. The micromixer consists of a mixing element, interdigitated microchannels, in the center of the substrate made of thermally grown silicon dioxide. The mixing element is housed within a stainless steel container. Each microchannel has a dimension of 30 μm in width and 100 μm in height. Two streams of reactants were delivered to the interdigital micromixer through two syringe pumps. Each stream was divided into many ultra-thin lamellae by microchannels, and fast diffusion took place as the lamellae left the microchannel chip in the direction perpendicular to the income streams. The ultra-thin lamellae tremendously improve the mass transport between the two reactants, such that rapid reaction occurs immediately at the outlet of the mixer. Figure 1B illustrates the principle of mixing via an interdigital micromixer [11].

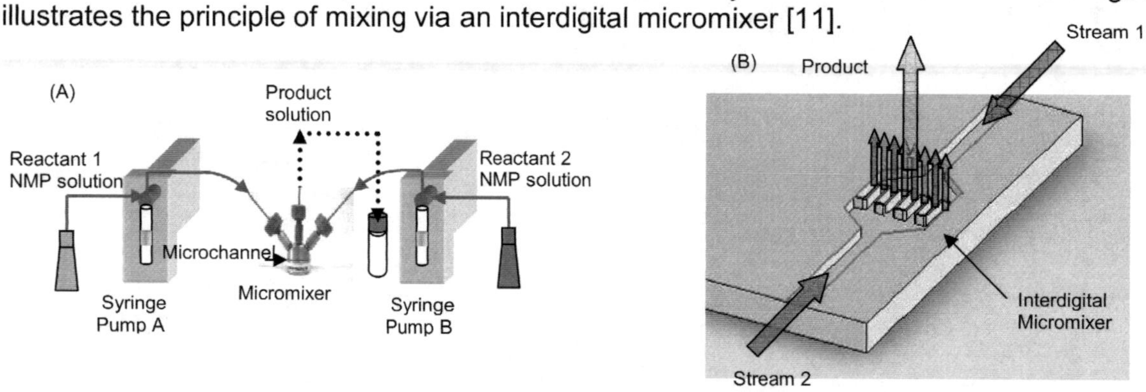

Figure 1. (A) A schematic diagram of the continuous flow micromixer. (B) Mixing via an interdigital micromixer.

General Methods:

All chemicals, except acetyl chloride (from Alfa Aesar), were purchased from TCI America, and were used as received. The interdigital micromixer was purchased from Institut für Mikrotechnik Mainz, Germany. A NMP solution of precursor and a NMP solution of reagent, thionyl chloride (for activation of precursor), building block (coupling reaction to synthesize dendrons) or core molecule (to synthesize dendrimers), were fed into the mixing element through the two micromixer inlets, respectively, by means of syringe pumps. Once the solution streams were introduced into the micromixer, each stream was divided by micro-scale channels into many thin substreams, which left the channels perpendicular to the direction of feed flows. In this manner, the feed flows were rapidly mixed by diffusion at the outlet of the micromixer. The mixed solution passed through the outlet of the micromixer at a flow rate of 0.052 cm^3/s and the tube (35cm long and 0.75 mm ID) connected with the outlet with an estimated mean residence time of 3 seconds. Subsequently, the adduct solution was collected, and poured into water, and the precipitate was collected and dried.

Synthesis of dendron G1:

Solutions of 3, 5-bis(4-aminophenoxy)-benzoic acid (building block) dissolved in NMP (concentration of 0.58 mol/L) and 3 equivalent of acetyl chloride dissolved in NMP (concentration of 1.74 mol/L) were introduced into microchannels by syringe pumps at a flowrate of 0.026 ml/s, respectively. The mixture was collected at the outlet and precipitated in water. The precipitate was collected and dried at 120 ^oC.

Synthesis of dendron G2:

Solutions of dendron G1 dissolved in NMP (concentration of 0.323 mol/L) and 1.04 equivalent of thionyl chloride dissolved in NMP (concentration of 0.336 mol/L) were introduced into the microchannels by syringe pumps at a flowrate of 0.026 ml/s, respectively. The intermediate was collected at outlet. Then solutions of the intermediate and 0.48 equivalent of building block which had been dissolved in NMP (concentration of 0.078 mol/L) were fed into the micromixer by syringe pumps at a flowrate of 0.026 ml/s, respectively. The mixture was collected at the outlet and precipitated in water. The precipitate was collected and dried at 120 ^oC.

In order to facilitate a quantitative comparison of synthesis results between the micromixer approach and the conventional batch approach, the batch approach was conducted using the same starting materials as were employed in the microreactor. To a 1.6 ml of solution of dendron G1 (0.517 mmol) dissolved in NMP, 1.04 equivalent of thionyl chloride was added at 0 ^oC under nitrogen and stirred for 20 minutes at 0 ^oC and for another 20 minutes at room temperature. Sequentially, 0.48 equivalent of building block 1 (0.248 mmol) was added to the solution and the reaction proceeded overnight. The mixture was precipitated in water. The precipitate was collected and dried at 120 ^oC.

Synthesis of dendron G3:

Solutions of dendron G2 (0.0613 mmol) dissolved in NMP (concentration of 0.153 mol/L) and 1.1 equivalent of thionyl chloride (0.0675 mmol) dissolved in NMP (concentration of 0.169 mol/L) were introduced into the microchannels by syringe pumps at a flowrate of 0.026 ml/s, respectively. The intermediate was collected at the outlet. Then solutions of the intermediate and 0.48 equivalent of building block which had been previously dissolved in NMP (concentration of 0.0368 mol/L) were fed into the micromixer by syringe pumps at a flowrate of 0.026 ml/s, respectively. The mixture was collected at the outlet and precipitated in water. The precipitate was collected and dried at 120 °C.

Synthesis of dendrimer G1:

Solutions of dendron G1 (0.2 mmol) dissolved in NMP (concentration of 0.2 mol/L) and 1.5 equivalent of thionyl chloride dissolved in NMP (concentration of 0.3 mol/L) were introduced into the microchannels by syringe pumps at a flowrate of 0.026 ml/s, respectively. The intermediate was collected at the outlet. Then solutions of the intermediate and 0.5 equivalent of core molecule which had been previously dissolved in NMP (concentration of 0.05 mol/L) were fed into the micromixer by syringe pumps at a flowrate of 0.026 ml/s, respectively. The mixture was collected at the outlet and precipitated in water. The precipitate was collected and dried at 120 °C.

Synthesis of dendrimer G2:

Solutions of dendron G2 (0.0885 mmol) dissolved in NMP (concentration of 0.0885 mol/L) and 1.5 equivalent of thionyl chloride (0.1328 mmol) dissolved in NMP (concentration of 0.1328 mol/L) were introduced into the microchannels by syringe pumps at a flowrate of 0.026 ml/s, respectively. The intermediate was collected at outlet. Then solutions of the intermediate and 0.5 equivalent of the core molecule which had been previously dissolved in NMP were fed into the micromixer by syringe pumps at a flowrate of 0.026 ml/s, respectively. The mixture was collected at the outlet and precipitated in water. The precipitate was collected and dried at 120 °C.

Measurement:

We used a combination of ^{1}H NMR spectroscopy to determine the purity and the structures of the products from the micromixer and from the conventional batch reactor. Syntheses from the micromixer provided comparatively pure product without measurable side product. On the other hand, the NMR spectrum of G2 dendron synthesized from the batch reactor shows some side products even after 20 hours of vigorous mixing (see Figure 3). The ^{1}H NMR spectra of dendrons G1, G2 and G3 synthesized via the micromixer are shown in Figure 2.

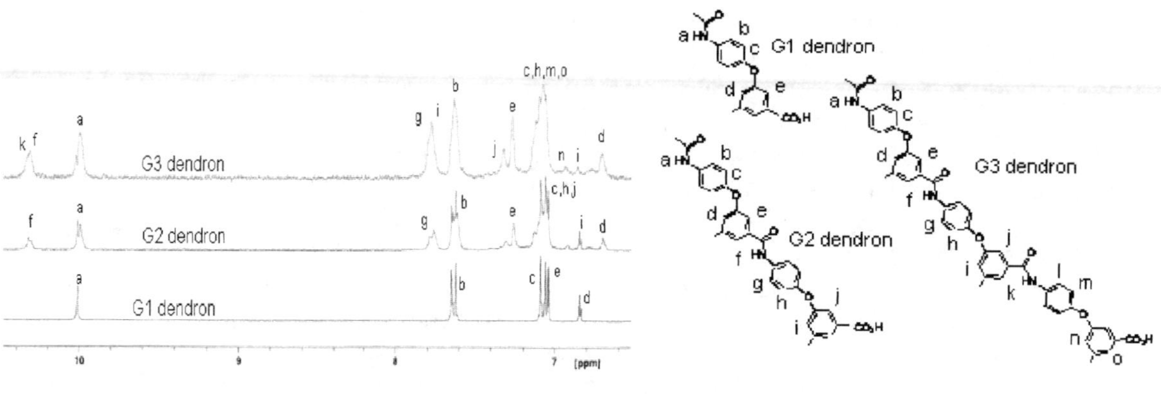

Figure 2. ^{1}H-NMR spectra of dendrons G1, G2 and G3 synthesized via a micromixer.

Figure 3 illustrates the comparison of dendron G2 synthesized via an interdigital micromixer and a flask respectively. Clearly, the dendron G2 synthesized via the micromixer is purer than that via the flask.

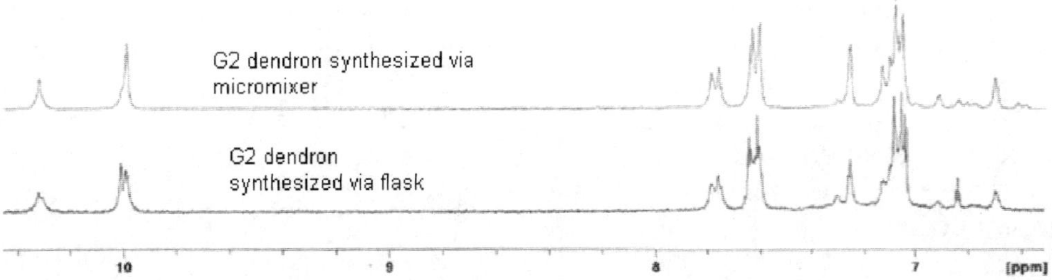

Figure 3. ^{1}H-NMR spectra of dendron G2 synthesized via a micromixer and a flask.

The ^{1}H-NMR spectrum of G1 dendrimer is compared with that for a G1 dendron in Figure 4. The signals assigned to the protons of position d shift from 6.84 to 6.71 ppm, position e from 7.04 to 7.25 ppm and the signals in 10.26 (f), 7.71 (g) and 6.98 (h) appear after the coupling reaction, indicating the formation of G1 dendrimer.

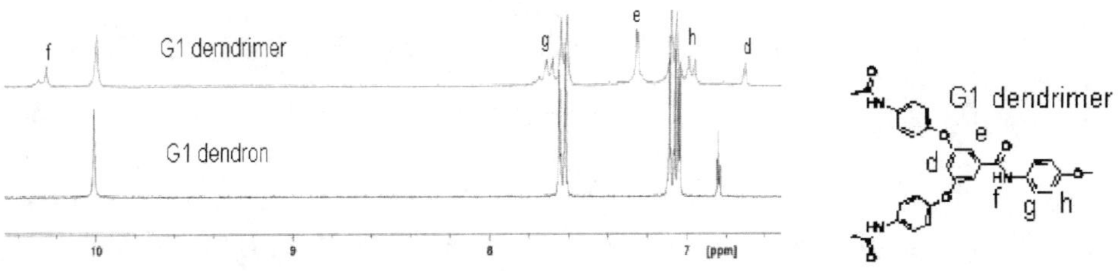

Figure 4. ^{1}H-NMR spectra of dendron G1 and dendrimer G1 synthesized via a micromixer.

Similarly, the ^{1}H-NMR spectrum of G2 dendrimer is compared with that for a G2 dendron in Figure 5. The signals assigned to the protons of position i shift from 6.85 to 6.80 ppm, and the signals in 7.73 (l) and 7.02 (m) appear after the coupling reaction, indicating the formation of the G2 dendrimer.

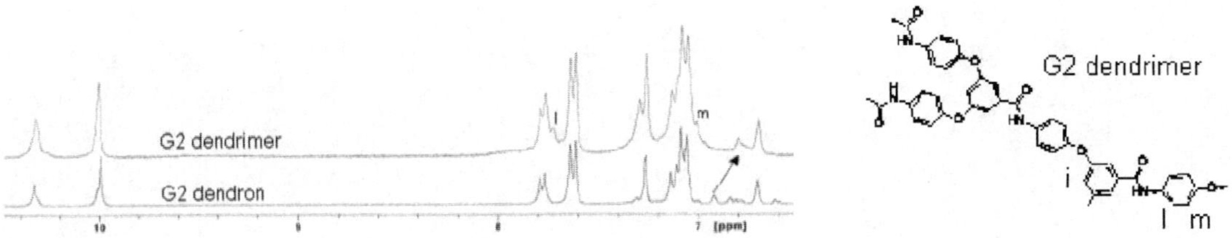

Figure 5. ^{1}H-NMR spectra of dendron G2 and dendrimer G2 synthesized via a micromixer.

Conclusions:

The micromixer approach to dendrimer synthesis using a convergent method demonstrates several advantages over the conventional batch process. The conventional reaction requires cooling and stirring, inert gas protection, and stepwise manual addition of subsequent reactants. All reactions conducted using the micromixer were performed continuously at ambient temperature without cooling or stirring, or the need for an inert atmosphere. A number of experiments were conducted using the micromixer and the results demonstrated good reproducibility. The most attractive advantage of the micromixer approach is that the residence time is 3 seconds versus several hours in a conventional batch reaction, yet yields remain high. Given the advantages of rapid, continuous flow, high yield and selectivity, and most importantly, the possibility of numbering-up the process for industrial production scale, microreactor-based synthesis appears to be a promising approach for dendrimer production.

Acknowledgment: Financial support of this research by the W. M. Keck Foundation is greatly appreciated.

References:

[1] D.A. Tomalia, K. Mardel, S.A. Henderson, G. Holan, R.Esfand, *Handbook of Nanoscience, Engineering, and Technology,* CRC Press LLC, **2003**, 20-1 - 20-34.
[2] S.M. Grayson, J.M.J. Frechet, *Chem. Rev.* **2001**, *101*, 3819-3868.
[3] D.A. Tomalia, H. Baker, J. Dewald, M. Hall, G. Kallos, S. Martin, J. Roeck, J. Ryder, P. Smith, *Polym. J.* **1985**, *17*, 117-132.
[4] B. Halford, *C&EN.* **2005**, *83*, 26-29.
[5] J.R. Baker, A.Quintana, L. Piehler, M. Banazak-Holl, D. Tomalia, *Biomedical Microdevices*, **2001**, *3*, 61-69.
[6] A.W. Bosman, H.M. Janssen, E.J. Meijer, *Chem. Rev.* **1999**, *99*, 1665-1688.
[7] V. Hessel, S. Hardt, H. Lőwe, *Chemical Micro Process Engineering*, Wiley-Vech, **2004**; pp. 74.
[8] C. Wille, R. Pfirmann, *Chemistry Today*, September, **2004**, 20.
[9] C.-H. Chang, S.-H. Liu, Y. Tennico, J.T. Rundel, V. T. Remcho, E. Blackwell, T. Tseng and B.K. Paul, *IMRET* **2005**.
[10] I. Washio, Y. Shibasaki, M. Ueda, *Macromolecules,* **2005**, *38*, 2237-2246.
[11] V. Hessel, H. Lőwe, F. Schőnfeld, *Chem. Eng. Sci.* **2005**, *60*, 2479-2501.

62d NMP Homopolymers and Two-Stages Copolymers Synthesis in Microtube Reactors: Improvement of the Macromolecular Architecture Control

Christophe Serra, Carine Rosenfeld, Cyril Brochon, and Georges Hadziioannou. Laboratoire d'Ingénierie des Polymères pour les Hautes Technologies, Ecole de Chimie Polymères et Matériaux, 25, Rue Becquerel, Strasbourg Cedex-2, F-67087, France
Microfluidic devices are now widely used in scientific fields such as biology, chemistry and chemical engineering. These devices revealed to be powerful tools to study highly exothermic, fast and diffusion-controlled chemical reactions. However, these microsystems only begin to be used within the framework of free radical and living free radical polymerizations. For the later, Atom Transfer Radical Polymerization (ATRP) in solution, Reversible Addition-Fragmentation Transfer (RAFT), Nitroxide-Mediated Polymerization (NMP) in miniemulsion and anion ring opening polymerization have been reported up to now. Here, we look at the NMP in solution of acrylic and styrenic monomers having very different behaviors in terms of their reactivity and exothermicity. We are also interested in the livingness of the polymer by studying the copolymerization of the homopolymer synthesized with a comonomer. Moreover, we look at the influence of the reactor (macro or micro scale reactor) on the control of the macromolecular architecture.

The first block is carried out either continuously in a stainless steel tube microreactor of 0.9 mm ID at 140°C in toluene with a TIPNO alcoxyamine as initiator or in sealed tubes (i.e. in batch process). The copolymerization is also carried out in a similar microreactor or in sealed tubes. However, for the continuous process, the viscous first block solution is mixed with the pure comonomer through a multi or bilamination micromixer (either an interdigital multilamination micromixer or a basic T-junction) prior to enter in the second microtube reactor. Several operating conditions like different reaction times, initial alcoxyamine concentrations and the presence of a rate accelerating agent were investigated.

Concerning the first homopolymer block, for the less reactive and exothermic monomer (styrenic) and for low molecular weights targeted, no influence was observed between the batch and the microtube reactor. The number-average molecular weight (Mn) increases linearly with monomer conversion and the polydispersity index (PDI) ranges from 1.1 to 1.4 for batch and continuous reactions, indicating that the controlled nature of the polymerization mechanism is maintained. However, for high molecular weights targeted, the microtube reactor allows to get lower PDIs than the batch reactor. For a highly reactive and exothermic monomer (acrylate), whatever the molecular weight targeted, the microtube reactor returned the lowest PDIs whereas he polydispersity indexes obtained in the batch reactor are higher than 1.5 giving evidences of the lost in the polymerization control. The presence of a rate accelerative agent like acetic anhydride allows to increase the monomer conversion by at least 20% without affecting significantly the controlled nature of the polymerization. Then, the use of an accelerative agent enables the synthesis of a copolymer by a continuous two-stages process in which the residual first monomer is not removed from the reactive medium.

Concerning the block copolymer, it was found that the interdigital multlamination micromixer allows on one hand a higher incorporation rate of the second comonomer onto the first block (higher overall monomer conversion) and on the second hand a lower overall PDI compared to the T-junction and the batch process.

In conclusion, it was demonstrated that the use of microfluidic devices (microtube reactor and micromixer) can significantly improve the control over the polymerization as well as the macromolecular architecture.

62e A Micro-Reactor for Preparing Uniform Molecularly Imprinted Polymer Beads

Mohammed Zourob, School of Materials Science,, The University of Manchester,, Grosvenor street, P.O. Box 88, Manchester,, United Kingdom, Nicholas Goddard, School of Chemical Engineering and Analytical Science, The University of Manchester, Manchester, The University of Manchester,, Grosvenor street, P.O. Box 88, Manchester,, United Kingdom, and Stephan Mohr, The University of Manchester,, Grosvenor street, P.O. Box 88, Manchester,, United Kingdom.

Molecular imprinting is an emerging technique for the preparation of synthetic polymers with high selectivity for a predetermined ligand. The most widely used method for preparing molecularly imprinted polymers is by bulk polymerization, followed by grinding to particles. These particles are irregular, polydisperse and usually include a large portion of wasted fine particulate material (estimated 70%). Extensive sieving and sedimentation is required to achieve a more narrow size distribution and to remove fine particles, making the method time and labour consuming. Methods for preparation MIPs in beaded form have been recently reviewed1, 2. They include: suspension polymerization in liquid perfluorocarbon 3 and mineral oil 4, multi-step swelling 5, 6 and grafting polymerization directly to a suitable support 7, 8. The synthesis of MIP in the form of spherical particles has numerous advantages in terms of the process and performance in their final applications.

In this study, uniform spherical molecularly imprinted polymer beads were prepared via controlled suspension polymerization in a spiral-shaped microchannel using mineral oil and perfluorocarbon liquid as continuous phases. Monodisperse droplets containing the monomers, template, initiator, and porogenic solvent were introduced into the microchannel, and particles of uniform size were produced by subsequent U.V. polymerization, quickly and without wasting polymer materials. The droplet/particle size was varied flexibly by changing the flow conditions in the microfluidic device. The diameter of the resulting products typically had a coefficient of variation (CV) below 2%. The specific binding sites that were created during the imprinting process were analysed via radioligand binding analysis. The molecularly imprinted microspheres produced in the liquid perfluorocarbon continuous phase had a higher binding capacity compared to the particles produced in the mineral oil continuous phase. The successful imprinting against a model compound using two very different continuous phases demonstrates the generality of this current simple approach.

Keywords: Molecularly imprinted polymers, microsystem.

1- Perez-Moral, N.; Mayes, A. G. Bioseparation 2001, 10, 287-299. 2- Mayes, A. G. In Molecularly Imprinted Polymers: Man Made Mimics of Antibodies and their Application in Analytical Chemistry; Selelrgren, B., Ed.; Elsevier Science B.V.: Amsterdam, 2001, pp 305-322.2- 3- Mayes, A.; Mosbach, K. Anal. Chem. 1996, 68, 3769-3774. 4- Kempe, H.; Kempe, M. Macromol. Rap. Comm. 2004, 25, 315-320. 5- Hosoya, K.; Yoshikazo, K.; Tanaka, N.; Kimata, K.; Araki, T.; Haginaka, J. Chem. Lett. 1994, 1437-1438. 6- Haginaka, J.; Sanbe, H. Anal. Chem. 2000, 72, 5206-5210. 7- Wulff, G.; Oberkobusch,

D.; Minarík, M. React. Polym. 1985, 3, 261-275. 8- Plunkett, S.; Arnold, F. J. Chromatogr. A 1995, 708, 19-29. 9- Glad, M.; Reinholdsson, P.; Mosbach, K. React. Polym. 1995, 25, 47-54.

62f Synthesis of Monodisperse Polymer Particles by Photopolymerization in Flow Micro-Reactors

Zhiqiang Gao, University of Kentucky, 161 F. Paul Anderson Tower, Lexington, KY 40506-0046 and Asit Ray, Chemical Engineering, University of Kentucky, 161 F. Paul Anderson Tower, Lexington, KY 40506-0046.

In recent years, microdroplet-based processes are finding increasing applications in existing and emerging technologies. In each of these cases, a solution microdroplet as a micro-reactor undergoes liquid to solid phase transformation. The property of the final solid phase depends on a number of time-dependent physical and chemical transformations occurring in the microdroplets. Highly monodisperse multiacrylate microspheres of diameters of 5-50 ìm, containing nanoparticles and additives, were prepared by UV-radiation photopolymerization. The microdroplets, with size fluctuations of less than 0.01%, with different monomer concentration were generated with a vibrating orifice aerosol generator (VOAG). The properties of the resulting particles (morphology, size, size distribution, and surface characteristics) were measured by scanning electron microscopy and optical microscopy. The resulting particles were monodisperse with a coefficient of variation (ratio of the standard deviation of the particles to the mean particle diameter) of 1% or less. The solid microparticles may be tuned in size by varying reaction procedures. The results of our experiments show that tailored microspheres (e.g., multicomponent, layered, and with nanoparticle inclusions) of highly reproducible size and physical characteristics can be produced, and the morphology of the final solid polymer particles can controlled by the initial droplet size and composition as well as through the reaction conditions. The results suggest possibility of preparing microparticles of various compounds with reproducible physical properties and functions through adding desired components, such as salt, additives and drugs et al. It is also possible to produce monodisperse multicomponent particles, spherical microcapsules, and micro-resonators

62g Synthesis of Polymer Beads Necklaces in a Microfluidic Device

Michel Bouquey[1], Christophe Serra[1], Laurent Prat[2], and Georges Hadziioannou[1]. (1) Laboratoire d'Ingénierie des Polymères pour les Hautes Technologies, Ecole de Chimie Polymères et Matériaux, 25, Rue Becquerel, Strasbourg Cedex-2, F-67087, France, (2) Laboratoire de Génie Chimique, CNRS UMR 5503 - INPT, 5 rue Paulin, Toulouse, 31106, France

Polymer particles synthesis has long been the focus of a very intensive research because from painting formulation to drugs delivery they cover a wide range of applications. So far, polymer particles were mainly prepared by either heterogeneous polymerization processes (emulsion, suspension, supercritical fluid) or by precipitation processes in a non solvent. Although, these two processes lead to polymer particules having a different size domain, they induce a large particle size distribution. Recently, microfluidic processes have been considered because of their unique capacity to generate billions of microdroplets with a very narrow size distribution. Indeed, if the microdroplets generated are polymerizable media, it is possible to obtain billions of polymer particles with well-defined characteristics like size, shape and morphology. The advantage is even greater considering the fact that the particle size can be simply adjusted by tuning the continuous phase and dispersed phase flow rates.

The work presented here aims at the synthesis of polymer beads necklaces. We are also interested to get necklaces with different compositions and structures like plain (a), alternated (b) or block (c) arrangement of two different polymer beads within the necklace (cf. figure, blue and red spheres represent two different polymer microparticles).

Necklaces are formed from polymer microparticles as follow: first, alternate formation of two different polymerizable microdroplets is promoted at a cross-junction; then the microdroplets are polymerized by free radical copolymerization using a photo or thermal initiator. By mixing a functional comonomer in each monomer phase, one obtains two differents polymer microparticles bearing on their surface an antagonistic chemical group capable to form a covalent bond by condensation reaction. Then, each reactive bead is individually and alternately confined in a channel in such a way that each microparticle get into contact with two other microparticles. Finally, the chemical reaction between the reactive groups is promoted resulting in a covalent bond between the microparticles and the formation of the polymer beads necklace. By varying the amount of crosslinking agent and the nature of the monomer (styrenic, acrylate), it is possible to adjust some of the physical properties of each beads like viscosity and rheological characteristics.

In this work, it was demonstrated that the use of microfluidic devices allow the synthesis of totally new polymer materials like polymer beads necklaces with a large range of morphologies and properties.

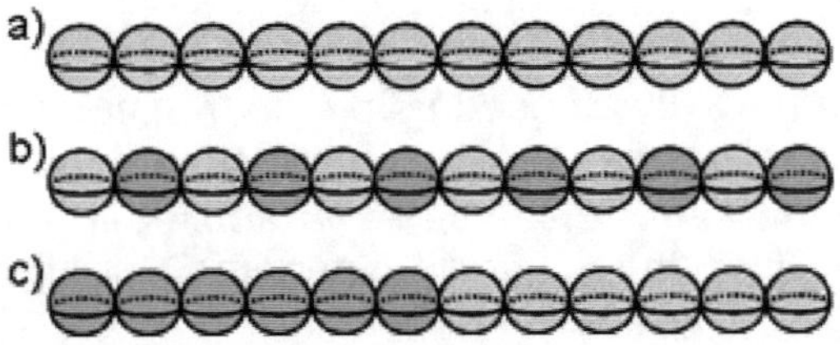

Nanoparticle-doped polymer minirods made by micro segmented flow

G.A. Groß, C. Hamann, J. Wagner, P.M. Günther, J.M. Köhler
Institute of Physics, Microreaction Technologies Department, Technische Universität Ilmenau, Germany

Abstract

A micro fluidic assembly was used for the formation of monomer segments separated by an immiscible perfluorinated aliphatic solvent. Segments were transformed in polymer rods by thermal polymerization. In this way uniform sized polystyrene rods with different diameters and length were formed. These rods could be modified by doping with metal nanoparticles and coupling reaction with diazonium ions.

Perfluorinated alkanes allow the generation of fluid segments composed of organic liquids in micro flow through arrangements. We used this technique in order to produce polymer rods. Therefore, pure monomers or monomer mixtures were used for the formation of micro fluid segments by injecting them into a carrier stream of a perfluorinated liquid. The volume of these fluid segments was controlled precisely by the flow rate of both immiscible liquids. Syringe pumps were applied for pulsation-free flow generation. The monomer segments were transferred into PTFE-tubes with various inner diameters (between 0.25 and 1 mm) and polymerized for 48 hours at 80 degrees (Celsius). The shape of the segments is determined by the inner diameter of the tube and the wetting behavior between the wall, the monomers and the carrier liquid. Segment size was adjusted by the dispensing volume from spherical droplets with the diameter of the tube up to cylindrical segments of some millimeter length.

Surface modified gold and silver nanoparticles were dispersed in the monomer solution only if the organic shell exposes hydrophobic groups. Therefore, metal nanoparticles were prepared by reduction of metal ions in a two phase water / toluene system with $NaBH_4$. Hexadecylamine (HDA) was used as phase transfer catalyst. HDA was displaced by alkylthiols in a subsequent reaction. After phase separation and purification, these particles were colloidal in the monomer dispersed and applied in the preparation of monomer micro fluid segments. After polymerization, nanoparticle-doped polymer micro rods were obtained.

Keywords

Micro reaction technology, polymer rods, nanoparticle, dye-doped, fluid segments, in-situ polymerization, metal composite

1. Introduction

Polymer particles are of broad interest for diverse applications. For example, they are used for the formation of different composed polymer materials in the plastic industry. Polymer particles are also in use as filling materials to adjust the physical properties of composite materials. Hence, there are many applications in the field of life sciences, for example filling material for separation columns[1] or as supports for enzymes or catalysts[2-5].

Chemically functionalized particles are of particular interest as supports for solid-phase synthesis.[6-10] They can be used for single bead synthesis procedures as well as for combinatorial syntheses by split-and-recombine methods.[11] In order to optimize the particle handling and to realize high loading capacities (per particle), the preparation of larger beads is important. A non-spherical shape is of particular interest for larger particles because the substance exchange by diffusion is faster, if the particles possess high surface to volume ratios, i.e. high aspect ratios. Therefore, cylindrical rod particles are desired for synthetical as well as separation purposes.

In principle, polymer synthesis particles can be handled in micro fluidic systems. A system for combinatorial chemistry in the 0.1 µl range using single polymer particles in microstructured silicon arrays was reported elsewhere.[12, 13]

Here, the principle of micro-segmented flow[14, 15] was applied for the formation of monomer segment, prior to polymerization. The preparation conditions were tuned in order to obtain polymer rods of various lengths and diameters. To prove the possibility of doping, nanoparticles and azo-dye chromophores were incorporated before and after polymerization.

2. Experimental arrangement, materials and methods

Commercially available PTFE tubing materials and fluid manifolds were applied for the micro fluidic setup. A PTFE T-junction was used as injector and connected with two syringe pumps. Here, the micro fluid segments were generated in this device. The first pump was used for the actuation of the perfluorinated separation liquid, the second for the monomer mixtures. Syringe pumps were controlled in an alternating order and uniform sized monomer segments were formed corresponding to the delivered volumes. The T-junction outlet was connected to a tubing coil which kept the monomer segment series. The syringe pump system[16] was control by a LabView application. Tubings with inner diameters between 0.25 mm and 1 mm were used in the polymerization experiments. The experimental set-up for the polymerization is shown in scheme 1.

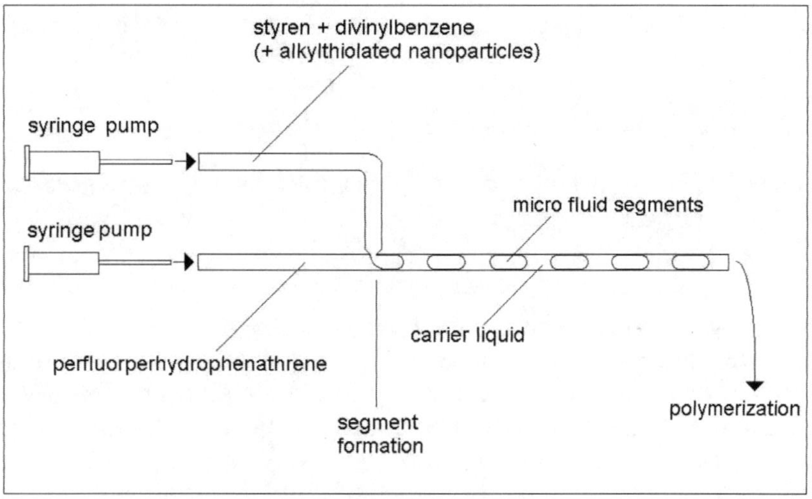

Scheme 1: Experimental assembly

Perfluorinated aliphatic solvents were applied as separation liquids due to their immiscibility with non-fluorinated hydrocarbons. Perfluorperhydrophenanthrene[17] was chosen as suitable separation medium for the monomer liquids. A mixture of styrene (95 % vol.) with divinylbenzene (5% vol.) as crosslinking agent was used for the polymerization. Additionally, N-methylpyrolidone (NMP) was added as porogeneous co-solvent. For initiation of the polymerization reaction, azobisisobutyronitrile (AIBN) was added in a concentration of 1.5 % (wt.) to the monomer mixture (see scheme 2) [18].

In each experimental run, one PTFE tubing coil with a length of about 2 m was filled with 100 – 300 monomer segments. The flow rate ratio between carrier liquid and monomer liquid was kept at unity (1:1). After filling, the tubing coils were closed and the monomer mixtures inside the segments were polymerized for 48 hours at 80 degrees Celsius in an inert gas atmosphere. During heating, the polymer rods were formed by curing of the monomers and kept the shape of the prior monomer segment. Uniform sized polymer rods with the dimensions of the monomer fluid segments were obtained. Received rods were released from the tubing coils easily by manually delivering a liquid with a connected syringe. Perfluorinated aliphatic solvent was filtered and recycled. Polymer rods were washed with dichloromethane and 2-propanole to remove remaining monomers.

Metal nanoparticles were synthesized following the method developed by Brust[19], but instead of tetraoctylammoniumbromide (TOAB) we used hexadecylamine (HDA) as phase transfer catalyst during reduction of tetrachloroaurate with $NaBH_4$ in an emulsion of toluene and water. The organic phase was separated and dodecylthiole (DDT) or octadecylthiole (ODT) was added. Yielded hydrophobic colloidal gold nanoparticles were purified by precipitation from toluene with ethanol. After centrifugation, the nanoparticles were obtained as brown residue with weak metallic lustre. Residual solvent was simply removed by vacuum drying at room temperature. Then, the lipophilic gold nanoparticles were dispersed in the monomer mixtures and applied for the polymerization reaction (see scheme 2).

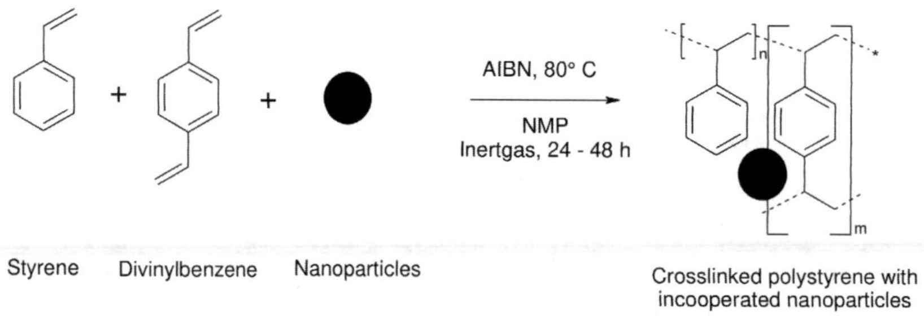

Scheme 2: Formation reaction of crosslinked polystyrene with incorporated nanoparticles

Azo-dye coupled polymer rods were prepared by direct coupling of the polystyrene rods with p-methoxyphenyl-p-diazoniumphenylamine (DAS-4) (see scheme 3). Therefore, approx. 100 mg polystyrene rods were swollen in dimethylformamide (DMF) and 10 mg of the diazonium salt (DAS-4) was added. The coupling reaction was conducted at 80 degrees Celsius within six hours. Received colored rods were washed subsequently with DMF, dichloromethane and 2-propanole three times and dried in streaming air.

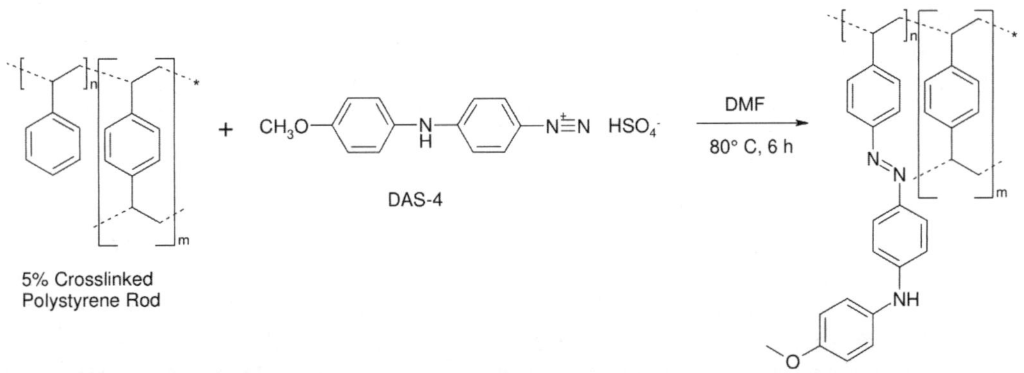

Scheme 3: Direct synthesis of polymere bound chromophoric groups by diazonium coupling reaction

Obtained polymer rods were characterized by light microscopy and by spectrophotometry using a fiber-based microoptical arrangement with an UV/VIS compact spectrometer.[20] A light emitting diode (450 - 650 nm) was used, as light source.

3. Results and discussion

Size and shape of the prepared polymer rods correspond to the shape of the monomer fluid segments. The length of the segments was adjusted by the precise dispensing of monomer volumes. Examples of polymer rods with different lengths are shown in picture 1. The volumes of injected monomer liquid in each segment can be controlled reproducibly. In result, the length of the formed rod can also be controlled precisely. If the fluid segments are formed with low monomer volumes, spherical or nearly spherical monomer segments were formed. Polymer spheres were received after polymerization instead of polymer rods (see picture 1, right hand side).

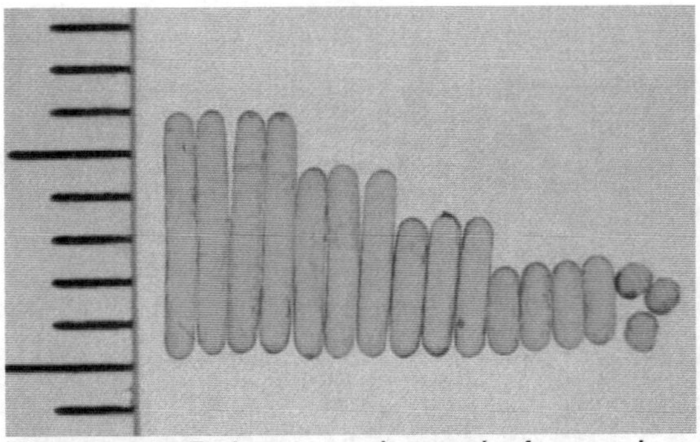

Picture 1: Polymer rods made from micro fluid segments with different length (tubing i.d.: 1.0 mm)

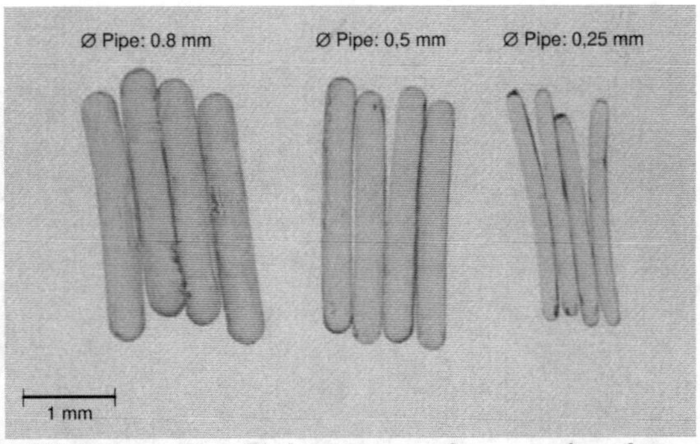

Picture 2: Polymer rods made from segments formed by different tubing diameters

The diameter of the polymer rods was adjusted, by varying the inner diameter of the PTFE tubes. For polymerization, tubes with internal diameters of 250 µm, 500 µm, 800 µm and 1 mm were applied. In result, polymer rods with different diameters were generated. Examples for three different diameters are given in picture 2. The slight bowing of some rods is caused by the radius of used tubing coils as well as the mechanical stress due to local shrinking during the washing and drying process. The aspect ratio of polymer rods (length / diameter ratio) was defined by the segment forming conditions. Different rod geometries can be formed by varying segment volumes and tubing diameters during segment formation and subsequent polymerization. Segment shape and size is completely controlled by surface tension. The polymerization after segment formation acts like a "molding" process which transfers the fluidic determined segment into a solid rod.

Nanoparticles with lipophilic surfaces were required for the preparation of nanoparticle doped polymer rods because of the lipophilic character of the monomer liquids. Therefore, colloidal gold particles were synthesized as described above. Normally, the colloidal gold particles are stabilized by polar ligands and electrostatic repulsion through excess charges on the surface. These ligands and the charges are responsible for the good solvatization in polar solvents. The electrostatic repulsion causes the high stability of the colloidal solution. Gold particles with lipophilic character are needed for the dispersion in styrene / divenylbenzene mixtures. In order to generate gold nanoparticles with a hydrophobic surface layer, original stabilizing ligands were substituted by alkylthiols. These thiols are well known for their high tendency to form self-assembled monolayers (SAM) on gold surfaces.[21] Appropriate gold nanoparticles were purified by precipitation and centrifugation.

After purification, the thioalkylated gold nanoparticles were dispersed in the monomer mixture by ultrasound. Hence, a suspension of metal particles in the monomer mixture was obtained. This suspension was transferred in one syringe and applied for the segment formation process. During polymerization, the dispersed nanoparticles are surrounded by the polymerization mixture. In result, a polymer matrix with incorporated nanoparticles was obtained. The polymer rods were released from the tube, washed and dried analogously to the pure polymer rods. The embedded nanoparticles enable the optical observation of the penetration of washing solvents during the initial washing steps. At suiting conditions, transparent polymer rods with a reddish color were obtained. This homogenous color indicates a homogenous dispersion of the gold nanoparticles inside the polymer matrix. But, only a small amount of the nanoparticles was entrapped in the crosslinked polymer matrix.

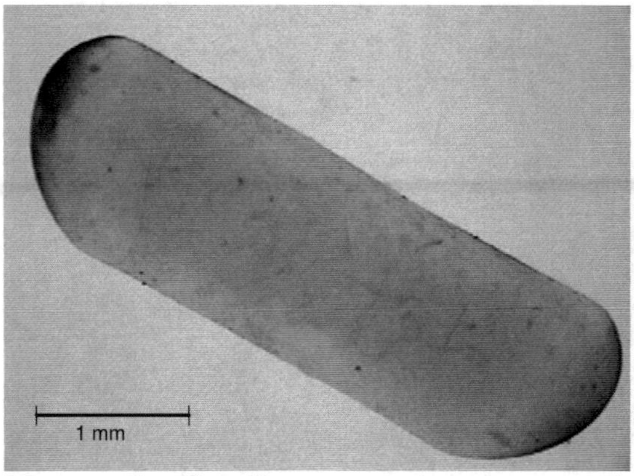

Picture 3: Swelling polymer rod. Nanoparticles are moved by the penetrationg solvent into the rod core.

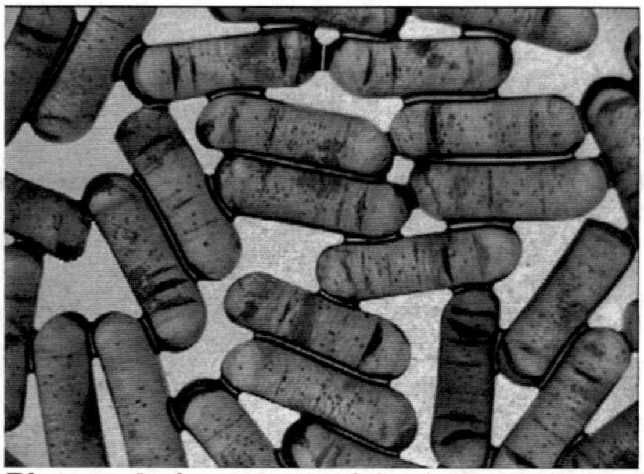

Picture 4: Au-nanoparticle doped polymer rods after removal of the moveable nanoparticles.

Most particles were moved by the permeation of a washing solvent during the polymer swelling. This movement was observed visually by the changing polymer color. At the beginning, the penetrating solvent moves the particles from the outer regions into the core (see picture 3). If the solvent permeates approx. 50 % of the rod, spherical domains with high nanoparticle concentrations are formed. In the following, these domains shrink to stripes until the rod is fully swollen (see picture 4). The intensity of these nanoparticle stripes gets lost within 10 - 30 min. and is caused by the normal diffusive movement of particles and the resulting washout effect. In subsequent shrinking and swelling steps the moveable particles were washed out completely and a weak reddish polymer was obtained. Remaining color indicates that a small amount of nanoparticles is entrapped in the polymer. Later, cracks were observed inside the polymer at the positions of the former nanoparticle stripes.

A concentration of aggregated nanoparticles was observed in the pole regions of the rods in some cases. Even, if agglomerated particles were used. The particles move considerably in the polymer matrix. In some examples a lowered concentration of particles was found in the outer region of the rod after washing. In other cases, the mechanical stress during drying of the rods cause crack regions. These are decorated by nanoparticle aggregates under further solvent treatment. Different types of nanoparticle distributions were observed inside the polymer rods as summarized in scheme 4.

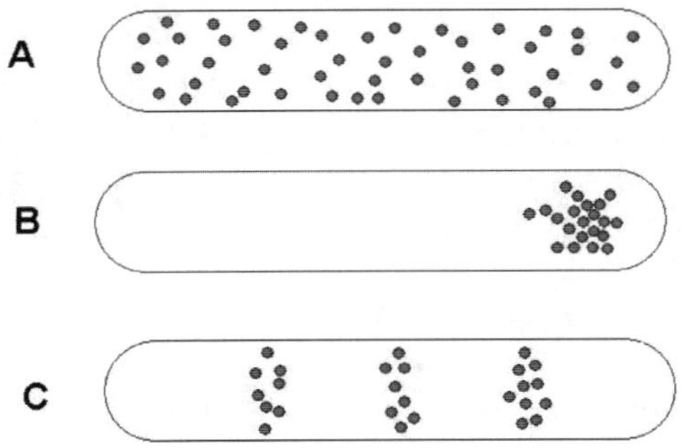

A: homogenous distribution of metal particles after polymerization
B: accumulation of aggregates of nanoparticles in one or two pole regions
C: strip-like structures

Scheme 4: Example for the observed nanoparticle distribution inside the rods

The nanoparticle-doped polymer rods were characterized by optical spectroscopy after washing. Therefore, a white light emitting diode (LED) was used for the spectral excitation of to the polymer rod. Transmitting light was analyzed by means of optical fibre and a compact spectrometer.[22] In result, the specific absorbance spectrum was taken from the transmitted light (see figure 1). The spectrum shows the expected absorption band around 525 nm, which is typical for the plasmon absorption of gold.[23] The shoulder at about 630 nm is probably related to the appearance of nanoparticle aggregates.

The introduction of dyes into the polymer matrix represents a second type of modification of the synthesized polymer rods. The idea was to couple the dyes covalently onto the polymer molecules in order to avoid a diffusive loss of dye molecules during the rinsing procedures. Dye derived polymers are of interest for chemical sensor applications.[24-26] If the sensor dye is covalently bound to the polymer, resulting rods can be added directly into the process fluids. The observed process parameter can by monitored in situ by the color of the rod. The interaction or influence of the reporting dye with the observed process should by minimal in that way.

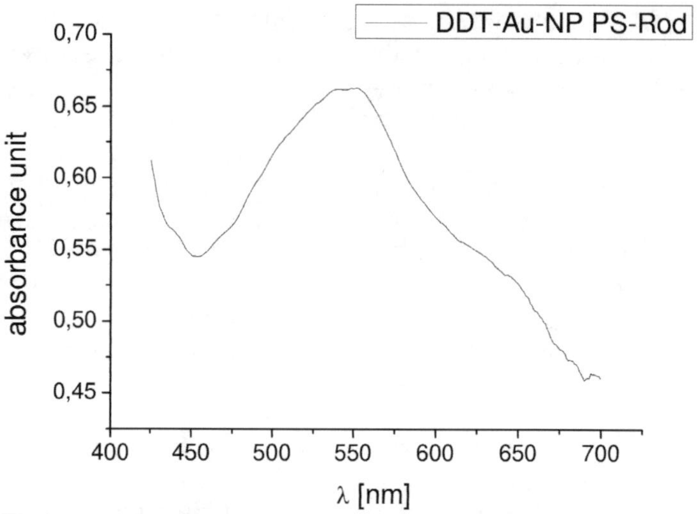

Figure 1: Qualitative absorbance spectrum of polymer-entrapped Au-nanoparticles

A very simple way for the covalent incorporation of dyes inside a polymer matrix consists in the generation of azo-dye chromophores by direct coupling of aromatic groups inside the polymer with diazoniums salts. The diazonium compound DAS-4 (p-methoxyphenyl-p-diazonium-phenylamin-hydrogensulfate) was chosen for coupling. The diazonium cations diffuse into the polymer matrix, if the polymer rods are swollen. Therefore, dimethylformamide (DMF) was applied. A reaction temperature of 80 degrees Celsius was used for the coupling with the aromatic cores of the polystyrene matrix, which are only slightly activated by the weak -I-effect of the aliphatic polymer backbone. The reaction scheme for the in-situ polymerization is given in scheme 3. After careful washing, derived polymer rods possess a homogeneous intensive dark-brown color (spectrum of a single rod is given in figure 2). The success of the coupling and the homogenous chromophore density inside the polymer matrix become evident by the visual comparison of untreated and azo-coupled polymer rods (see picture 5). The chromophore intensity of the rods was easily controlled by varying the concentration of dye groups by different diazoniumsalt concentrations and appropriate reaction times.

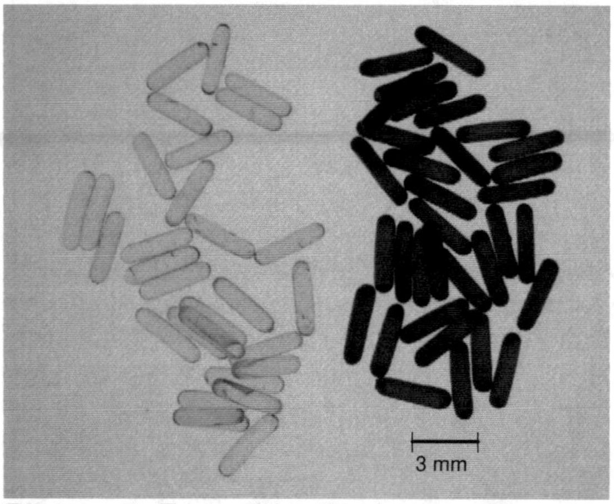

Picture 5: Ensemble of polymer rods before (left) an after (right) azo-coupling.

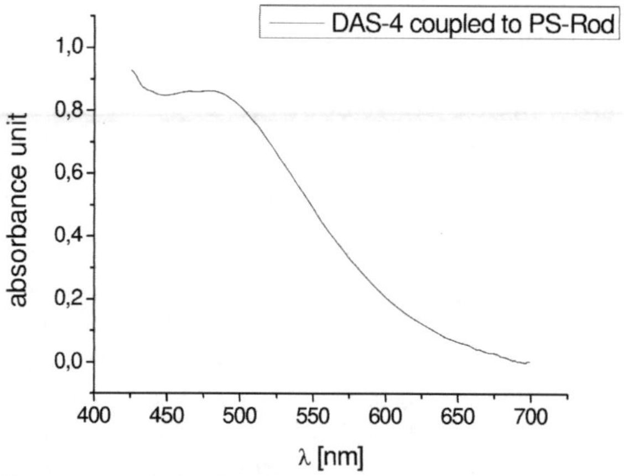

Figure 2: Absorbance spectrum of a polymere rod after coupling of the diazonium ion (DAS-4).

Summarizing conclusions

Polymer rods were prepared in a micro flow-through system by the application of the segmented-flow technique. Therefore, micro fluid segments of the monomer-containing reaction mixture were generated in a carrier stream of perfluorinated hydrocarbons of low vapor pressure. The length and the diameter of the polymer rods were defined by the size of generated fluid segments and depends on the segment forming conditions. The fluidic determined shape of the segments can be "molded" in solid polymer rods.

The polymer rods were modified by the introduction of nanoparticles before polymerization and the synthesis of chromophoric groups after polymerization. The incorporation of surface-modified gold nanoparticles was succeeded by dispersing them in the monomer solution before segmentation. The dye was formed after polymerization in situ by azo-coupling. The obtained polymer rods were characterized by light microscopy and micro spectrophotometry.

Acknowledgement

Finical support from the Deutsche Bundesstiftung Umwelt (DBU) is gratefully acknowledged.

References

[1] C. W. Huck, R. Bakry, G. K. Bonn, *Engineering in Life Sciences* **2005**, *5*, 431.
[2] S. Brocchini, *Advanced Drug Delivery Reviews* **2001**, *53*, 123.
[3] R. Reents, D. A. Jeyaraj, H. Waldmann, *Advanced Synthesis & Catalysis* **2001**, *343*, 501.
[4] H. Fenniri, *Current Medicinal Chemistry* **1996**, *3*, 343.
[5] E. Bayer, *Angewandte Chemie-International Edition in English* **1991**, *30*, 113.
[6] A. Ganesan, *Mini-Reviews in Medicinal Chemistry* **2006**, *6*, 3.
[7] M. Delgado, K. D. Janda, *Current Organic Chemistry* **2002**, *6*, 1031.
[8] S. V. Ley, I. R. Baxendale, G. Brusotti, M. Caldarelli, A. Massi, M. Nesi, *Farmaco* **2002**, *57*, 321.
[9] A. J. Mendonca, X. Y. Xiao, *Medicinal Research Reviews* **1999**, *19*, 451.
[10] D. H. Drewry, D. M. Coe, S. Poon, *Medicinal Research Reviews* **1999**, *19*, 97.
[11] A. Furka, F. Sebestyen, M. Asgedom, G. Dibo, *International Journal of Peptide and Protein Research* **1991**, *37*, 487.
[12] G. Mayer, A. Schober, J. M. Kohler, *Reviews in Molecular Biotechnology* **2001**, *82*, 137.
[13] G. A. Groß, G. Mayer, J. Albert, D. Riester, J. Osterodt, H. Wurziger, A. Schober, *Angewandte Chemie-International Edition* **2006**, *accepted.*
[14] P. M. Gunther, F. Moller, T. Henkel, J. M. Kohler, G. A. Gross, *Chemical Engineering & Technology* **2005**, *28*, 520.
[15] K. S. Lam, A. L. Lehman, A. M. Song, N. Doan, A. M. Enstrom, J. Maxwell, R. W. Liu, in *Combinatorial Chemistry, Pt B, Vol. 369*, Academic Press Inc, San Diego, **2003**, pp. 298.
[16] CeDOSYS, *Cetoni-GmbH; Korbusen (Germany), www.cetoni.de.*
[17] PP11, *F2 Chemicals Ltd., Lancashire (United Kingdom).*
[18] D. C. Sherrington, *Journal of Polymer Science Part a-Polymer Chemistry* **2001**, *39*, 2364.
[19] M. Brust, M. Walker, D. Bethell, D. J. Schiffrin, R. Whyman, *Journal of the Chemical Society-Chemical Communications* **1994**, 801.
[20] J. R. Burns, C. Ramshaw, *Lab on a Chip* **2001**, *1*, 10.
[21] Y. Joseph, I. Besnard, M. Rosenberger, B. Guse, H. G. Nothofer, J. M. Wessels, U. Wild, A. Knop-Gericke, D. S. Su, R. Schlogl, A. Yasuda, T. Vossmeyer, *Journal of Physical Chemistry B* **2003**, *107*, 7406.
[22] USB2000, *OceanOptics Inc., Duiven (Netherlands).*
[23] U. Kreibig, L. Genzel, *Surface Science* **1985**, *156*, 678.
[24] M. E. Diaz-Garcia, G. Pina-Luis, I. A. Rivero, *Trac-Trends in Analytical Chemistry* **2006**, *25*, 112.
[25] G. J. Mohr, *Chemistry-a European Journal* **2004**, *10*, 1082.
[26] G. J. Mohr, F. Lehmann, U. W. Grummt, U. E. Spichiger Keller, *Analytica Chimica Acta* **1997**, *344*, 215.

Colloidal Low-Temperature Synthesis of Zinc Oxide Nanoparticles

Eun Kyung Na[1], *Mingsong Wang*[1], *Eui Jung Kim*[1] and *Sung Hong Hahn*[2], (1)*Department of Chemical Engineering, University of Ulsan, Ulsan 680-749, South Korea*, (2)*Department of Physics, University of Ulsan, Ulsan 680-749, South Korea*

ABSTRACT

We have synthesized ZnO nanoparticles from anhydrous $ZnCl_2$ and urea at low temperatures by a colloidal chemistry method. The ZnO nanoparticles synthesized had a hexagonal structure and were uniform in both shape and size. TG results showed that there is no significant weight loss above 500°C. According to PL results, the synthesized ZnO particles may be used as an efficient blue-green luminescent material.

INTRODUCTION

Zinc oxide (ZnO) has found its important applications in optoelectronic devices [1-3], catalysts [4,5], sensors [6,7], and varistors [8-10]. Because of its wide band gap of 3.37 eV, ZnO has gained attraction as a potential material for low-voltage luminescence and short wavelength applications [1]. In addition, ZnO has efficient exciton emission at room temperature due to its high exciton binding energy of 60 meV [11]. ZnO powders have been prepared by several methods including sol-gel [12-14], gas-phase condensation [15], and liquid-phase chemical condensation [16-18]. There are only a few cases where pure hexagonal ZnO nanoparticles were obtained. In this work, ZnO nanoparticles were synthesized from anhydrous $ZnCl_2$ and urea at low temperatures by a simple colloidal chemistry method [19].

EXPERIMENTAL

The starting material used in the present study was anhydrous $ZnCl_2$. Ethylene glycol (EG) was used as the reaction medium for synthesizing ZnO nanoparticles. Tetramethylammonium hydroxide (TMAH) was used to adjust the pH of the reaction solution. Water used in the experiments was doubly distilled and deionized. Typically, 7.0 mmol of anhydrous $ZnCl_2$ was dissolved into 50 mL of EG (solution 1); 7.0 mmol of urea was separately dissolved into another 50 mL of EG (solution 2). Solution 2 is then injected into solution 1. 14.0 mmol of TMAH was dissolved into the mixed solution under vigorous stirring. The reaction solution was heated to 150°C and higher, resulting in a milky white solution. The ZnO particles formed were separated in a Hanil HA-500 centrifuge at 10,000 rpm for 10 min and were then washed with water, ethanol, and acetone consecutively to remove organics from the particles. The particles were then dried in vacuum at room temperature for 6 h.

A thermal behavior of the ZnO particles synthesized was examined by thermogravimetry (TG) using a Rhometric STA1500 TG thermal analyzer. Powder X-ray diffraction (XRD) was used for crystal phase identification. XRD patterns were obtained with a Philips PW3710 diffractometer using Cu$K\alpha$ radiation at 30 kV and 20 mA. The particle size and morphology were analyzed by scanning electron microscopy (SEM). The chemical structure of the particles was examined using an ATI Mattson Genesis Series Fourier transform infrared (FTIR) spectrophotometer. Photoluminescence (PL) measurements were performed using a SPEX 750M PL spectrometer. Samples were excited by employing a 325 nm He–Cd laser with an output of 4 mW power at room temperature.

RESULTS AND DISCUSSION

TMAH is a strong organic alkali, which can be used to adjust pH values. The advantage of TMAH compared to a strong inorganic alkali is that it does not contaminate metal oxides with alkali metal cations, which would affect the ohmic conductivity of the oxide material [20]. No ZnO particles were formed without TMAH in this work.

Fig. 1 illustrates the XRD patterns of the particles synthesized at different temperatures from 150°C to 194°C. The XRD data of the particles synthesized indicate that the hexagonal ZnO phase is formed at temperature as low as 150°C. The peak intensities are increased with increasing temperature, indicating an improvement in crystallinity. They are also increased with increasing time at a given temperature of 150°C. Of course, the existence of cubic ZnO phase from XRD patterns alone cannot be excluded because of the large similarity in the structures between cubic and hexagonal ZnO.

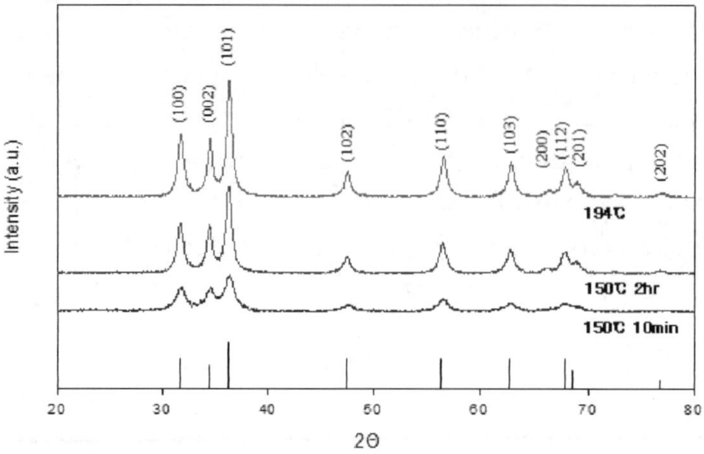

Fig. 1. XRD spectra for ZnO particles prepared at various temperatures.

SEM micrograph of ZnO nanoparticles prepared at 150°C is shown in Fig. 2. The as-prepared powders are composed of the secondary particles of 60-80 nm in size. They are uniform in both shape and size. The crystalline primary particles are observed in the agglomerated secondary particles with a spherical shape. The size of the primary particles is about 20 nm.

Fig. 2. SEM micrograph of ZnO particles prepared at 150°C.

Fig. 3 shows TG data for the ZnO particles prepared at 150°C. For thermal analysis the sample was heated in nitrogen at a heating rate of 50°C/min. The volumetric flow rate of nitrogen was 40 mL/min. The TG curve indicate that the major weight loss occurs below 500°C and no significant change in weight occurs with further increase in temperature above 500°C.

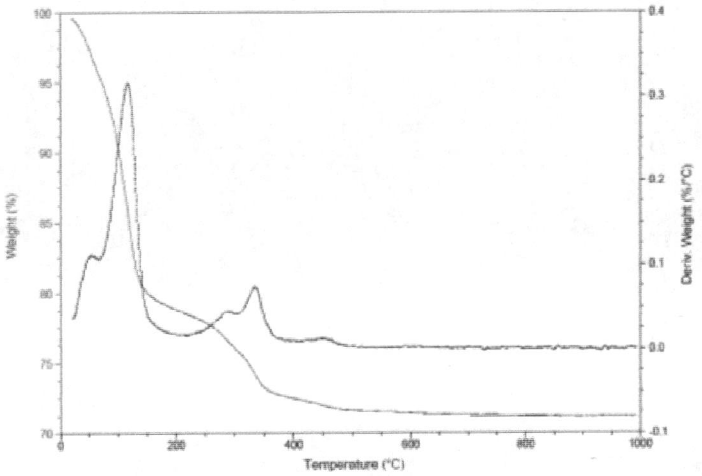

Fig. 3. TG for ZnO particles prepared at 150°C.

Fig. 4 shows FTIR spectra for the ZnO particles prepared at 150°C. In the figure the absorption peaks appearing near 1620 and 3400 cm^{-1} is relevant to an O-H group. The peaks appearing at 2900 cm^{-1} and 900-1300 cm^{-1} are due to organic bonds. Zn-O-Zn bonds appeared in the range of 400-600 cm^{-1}.

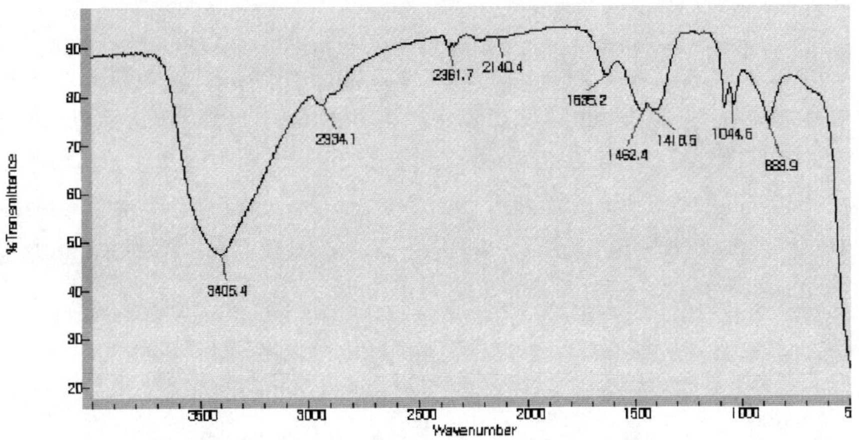

Fig. 4. FTIR spectra for ZnO particles prepared at 150°C.

Fig. 5 shows the PL spectra of the as-prepared ZnO particles measured at room temperature. One broad band of luminescent peak appears in the spectra. The band consists of blue, green, and yellow luminescence ranging from 450 to 600 nm. Intrinsic defects including O_i^- and V_O^+ produced may serve as deep acceptor level and deep donor level, respectively, which are responsible for the emission in the visible range.

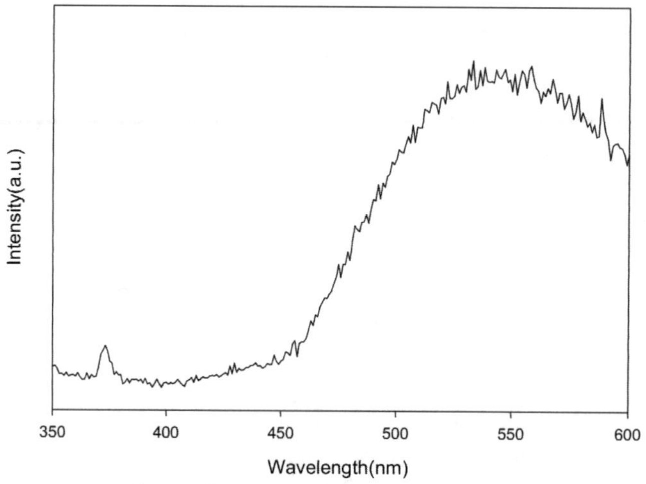

Fig. 5. Photoluminescence spectra of ZnO particles prepared at 150°C.

CONCLUSION

ZnO nanoparticles have been prepared from anhydrous $ZnCl_2$ and urea at low temperatures by a colloidal chemistry method. No ZnO particles were formed without TMAH. The ZnO hexagonal phase was formed at temperature as low as 150°C. The ZnO particles prepared were uniform in both shape and size. The ZnO particles synthesized may be used as an efficient blue-green luminescent material.

REFERENCE

[1] E.M. Wong, P.C. Searson, Appl. Phys. Lett. 74 (1999) 2939.

[2] J.C. Johnson, H. Yan, R.D. Schaller, L.H. Haber, R.J. Saykally, P. Yang, J. Phys. Chem. B 105 (2001) 11387.

[3] K. Govender, D.S. Boyle, P. O'Brien, D. Binks, D. West, D. Coleman, AdV. Mater. 14 (2002) 1221.
[4] P.V. Kamat, R. Huehn, R. Nicolaescu, J. Phys. Chem. B 106 (2002) 788.
[5] G. Ramakrishna, H.N. Ghosh, Langmuir 19 (2003) 3006.
[6] L.F. Dong, Z.L. Cui, Z.K. Zhang, Nanostruct. Mater. 8 (1997) 815.
[7] S. Roy, S. Basu, Bull. Mater. Sci. 25 (2002) 513.
[8] Y. Suzuoki, A. Ohki, T. Mizutani, J. Phys. D: Appl. Phys. 20 (1987) 511.
[9] T.K. Gupta, J. Am. Ceram. Soc. 73 (1990) 1817.
[10] N. Horio, M. Hiramatsu, M. Nawata, K. Lmaeda, T. Torii, Vacuum 51 (1998) 719.
[11] K. Hummer, Phys. Stat. Sol. 56 (1973) 249.
[12] N.I. Kovtyukhova, E.V. Buzaneva, C.C. Waraksa, T.E. Mallouk, Mater. Sci. Eng. B 69–70 (2000) 411.
[13] R. Vacassy, S.M. Scholz, J. Dutta, C.J.G. Plummer, R. Houriet, H. Hofmann, J. Am. Ceram. Soc. 81 (1998) 2699.
[14] B. Bhattacharjee, D. Ganguli, S. Chaudhuri, A.K. Pal, Thin Solid Films 422 (2002) 98.
[15] J.C. Sanchez-Lopez, A. Fernandez, Thin Solid Films 317 (1998) 497.
[16] J.F. Suyver, S.F. Wuister, J.J. Kelly, A. Meijerink, Nano Lett. 1 (2001) 429.
[17] S. Wageh, Zhao Su Ling, Xu Xu-Rong, J. Crystal Growth 255 (2003) 332.
[18] Heesun Yang, P.H. Holloway, J. Appl. Phys. 93 (2003) 586.
[19] Y. Zhao, Y. Zhang, H. Zhu, G.C. Hadjipanayis, J.Q. Xiao, J. Am. Chem. Soc. 115 (1993) 8706.
[20] M. Ristić, S. Musić, M. Ivanda, S. Popović, J. Alloys Compd. 397 (2005) L1–L4.

71m Comparative Investigations of Dispersion Formation in Different Types of Micromixers

Patrick Löb, Volker Hessel, Stephan Hollomotz, and Yong Men. Chemical Process Engineering, Institut für Mikrotechnik Mainz GmbH, Carl-Zeiss-Straße 18-20, Mainz, D-55129, Germany

Different types of micromixers have been applied for l/l-dispersion and their potential for this purpose have been demonstrated in recent years. As advantages have been addressed e.g. a narrower droplet size distribution compared to conventional process (e.g. A. Kawai et al. [1]), numbering-up as scale-up concept while keeping product quality (e.g. A. Kawai et al. [1]), predictability of droplet size (e.g. S. Hardt et al. [2]), higher energy efficiency compared to conventional process (e.g. T. Bayer et al. [3]), high absolute energy efficiency (e.g. S. Sugiura et al. [4-5]) and production on demand (e.g. Schwesinger et al. [6]). Certainly, not all these potentials can be exploited by now for selected applications at one time especially when higher flow rates have to be considered. Beside the in literature documented experience, it is difficult at the moment to select the best suited mixer for a specific application. Mainly this is due to the fact that performance data of different mixers are often not comparable and that performance data are often incomplete. Comparability would require information on aspects like mean droplet size, width of droplet size distribution, throughput per unit, numbering-up concept, flow rate ratio necessary to obtain required droplet size, specific energy input and predictability of droplet size and mechanism of dispersion formation. Based on this background we performed comparative studies using different types of micromixers out of the range of IMMxs micromixers (Slit Interdigital Micro Mixer Version 2, Caterpillar Micro Mixer and Starlaminator Micro Mixer) representing different functional principles with heptane/water (SDS) as model system. The covered mixer types are parallel lamination micromixers, serial lamination (split-and-recombine) micromixers and injection micromixers (see the recent reviews by N.-T. Nguyen [7] and V. Hessel [8] for the classification of micromixers). Mean droplet size, coefficient of variation and energy input as function of flow rate were determined for all the mixers and compared. Furthermore, new investigations were performed to determine the influence of number of mixing steps and channel dimension in the case of the Caterpillar Micro Mixer on mean droplet size. Together with former experiments dealing with the Slit Interdigital Micro Mixer [9-11], this will allow a comparison of mixer performance on a broad basis and facilitate future equipment selection.

[1]A. Kawai, T. Futami, H. Kiriya, K. Katayama, K. Nishiawa, in Y. Baba et al. (eds.), Micro Total Analysis Systems 2002, Volume 1, 368-370, Kluwer Academic Publishers, the Netherlands, 2002. [2]S. Hardt, F. Schönfeld, F. Weise, C. Hofmann, W. Ehrfeld, in Technical Proc. 2001 Int. Conf. on Modeling and Simulation of Microsystems MSM 2001, Hilton Oceanfront Resort SC, USA, March 19-21, 2001, 223. [3]T. Bayer, H. Heinichen, T. Natelberg, in Proceedings of the 4th International Conference on Microreaction Technology, IMRET 4, 5-9 March 2000, pp. 167-173, AIChE Pub., Atlanta, USA (2000). [4]S. Sugiura, M. Nakajima, S. Iwamoto, M. Seki, Langmuir, 17 (2001) 5562-5566. [5]S. Sugiura, M. Nakajima, M. Seki, Langmuir 18 (2002) 10, 3854-3859. [6]N. Schwesinger, T. Frank, H. Wurmus, J. Micromech. Microeng. 6 (1996) 99-

102. [7]N.-T. Nguyen, Z. Wu, J. Microm. Microeng. 15 (2005) R1-R16. [8]V. Hessel, H. Löwe, F. Schönfeld, Chem. Eng. Sci. 60 (2005) 2479-2501. [9]H. Pennemann, S. Hardt, V. Hessel, P. Löb, F. Weise, Chem. Eng. Tech. 28, 4 (2005) 501-508. [10]V. Haverkamp, W. Ehrfeld, K. Gebauer, V. Hessel, H. Löwe, T. Richter, C. Wille, Fresenius J. Anal. Chem., 364 (1999) 617-624. [11]P. Löb, H. Pennemann, V. Hessel, Y. Men, Chem. Eng. Sci., Fluid Mixing 8 International Conference 2006, Special Issue,accepted for publication.

Immobilization of enzymes on microchannel surface through cross-linking polymerization

Takeshi Honda,[a] Masaya Miyazaki,*[ab] Hiroyuki Nakamura[a] and Hideaki Maeda*[ab]

[1]*National Institute of Advanced Industrial Science and Technology (AIST), Tosu, Saga 841-0052, Japan*
[2]*Department of Molecular and Materials Sciences, Interdisciplinary Graduate School of Engineering Sciences, Kyushu University, Kasuga, Fukuoka 816–8580, Japan*

Abstract

The enzyme-microreactor has a strong potential for applications such as biotechnological syntheses and analyses. At present, interesting microreactors with various enzymes have been reported. However, the developed methods of microreactor preparations require high-level techniques and/or multi-step procedure which could lead to low cost performance. Also, disposability of the reactor is considered essential in bioanalyses. Simplifying the procedure of enzyme-immobilization in microreactor has been attractive. This is achievable by concertedly utilizing the reaction for enzyme-immobilization and the characteristic of microfluidics. We have designed an enzyme-immobilized microreactor possessing a substrate membrane that covers the internal surface of the micro-channel. The cylindrical membrane is composed of cross-linked polymerized enzyme product. This carrier-free method based on enzyme-cross-linked aggregation (CLEA) realized a facile and inexpensive preparation of an enzyme-immobilized microreactor. In this study, α-chymotrypsin was used as a model enzyme for protein digestion. Recently, microreactors capable of protein digestion has been attractive in proteomics studies. By utilizing the optimized condition in CLEA formation, α-chymotrypsin-immobilized tubing microreactor was successfully prepared by cross-linking polymerization in a concentric laminar flow. The resulting microreactor showed higher stability against heat and organic solvent than those of the free enzyme. These results show that this preparation method leads to a flexible technology platform for screening and designing a potential enzyme-microreactor for broad applications.

Introduction

Microreactors are potentially powerful tools in chemistry and biotechnology fields.[1] The excellent performance of microreaction systems is achieved by rapid heat and mass transfers, and larger surface/interface area. These attractive features are favorable to catalytic reactions through catalysts immobilized on the reactor.[2] Enzyme is one of the catalysts, which is useful in substance production and has high potential for analysis applications. There have been demands for enzymatic microreaction devices in several fields, especially for high-throughput biotechnological syntheses and bioanalyses.[3] Various enzyme-immobilized microreactors have been developed and have shown high-performance. Nano-structure fabrication and direct chemical modification on the micro-channel surface are the useful technology for preparation of such microreactor.[4] We have also developed modified sol-gel techniques for surface modification of micro-channel and their application for preparation of microreactor.[5] However, these preparation methods require high-level techniques and multi-step procedure, leading to low cost performance. Microreactors for on-demand bioanalyses and bio-syntheses should be inexpensively prepared without difficulty because such reactors are generally preferable due to its disposable nature. We have designed an enzyme-immobilized microreactor possessing a substrate membrane covering the internal surface of the micro-channel. The cylindrical membrane is composed of cross-linked polymerized enzyme product. This microreactor was simply and inexpensively prepared by concertedly utilizing the reaction for membrane-formation and microfluid characteristics. We used a product polymerizing enzyme as the membrane-substrate in order to achieve the more facile preparation method. The membrane was based on a cross-linking enzyme aggregate (CLEA) formed using cross-linker with aldehyde groups which react to amino groups of the enzyme.[6]

Results and Discussion

To fabricate an inexpensive microreactor, a tubular substrate requiring no micro-fabrication was selected. Commercially available PTFE tube is being used as a standardized equipment in various analytical apparatus. In the preparation of chymotrypsin-

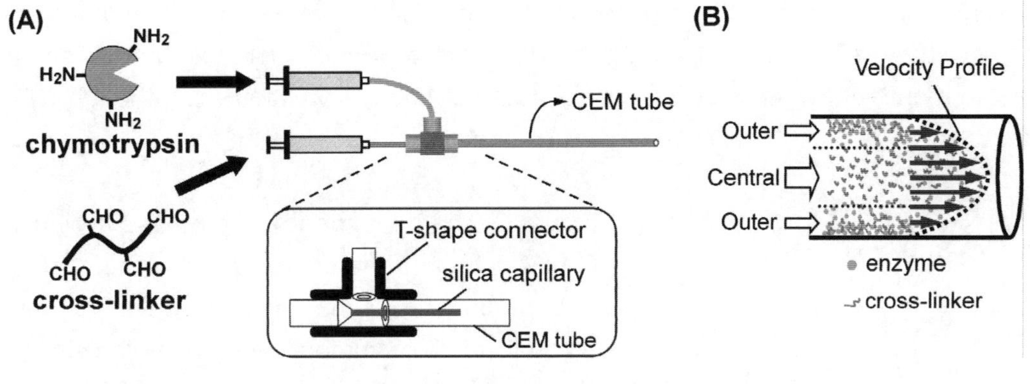

Fig. 1. (A) Illustration of preparation system of CLEA-based microreactor (CEM). (B) Possible

mechanism of polymerization process of enzyme and cross-linker reagent in a microchannel.

CEM, the following reagents were used: enzyme (20 mg ml^{-1}), and cross-linker containing glutaraldehyde (GA) and paraformaldehyde (PA) at various concentrations. The cross-linking polymerization in a PTFE tube as a CEM substrate (hereafter described as CEM-tube) was performed in a concentric laminar flow consisting of two streams of the cross-linker and enzyme. The preparation system is shown in Figure 1A. To form a stable laminar flow, the proper flow rates should be selected for each solution. We set the pumping rate of cross-linker (corresponding to a central stream) to 12.5-times higher value than that of the enzyme (outer stream). Formation of the concentric laminar flow was confirmed by confocal laser scanning microscopy (data not shown). Calculating from the pumping rates and volumes of central and outer streams in the concentric laminar flow, residence time of the central stream was comparatibly shorter (0.3 min) than that of the outer stream (19.6 min). This residence time of the enzyme solution would be sufficient reaction time for CLEA formation because CLEA formationcould be completed within 10 min in a batchwise system (data not shown). When enzyme molecules diffusing into the central stream polymerize at the central region, the polymerizing product would be incapable of obtaining enough reaction time for maturation into an insoluble aggregation due to the greater linear velocity compared to the outer flow velocity (Figure 1B). In addition, residence time of the chymotrypsin on the PTFE wall surface would be extended due to its adsorption to the hydrophobic surface.[7] These factors would lead to promotion of the enzyme-polymer growth at the vicinity of the inner-wall of the tube as compared to that of the central region of the microchannel.

It is well known that enzymes are deactivated by over-reactivity of chemical modifiers including cross-linkers. Also, such reactivity and a relative amount of the cross-linker for protein are linked to the degree of immobilization. The CEM prepared by using GA only or PA only as cross-linkers showed unsuccessful membrane-formation and/or relatively lower protease activity. Using PA (4 - 16%) also led to a similar result. The mechanical strength of the obtained membrane was relatively weak (data not shown). In contrast, although GA (0.5 - 2%) alone

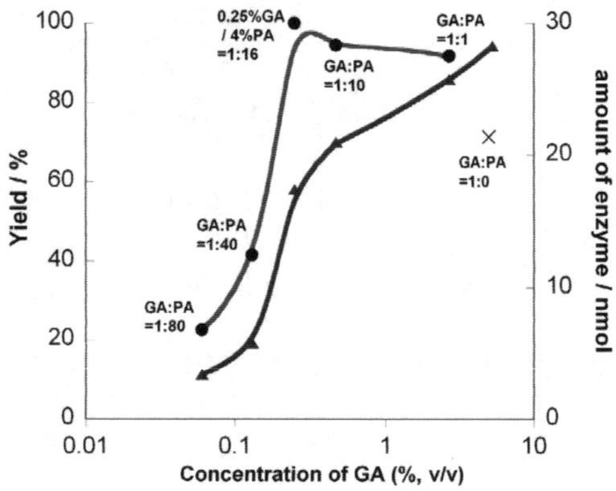

Figure 2. The dependency of protease activity on the GA/PA ratio and the concentration of total aldehyde. The graph shows the reaction yields (●) and the amount of the immobilized chymotrypsin (▲) of CEMs. Preparation was used at GA/PA with various combination ratios. The volume rate of aldehyde

reagents (GA and PA) was kept constant against total volume of reaction solution. There was no success in microreactor preparation at 1:0 of GA/PA ratio (5.25% GA) due to rapid tube obstruction (×). Horizontal axes indicate logarithmic plots of concentration of GA included in the reaction mixture. All enzymatic reactions were performed for 3 min (4 μl min^{-1} of flow rate) at 37°C.

had the strong ability of CLEA-formation and high protease activity by the strictly controlling the reaction time, it was very difficult to control enzyme-polymerization in a micro-space due to rapid aggregation of enzymes. This procedure frequently led to pronounced evenness of the membrane and tube obstruction, and resulted in a very low reproducibility in enzyme immobilization. We have tried to reinforce the mechanical strength of the CEM by addition of GA in preparation system using PA only. To determine the optimal condition for both effective immobilization and high residual activity, we examined the influence of the cross-linker, especially the GA/PA ratio and the concentration of total aldehyde, on the chymotrypsin-CEM. In Figure 2, the protease activity decreased with decrease in the amount of GA due to a decrease in the formed CLEA which lead to unsuccessful membrane formation. In contrast, higher amount of GA tended to promote evenness of the membrane and tube obstruction due to rapid aggregation. Also, the activities decreased despite the increase in the amount of the immobilized enzyme. This result could be attributed to inactivation of chymotrypsin by over-reactivity of GA. It was found that the GA/PA ratio (0.25%/4%PA) is most suitable for microreactor preparation. These results indicate that the combining characteristics of GA and PA are useful for effective formation of enzyme-membrane in a microchannel, and the ratio and concentration of GA/PA are very important for preparation of the CEM with high activity.

The CEM prepared using 0.25%GA/4%PA was dried by a vacuum pump. Figure 3 shows CCD images of the cylindrical enzyme-membrane (dry state) exposed from the PTFE tube and the sectional view of the CEM. The cylindrical membrane was shrunk by drying-out, and the thickness of the obtained membrane was 30 – 65 μm. The amount of protein content in the

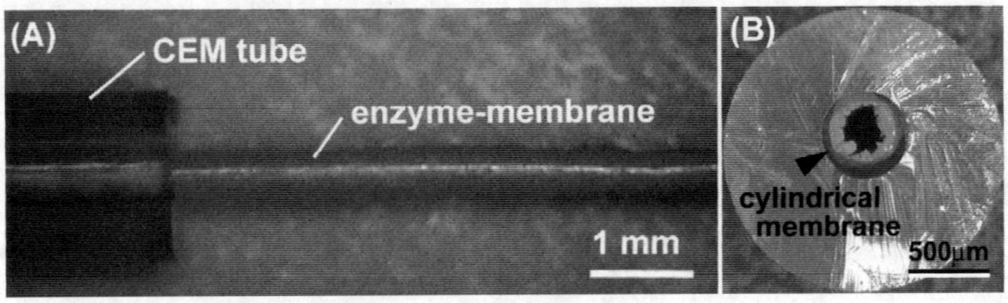

Figure 3. CCD images of (A) cylindrical enzyme-membrane (dry state) exposed from PTFE tube, which forms on the inner wall of the tube and (B) the sectional view of the obtained CEM.

membrane was 400 μg (16 nmol) in a microreactor. When a suspension of CLEA prepared in batchwise system was passed through a PTFE tube, no polymer product of membrane state was observed on the inner wall (data not shown). This result indicates that polymerizing in a microfluid enables the enzyme-polymer product to form the membrane on the inner wall of the CEM tube.

Next, we examined the reaction efficiency using the obtained CEM. Hydrolysis of 1 mM GPNA was used for evaluation. In the CEM, the hydrolysis reaction was completed at a flow rate of 4 μl min^{-1}, which yields a residence time of 3 min. Furthermore, the enzymatic reaction performed at various concentration of GPNA, and the kinetic constants K_m (0.5 mM) and V_{max} (0.8 mM min^{-1}) were calculated from Lineweaver-Burke plot (Fig. 4A). There was no significant difference on the hydrolysis efficiency compared to that of solution-phase batchwise reaction at the same molar ratio of enzyme (3 mg ml^{-1})/substrate, which took 170 sec to complete the reaction. On the other hand, CEM was found to have advantages of higher stability compared to the free enzyme. The present CEM was durable for at least 5 days while the hydrolytic activity of the free chymotrypsin solution became progressively low (Fig. 4B). In addition, this microreactor could maintain the hydrolysis yield at 70 % and above in a continuous flow (4 μl min^{-1}) of substrate solution for 5 days. A collapse of the conformation and self-digestion are well-known to be responsible to inactivation of chymotrypsin. Therefore, the conformation rigidified by immobilization through cross-linking would allow chymotrypsin to

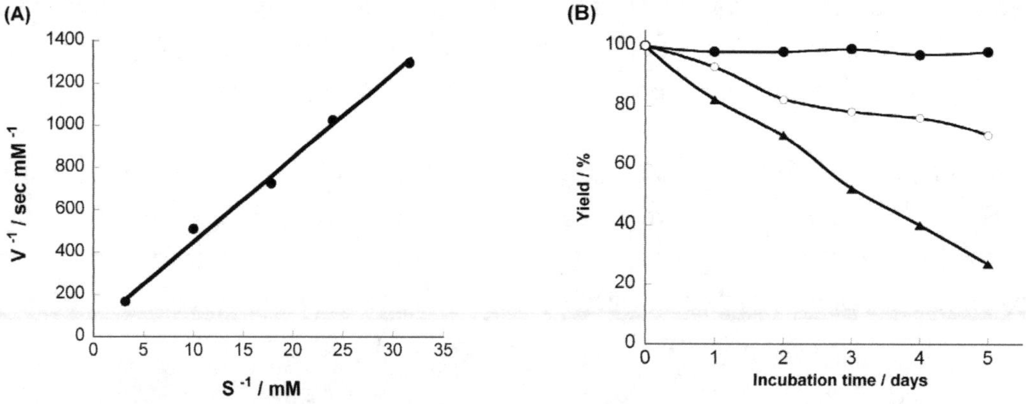

Figure 4. (A) A Lineweaver-Burke plot was built from the kinetic data of the obtained CEM. (B) Durability of enzyme catalysts. The graph shows comparison of reaction yield of microreactor (●) and the solution-phase reaction (▲). The CEM or free enzyme solutions were incubated for various times at 37 °C, and then these protease activities against 1 mM GPNA were measured. In the CEM, the relative activity was measured when continuously flowing 1 mM GPNA at 4 μl min^{-1} (○).

avoid such collapse and self-digestion.[8] High stability of the enzymatic activity for protein digestion should be an important feature of the enzyme-reactors used in proteomics study because digestion is often carried out under denature conditions to achieve an efficient and rapid digestion of protein substrates.[9] Also, the CEM showed resistance to heat and chemical denaturant like urea and organic solvents (dimethyl sulfoxide:DMSO and dimethyl formamide:DMF) as compared to free chymotrypsin (Fig. 5). This microractor demonstrated high stability and utility for hydrolysis under denaturating conditions. Our data were consistent with previous reports using GA cross-linking method which provide enzymes with high stability under denaturating conditions.[10] This effect was also considered to result from the rigid conformation and avoidance of autolysis of the immobilized enzyme.[8] An application of enzyme-immobilizing microreactor for protein digestion was examined using known protein substrates. This reaction was performed at 70 μg ml^{-1} myoglobin solution at 50 °C affording a residence time of 10 min to give digestion yields of 80%. In comparison, the digestion of the same proteins by free protease according to the conventional procedure[11] (enzyme-substrate protein mass ratio 1:50) took several hours to reach the same digestion yield (data not shown). These results showed the high potential of this microreactor for enzymatic digestions.

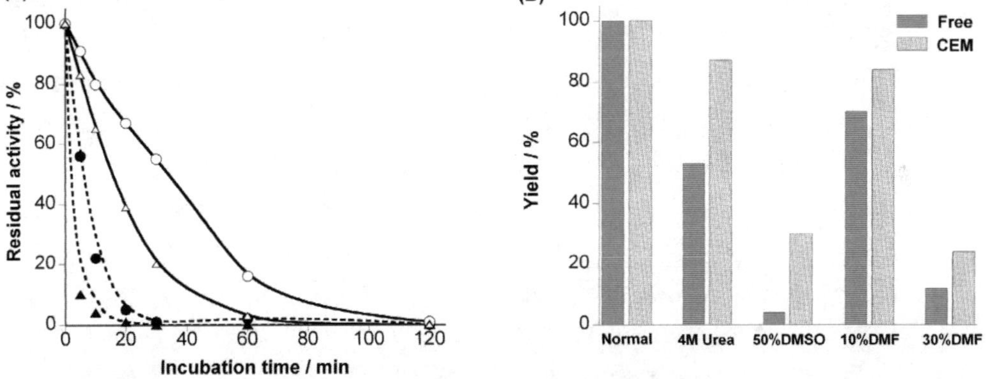

Figure 5. (A) Temperature dependence of the enzymatic activity of the CEM (solid lines) and free acylase (dashed lines) at 60°C (circles) and 80°C (triangles). The activities without pre-incubation were assigned values of 100%. (B) Effect of urea and organic solvents on the enzymatic activity of the CEM and the free enzyme. The activities without organic solvent were assigned values of 100%.

In summary, we have developed an enzyme-immobilizing microreactor by cross-linking polymerization. This method may also be of interest in rapid development and screening of various micro-analysis system involving the immobilized biomolecules such as enzyme, antibody, glycoprotein, and other functional proteins.

Experimental Section

Materials: Commercially available PTFE tubes (500 μm i. d., 1.59 mm o. d.) and heat-shrink tubing were obtained from Flon Chemical Inc. (Osaka, Japan). GL Science Co., Ltd. (Tokyo, Japan) provided a silica-fused microcapillary (100 μm i. d., 350 μm o. d.) and a stainless-steel T-shaped connector (Union Tee, SUS-316). GA, PA, α-chymotrypsin, and other reagents for amino acid analysis were obtained from Wako Pure Chemical Ind., Ltd (Osaka, Japan). Sodium cyanoborohydride (NaCNBH3), GPNA, and other reagents for buffer were obtained from Sigma–Aldrich Co. (St. Louis, MO, USA).

α-chymotrypsin-CEM preparation using a laminar flow system: The CEM preparation system was assembled as shown in Figure 1. A fused silica microcapillary (5 cm length) and a PTFE tube (5 cm length) for sending reagents and a PTFE tube (6 cm length) corresponding to CEM tube were attached to a T-shaped connector as shown in Figure 1. In order to form a concentric laminar flow in the CEM preparation, the silica capillary was set in the T-shape connector located at the concentric position of the CEM tube. All solutions were filtered and charged into 1 ml plastic syringes (Terumo Co., Tokyo, Japan). The cross-linker and enzyme solutions were supplied in the CEM tube through the capillary and another inlet of the T-shape connector using a PTFE tube, respectively. Transport of solution was performed at different pumping rates (6.25 μl min-1 for cross-linker and 0.5 μl min-1 for enzyme) by Harvard 11 Pico Plus Syringe Pump (Harvard Apparatus, Inc., Holliston, MA, USA). After polymerization was performed for 3 hours, the obtained CEM was rinsed with Tris-HCl (1 M, pH 9.0), which simultaneously quenched active aldehyde groups remaining on the formed membrane. In order to reduce the resulting Schiff base, the CEM was treated with $NaCNBH_3$ in borate buffer (50 mM, pH 9.0). Then, the CEM was well washed by PBS (pH 7.5). All preparations were performed at 4°C. An enzyme-polymerizing membrane of the obtained CEM was observed using a color CCD digital microscope system (HI-SCOPE DH-2700; Hirox Co., Ltd, Tokyo, Japan).

Estimation of the amount of enzyme: In order to estimate the amount of the immobilized enzyme in the CLEA, quantitative amino acid analysis was carried out. The obtained membrane was hydrolyzed by HCl (6 M) at 110°C for 24 h. The hydrolyzed product was reacted to phenyl isothiocyanate (PITC) and the PITC-labeled amino acids were analyzed by HPLC system using a Wakosil-PTC (4.0 mm x 200 mm, Wako Pure Chemical Ind., Ltd). The protein content was quantified by amino acid analysis using the standard PITC procedure. The amount of the immobilized α-chymotrypsin was calculated by comparison with the amino acid composition of the α-chymotrypsin.

Enzymatic hydrolysis of GPNA: Chymotrypsin activity was estimated using the hydrolytic assay of GPNA. The substrate was dissolved in 50 mM phosphate buffered saline (PBS, pH 7.5). In a batchwise system using free enzyme, an equal amount of substrate solution was mixed with α-chymotrypsin dissolved in PBS by vigorous stirring at 3,000 rpm in a micro-test tube. In the CEM system, substrate solution was charged into a 1 ml syringe, and supplied to the tubing CEM using a syringe pump. Standard enzymatic reactions were performed at 37°C in a thermostated incubator. The reaction product was analyzed by HPLC system using a Wakosil C18AR column (3.0 x 250 mm, Wako Pure Chemical Ind., Ltd). The reaction yields were calculated from the peak area calibrated with the standard compound. The K_m and V_{max} were computed by fitting the data to the Michaelis-Menten equation using a public domain program, GNUPLOT.

References

[1] V. Hessel, S. Hardt, H. Lowe, Chemical Micro Process Engineering, Wiley-VCH, Weinheim, **2004.**

[2] a) K. Kusakabe, D. Miyagawa, Y. Gu, H. Maeda, S. Morooka, J. Chem. Eng. Jpn. **2001**, 34, 441-443; b) C. de Bellefon, N. Pestre, T. Lamouille, P. Grenouillet, V. Hessel, Adv. Synth. Catal. **2003**, 345, 190-193; c) X. Li, H. Wang, K. Inoue, M. Uehara, H. Nakamura, M. Miyazaki, E. Abe, H. Maeda, Chem. Commun. **2003**, 964-965.

[3] N. Lion, F. Reymond, H. H. Giraut, J. S. Rossier, Curr. Opin. Biotechnol. **2004**, 15, 31-37.

[4] a) J. Drott, E. Rosengren, K. Lindstrom, T. Laurell, Mikrochim. Acta **1999**, 131, 115-120; b) M. Bengtsson, S. Ekstrom, G. Marko-Varga, T. Laurell, Talanta **2002**, 56, 341-353; c) H. Qu, H. Wang, Y. Huang, W. Zhong, H. Lu, J. Kong, P. Yang, B. Liu, Anal. Chem. **2004**, 76, 6426-6433.

[5] a) M. Miyazaki, J. Kaneno, M. Uehara, H. Fujii, M., Shimizu, H. Maeda, Chem. Commun. **2003**, 648-649.

[6] L. Cao, L. Langen, R. A. Sheldon, Curr. Opin. Biotechnol. **2003**, 14, 387-394.

[7] a) M. Morra, C. Cassinelli, J. Biomed. Mater. Res. **1995**, 29, 39; (b) S. Noinville, M. Revault, and M.-H. Baron, Biopolymers, **2002**, 67, 323.

[8] M. N. Gupta, B. Mattiasson, In Methods of Biochemical Analysis: Bioanalytical Applications of Enzymes, vol. 15. John Wiley: New Jersey; 1-34.

[9] A. K. Palm, M. V. Novotny, Rapid Commun. Mass Spectrom., **2004**, 18, 1374.

[10] N. L. S. Clair, M. A. Navia, J. Am. Chem. Soc., **1992**, 114, 7314.

[11] D. S. Peterson, T. Rohr, F. Svec, J. M. J. Frechet, J. Proteome Res., **2002**, 1, 563.

71q Effect of Size Constraints on Integration Potential in Logistical Fuels Processing

Rose M. Cummings[1], Kristin H. McGlocklin[1], **Mario R. Eden**[1], and Bruce J. Tatarchuk[2]. (1) Department of Chemical Engineering, Auburn University, 230 Ross Hall, Auburn University, AL 36849-5127, (2) Center for Microfibrous Materials Manufacturing, Auburn University, 230 Ross Hall, Auburn University, AL 36849-5127
The theoretical energy efficiency of a fuel cell system is approximately three times higher than a combustion engine based generator, thus it would provide substantial savings if alternative means of power production could be developed. Recent efforts have been focused on reforming existing logistical fuels, e.g. diesel or JP-8 for use in fuel cell systems. This is particularly important for military applications, as it would allow for the US armed forces to move towards using one single logistical fuel. To meet these ends the Center for Microfibrous Materials Manufacturing (CM3) at Auburn University has developed a bench scale test bed for investigating running a portable radar system of a PEM fuel cell stack by producing high purity hydrogen from steam reforming (SR) of JP-8. In principle, a PEM fuel cell system consists of the fuel processing section and the fuel cell itself, with the former being the reformer and post-combustion cleanup steps. Such systems inherently possess tremendous integration potential, not just limited to recycling unused material, but also in terms of energy recovery. Process integration techniques can be employed to realize this potential by providing global process insights and identifying overall process performance targets. Current methods for resource management such as thermal and mass pinch analyses are aimed at processing facilities, i.e. stationary plants, where the overall goal is to balance reductions in operating cost against increased capital investments to maximize profitability. For a certain class of problems however, conventional pinch analyses fail to adequately address the resource management problems. For compact and/or mobile applications the deciding factor is not simply the resource utilization level or cost of equipment, but is often a trade-off between the achievable resource utilization and the weight and/or volume of the equipment. In this contribution we study thermal management and resource conservation strategies in logistical fuels processing for mobile applications. Based on performance data obtained from the test bed, models have been developed for three reformation strategies, i.e. steam reforming (SR), partial oxidation (POX) and auto-thermal reforming (ATR). A comprehensive efficiency analysis of hydrogen production from logistical fuels also requires an evaluation of effects of changing the fuel itself. Therefore the fuel processing system for each reformation strategy has been modeled for a variety of fuels. This contribution will illustrate the results of a process integration analysis of different reforming strategies of various fuels. To further increase the integration potential, the use of heat pipe technology has been investigated. A simple, systematic method for targeting of optimum heat pipe usage in heat exchange networks has been developed to augment existing thermal pinch analysis methods.

71b Effective Vaccine Delivery System Using Biodegradable Nanoparticles

Ik Joong Kang and Bum Gil Kim. Department of Chemical & Bio Engineering, Kyungwon University, Sungnam, South Korea

All of the infants have to be prescribed by three consecutive injection of hepatitis vaccine just after his birth. Adults are also prescribed by an intra muscular injection every five years. Recently, it is very mortal that people over 8 % get infected with chronic hepatitis as increasing trend and consequently leaded to the death because of a complication arises of cirrhosis and liver cancer. Hepatitis virus(Hv) is the liver affinity DNA virus as a kind of hepadna virus has 42 nm diameter and no cell denaturalization phenomenon. Outer membrane of the virus consist of hepatitis surface antigen(HBsAg) and center is composed of HV DNA and hepatitis B core antigen(HBcAg). There are four genes of S, C, X, P in HV DNA. S gene has the pre-S region composed of pre-S1 and pre-S2 as an encoding gene of surface antigen. C gene is in the precore region for the encoding of core protein(HBcAg). In this study biodegradable polymer's derivatives was synthesised, made to nano particles as a supporter of vaccine, and analyzed by FT-IR, HPLC, atomic analyzer , and SEM for the development of effective drug delivery system. The character of vaccine delivery body as its size and surface polarity was analyzed by AFM, Zeta potential, and ELS.

Flow Regimes in a Catalyst Trap Microreactor

H. Gadre, S. McGovern, and R.S. Besser
Department of Chemical, Biomedical, and Materials Engineering
Stevens Institute of Technology
Hoboken, NJ 07030
USA.
Contact: hgadre@stevens.edu, 201-216-5257

In this paper, we propose the use of a novel catalyst trap microreactor to address the issues of multiphase mixing in microreactors. Multiphase mixing plays a significant role in multiphase reactions, which are typically carried out on a large scale. An extensive review of the multiphase reactions carried out in the pharmaceutical and fine chemicals industries is provided by Mills and Chaudari (1). Multiphase mixing has not been investigated extensively in microreactors (2). Much of the work in this area has been limited to a single micro channel or a capillary reactor (3-5). Observing the flow patterns and subsequently achieving efficient flow behavior can address the challenges associated with the multiphase mixing in microreactors. The realization of efficient multiphase reactions in microreactors would significantly benefit chemical process miniaturization in these industries.

Two-phase gas-liquid flow patterns in microchannels or capillary reactors have been explored in the past (6-9). For our convenience, we consider four different flow regimes, as visualized in Figure 1, namely (a, b) bubble flow, (c, d) Taylor or slug flow, (f) churn flow and (h) annular or trickle flow.

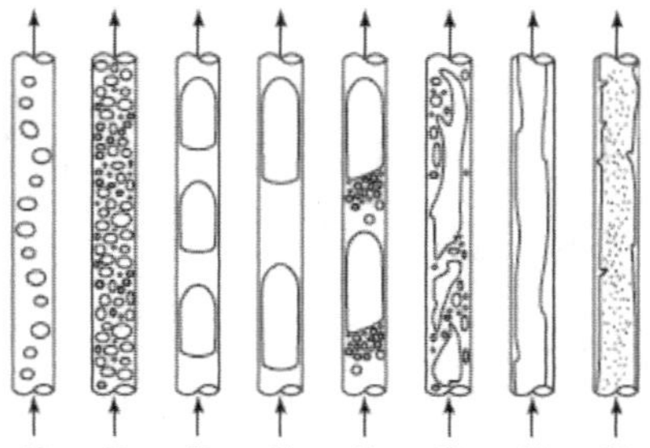

Figure 1: Flow Regimes: a, b: bubbly flow; c, d: segmented (Taylor) flow; e: transitional churn flow; f: churn flow g: film flow; h: annular flow

It is well known that in conventional reactors the slug flow regime typically gives the best gas-liquid mass transport in two-phase flow by constantly refreshing the gas-liquid interfacial area and because of the turbulence generated by the liquid slug. However, in the multiphase scenario in microreactors, it might very well be disadvantageous because much of the catalyst would be in contact with gas instead of liquid. Bubble flow, in contrast, has essentially all of the catalyst in contact with the liquid and there is higher gas-liquid interfacial area, although less turbulence to drive gas diffusion. So, if we could eliminate gas-liquid diffusion as a limitation, bubble flow may be the best option because of its high liquid to solid interfacial area. Hence, creating such a flow pattern in the microreactor could address the problem of multiphase mixing in microreactors. In order to study this possibility, three different novel catalyst trap microreactors were designed, fabricated, and characterized.

The reactor was designed keeping in mind the flow configuration to be achieved and also to maintain low transport resistance needed to enhance multiphase mixing. The dimensions of the reactor were based on the intrinsic kinetic model available in the literature. The reactor incorporates a cross flow configuration in which the gas diffuses into the liquid channel through a slotted wall. The gas diffuses at a perpendicular angle to the flow of the liquid solvent and hence the cross flow configuration. The liquid channel itself has an array of posts or catalyst traps, wherein the catalyst is trapped. Figure 2 portrays the concept of the catalyst trap microreactor design.

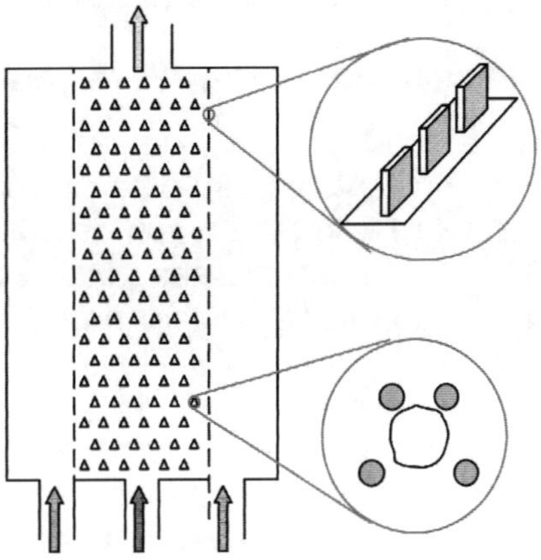

Figure 2: Reactor Design

The reaction of liquid-phase hydrogenation of o-nitroanisole to o-anisidine in the presence of methanol as an inert solvent and a Pt catalyst supported on carbon was used for the experiments. Since the flow patterns are not limited by the reaction, the study described here was carried out without the presence of catalyst. Based on the reactor model, a size of 20 mm X 18 mm was chosen for the liquid channel. The depth of the channel was calculated to be 350 μm in order to achieve a noticeable conversion at reasonable experimental flow rates around 0.05-0.1 mL/min for the above-stated reaction. Two separate gas channels were proposed in order to maximize the gas-liquid interfacial area so that gas diffusion into the liquid channel is not a limiting factor for the reaction. The liquid channel had dimensions of 20 mm x 18 mm and 350 μm in depth. There were 14,082 catalyst traps in the liquid channel. Figure 2 is a schematic of the liquid channel with the catalyst traps.

A catalyst trap is a trapezoidal arrangement of four posts designed in such a way that it can hold a catalyst particle in the desired size range. Smaller particle sizes intended for this reactor (35-50 μm), which can be prepared by grinding and sieving operations, are attractive for screening tests or small scale production due to lower internal mass transport resistance compared to larger particles. Figure 3 is an SEM image of the catalyst trap with catalyst particles.

Figure 3: SEM image of Catalyst Trap and Microreactor.

The catalyst traps were spaced 75 µm from each other in the liquid channel. This distance was chosen to ease the catalyst loading into the channel. Apart from these parameters a significant amount of area on the un-etched substrate portion was left to increase the anodic bonding contact area. This in turn increased the bonding strength of the glass cover to the chip, in order that the chip could tolerate higher pressure. A width of 2 mm was left on all sides for bonding.

Three different reactors were designed based on the above-mentioned parameters. In creating design variations, catalyst trap spacing and the liquid channel dimensions were manipulated in order to observe the effects of these parameters on the flow configurations in the reactor. The first reactor incorporated all the basic parameters as stated above such as 20 x 18 mm liquid channel with 300 µm depth and a spacing of 75 µm between the traps. This reactor will be called as 75 µm spacing wide channel reactor for future reference.

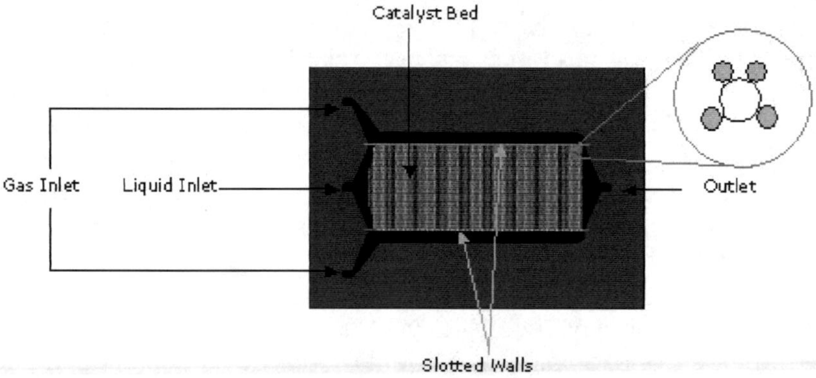

Figure 4: Auto Cad Diagram of the narrow channel reactor

The second reactor differed only in the trap spacing, which was increased to 150 µm, resulting in smaller total number of traps since the overall channel size was unchanged. This was done in order to explore the effect of trap spacing on catalyst loading and on the flow pattern. This reactor is named as 150 µm spacing wide channel reactor for further reference. The final reactor was quite different in terms of the dimensions of the liquid channel. This reactor had the same channel length of 20 mm, yet with the width reduced to 10 mm, with a view toward observing effects on the flow pattern. The trap spacing in this reactor was maintained at 150 μm. This reactor is the narrow channel reactor. Figure 4 is a schematic of the narrow channel reactor.

The reactors were fabricated at New Jersey Nanotechnology Consortium (NJNC)-Bell Labs in Murray Hill, New Jersey using the standard techniques involving photolithography and etching. Figure 3 shows SEM images of the reactor after the fabrication step.

Visual observations were obtained using a high-speed ccd camera. Typical photographs and sketches of the observed flow patterns, obtained at the middle of the microreactor, are shown in the figures below. These patterns varied according to the size and shape of the microreactor, i.e., the transition from the liquid dominated flow to gas dominated flow varied in each of the three different microreactors. In general, the observed flow patterns can be described as follows.

(a) Liquid-Dominated Flow: This flow pattern occurs at a relatively low gas velocity. The flow pattern is characterized by a continuous flow of gas through the slotted wall and into the channel forming a pattern, which is reproduced after an interval of time of 5-10 seconds. The amount of liquid in this flow pattern is more dominating than the amount of gas and hence its name

(b) Gas-Dominated Flow: This flow pattern occurs at a relatively higher gas velocity. A more gas dominating flow into the channel characterizes the flow pattern. The gas subdues the liquid flow in this case and hence the name the gas dominated flow.

(c) Transitional flow: This flow occurred in the region between the liquid dominated and the gas dominated flow. Neither gas nor liquid was dominating in this flow. It was characterized by a more continuous flow of gas into the channel. All the three flows are summarized in Figure 5.

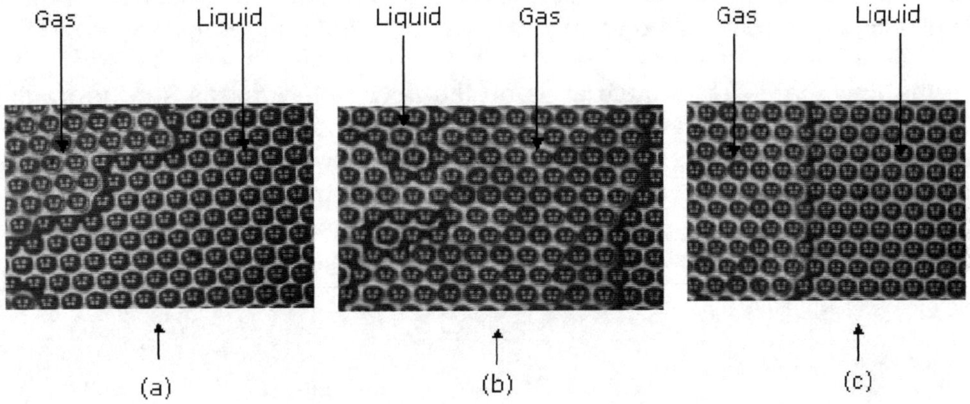

Figure 5: (a) Image of Liquid Dominated Flow; (b) Image of the Gas-Dominated Flow; (c) Image of the Transitional Flow.

The flow maps for all three different reactors show all these three different flow patterns. The following figure displays the flow pattern maps obtained in the present study for the three different microreactors.

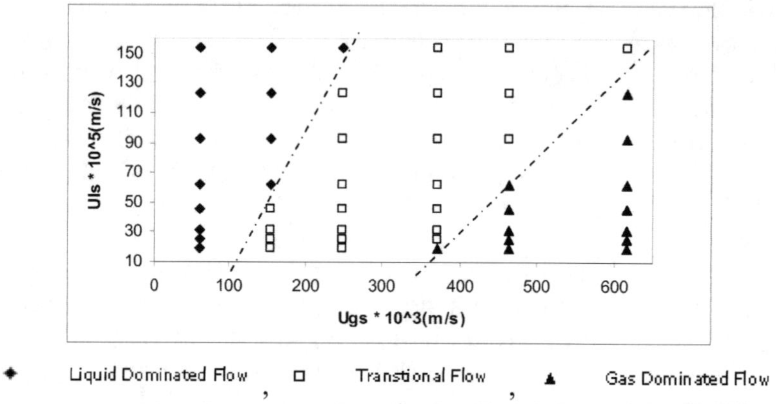

Figure 6: Flow patterns and flow pattern transition lines for 75 μm spacing wide channel reactor.

Figure 6 is the flow pattern map for the 75 μm -wide channel reactor. All the experiments were carried out at 30° c, below 90 psia. The liquid flow rates were varied between 0.06 to 0.5 mL/min and the gas flow rates were between 2 and 20 sccm. The coordinates of the flow pattern maps are gas and liquid superficial velocities. Gas and liquid superficial velocities were calculated taking into account the channel parameters. The different flow patterns liquid-dominated flow, gas-dominated flow and the transitional flow are all depicted on the flow maps. The dotted lines on the flow pattern maps are the transition lines from one pattern to another starting from the liquid dominated flow to transitional flow and from transitional flow to gas dominated flow. The transitional flow covered the major portion of the flow pattern map showing that it was feasible to access the more ideal transitional flow pattern in this particular microreactor. Figure 7 is the flow pattern map for the second kind of reactor i.e. the wide channel reactor with 150 micron catalyst post spacing.

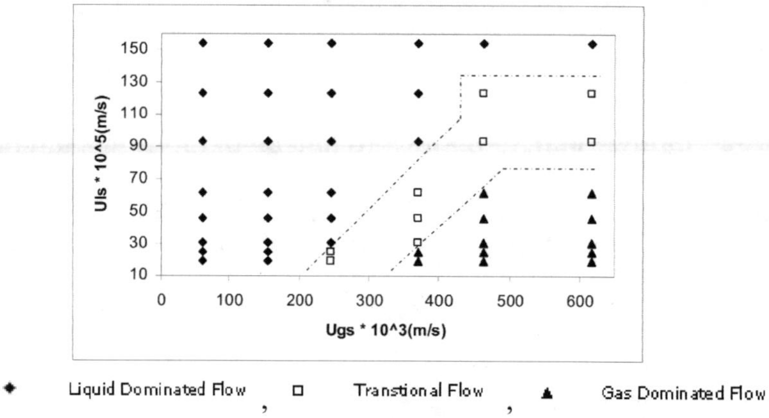

Figure 7: Flow patterns and flow pattern transformation lines for 150 μm spacing wide channel reactor.

The experimental conditions, liquid and the gas flow rates all remained the same in this experiment. The major difference being the distance between each catalyst post, which was 75 micron in the earlier case, in this microreactor the spacing was maintained at 150 μm. This variation in the design was done in order to see the effect of spacing on the catalyst loading and also the effect on the flow pattern if any. As expected there was a lot of difference between the flow patterns of the reactor with 75 μm spacing. One of the major disadvantages of this particular microreactor was that there was more of the liquid- dominated flow, which meant that even though all of the catalyst posts were encapsulated with liquid there was significantly less amount of gas diffusion into the reactor channel. This left many of the catalyst traps unused. The catalyst loading was also less, as there were less catalyst traps as a result of the higher spacing between them. However, loading the catalyst in this reactor was faster and more effective because of the higher spacing between the catalyst traps.

The same experimental conditions were maintained for the third reactor, i.e., the narrow channel reactor. Because of the narrow channel width, anything above the liquid flow rate of 0.3 mL/min would not allow the gas to diffuse through the slotted wall and hence the liquid flow rates were maintained between 0.06 to 0.3 mL/min in this case. The flow patterns and the flow behavior in the narrow channel reactor can be seen in Figure 8.

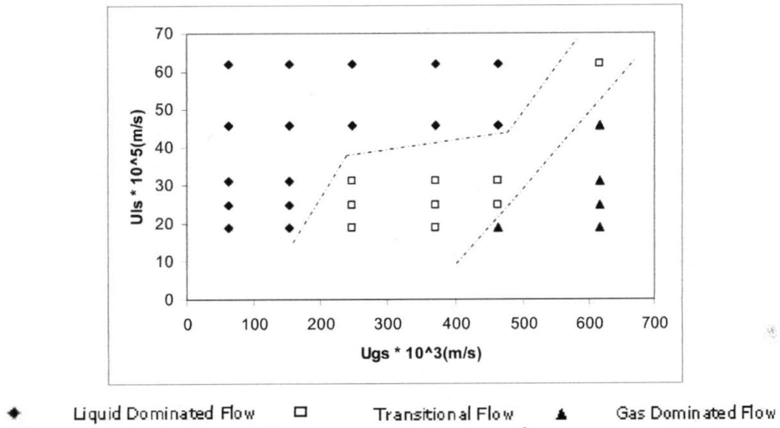

Figure 8: Flow patterns and flow pattern transition lines for the narrow channel reactor.

All three-flow patterns were observed in this reactor as well. Since the width of the channel was narrow, the catalyst loading in this type of reactor was less when compared with the other two reactors. Liquid was more dominating in this particular reactor, i.e., due to the narrow channel width the liquid flow rate was dominating the flow throughout the channel. Even in this reactor there was more of the liquid dominated flow than the other two flows. The significant difference in this reactor was the amount of gas-dominated flow pattern in this reactor, which was less compared to the other two reactors. And again the narrow channel width of the reactor can account for this.

Hence, it can be said that the 75-μm wide channel reactor possesses preferable flow characteristics for the multiphase reaction than the other reactors because more of the transitional flow configuration can be, which in turn would significantly assist the multiphase mass transfer by creating high mass transfer interfacial area.

References:

1. P.L.Mills, R.V. Chaudhari, 1997. Catal. Today 37, 367-404.
2. N Doku George, Verboom Willem, Reinhoudt David N. and Albert van den Berg. 2005. Tetrahedron 61, 2733.
3. D. Barnea, Y. Luninski, Y. Taitel, 1983. Can.J.Chem.Engng. 61, 617-620.
4. M. Suo, P.Griffith, 1964. J. Basic Engng. 86, 576.
5. K. Mishima, T. Hibiki, 1996. Int. J. Multiphase Flow 22, 703.
6. K.A. Triplett, S.M. Ghiaasiaan, S.I. Abdel-Khalik, D.L. Sadowski 1999. Int.J. Multiphase Flow 23, 377.
7. M.K.Akbar, D.A. Plummer, S.M. Ghiaasiaan. 2003. Int. J. Multiphase Flow 29, 855.
8. T. Kreutzer Michiel, Kapteijn Freek, A. Moulijn Jacob, J. Heiszwolf Johan, 2005. Chem Eng. Sci 60, 5895.
9. W.L. Chen, M.C. Twu, C.Pan, 2002. Int. J. Multiphase Flow 28, 1235.

Formation and fluorimetric characterization of micelles in a micro-flow through system with static micro mixer

10. März 2006 micell4

Michael Schuch, G. Alexander Groß, J. Michael Köhler

Technische Universität Ilmenau, Institute für Physik, Stiftungslehrstuhl der Deutschen Bundesstiftung Umwelt, Fachgebiet Phys.Chem./Mikroreaktionstechnik, PF 100565, D-98684 Ilmenau, Germany, Tel. - (49) - 3677-693655

Summary

The formation and behaviour of micelles of sodium dodecylsulfate in water by use of a static micro mixer were studied. Trisbipyridylruthenium(II) was applied as indicator dye, 9-methylanthracene was used for fluorescence quenching. All experiments were carried out by a micro fluid arrangement with three syringe pumps, a 2+1 two-step static micro mixer (IPHT Jena) and a on-line micro fluorimetry including a luminescence diode for excitation, a blue glass filter (BG 7, Linos), two edge filters (RG 630, Linos) and a photo counting module (MP 900, Perkin Elmer). It was possible to measure the fluorescence inside the PTFE tube (inner diameter 0.5 mm) directly. A linear dependence of fluorescence intensity from dye concentration was observed in absence of quencher and surfactant as expected. An aggregation number of about 62 was found in the flow rate range between 300 and 800 :l/min. The fluorescence intensity increases slightly, but significant with increasing flow rate, if no quencher is present. In the presence of quencher, the fluorescence intensity decreases with decreasing surfactant concentration and with enhanced flow rate. The strength of the flow rate effect on the fluorescence increases with decreasing surfactant concentration. The size of micelles was determined in micro channels by the micro fluorimetric method in analogy to the conventional system. The micelles extract the quencher from the solution and lower, this way, the quenching effect. The size of micelles was estimated and it could be shown, that the flow rate has only low effect on the aggregation number at the investigated flow rates. The effect of flow rate and surfactant concentration on the fluorescence in the presence of quencher was interpreted as a shift in the micelle concentration due to the shear forces. It is expected, that the fluorescence intensity is lowered, if more quencher molecules are molecular disperse distributed inside the solution. Obviously, the lowered fluorescence intensity at higher flow rates suggests a reduction of the micelle density causing an increase of quencher concentration outside the micelles.

Keywords

microreaction technology, micelles, aggregation number, shear stress, laminar flow, fluorescence quenching

1. Introduction

The first micro reactors were mainly constructed for the processing of homogenous phases. The handling of homogenous phases in microreactors is less demanding than that of heterogeneous systems in micro reactors. Although heterogenous systems are of great importance for many chemical and biological applications, they are avoided frequently, if micro reactors are used. The generation or introduction of heterogenous systems in micro fluidic systems is accompanied with disturbances in many cases. Such perturbations include precipitations, increasing flow resistance, sedimentation, particle adsorption at the wall surfaces and blocking. Gas bubbles cause elastic answering behaviour of the micro fluidic system against pressure changes by the compression and expansion of the gas.

Despite these problems, the application of micro reactors for heterogenous chemical systems attracted more and more interest during the last years. Liquid/liquid two phase systems and liquid/liquid gaseous multi phase systems were introduced in order to generate ideal plug flow behaviour in micro fluidic arrangements for fast serial processing [1-8]. Liquid/liquid heterogenous systems and particular micro segmented-flow systems are of interest for chemical [9-15] as well as for biological applications [16-18]. Beside liquid interfaces, solids can also be handled in micro reactors. It was shown, that dielectric micro and nanoparticles can be generated in micro flow-through systems [19-22]. Metal nanoparticles are of particular interest concerning labelling of biomolecules and biochips [23-24]. The formation of metal nanoparticles and the growth of metal nano seeds by reduction of metal salt solutions in micro reactors succeeded, too [25-27]. Micro mixers were used for the synthesis of polymers by emulsion polymerization and yield polymer nano particles 28, 29].
The control of interface states is a critical parameter for a successful handling of heterogenous systems in micro reactors. The properties of solid interfaces can be modified by deposition of thin organic or inorganic films as well as by the immobilization of molecules. Liquid surfaces and liquid/liquid interfaces can easily be modified by the introduction of surfactants. Surfactants stabilize the liquid droplets in emulsions and can also act as stabilizers for suspensions and colloidal solutions. The addition of surfactants influences the formation of nano particles [30, 31] and can be used for the generation of metal nanorods in a micro fluid synthesis [32].
Surfactants can form micelles as a particular type of nanoscale objects above a characteristic concentration, the critical micelle concentration cmc [33]. Micelles are supramolecular objects forming by self assembling of amphiphilic molecules. The single micelles are in a thermodynamic equilibrium with surfactant molecules molecular disperse distributed in solution. The exchange rates are high. Thus, micelles must be understood as highly dynamical objects [34]. The formation of micelles is expected in all cases of higher surfactant concentrations in micro reactors, and, therefore, it is of great importance for all heterogenous systems manipulated in micro reactors under addition of surfactants. Although, the application of surfactants is of great interest for micro fluidics, low attention was paid to the behaviour of micelles under micro fluidic conditions. In this work, the formation of micelles in a micro fluidic system was investigated by means of a fluorimetric method. Shear stress and the influence of flow rate on the optical response related to the micellar state were studied.

2. Experimental

2.1. Chemicals and micelle formation

All used chemicals were pA. graded. Sodium dodecylsulfate (SDS) was applied as model substance, because it represents a frequently applied anionic surfactant. It is known, that micelles of SDS in water possess an electrical excess charge. For the ratio between positive counter ions and the number of SDS molecules inside a micelle, numbers between 0.46 and 0.86 were reported [35]. This value depends on the size and chemical structure of the charged or polar head group and the length and mobility of the hydrophobic tail group and , in particular on the properties of the solvent and the kind and concentration of other ions contained. The excess charge of the micelles is mainly responsible for the repulsion between the molecular ensembles and therefore for the thermodynamical stabilization of the whole system.

Ionic surfactants like SDS preferably form smaller micelles. The aggregation number is normally in the range between 10 and 100, whereas non-ionic surfactants can form micelles of about 10^4 molecules. Mean aggregation numbers between 60 and 83 are reported for SDS in water, for example [35, 36].

2.2. Micro fluidic arrangement

It was the aim to characterize the micellar solution inside a tube capillary immediately after mixing of the components. Therefore, a two step static micro mixer was applied. Static micro mixers are widely accepted [37, 38]. The applied mixer is constructed based on a fluidic lamination principle based on several split and recombine units. The channel structures are prepared photo lithographically using Si and glass substrates and applying wet etching technology. A three chip sandwich technology including two glass chips at the bottom and the top and a silicon chip between them was applied in order to get a good splitting and reshaping behaviour at the fluid stream. The construction and preparation technology of the micro mixer class was described previously [39].
Here, a two step mixer was applied: Two liquids were brought together at first, and a third liquid was added by in a second mixing step. Syringe pumps were used for fluidic actuation.
The outlet of the mixer was directly connected with the PTFE tube, in which the characterization occurred. The glass fibres of the measurement arrangement were placed directly behind the outlet of the mixer (Fig. 1).

2.3. Measurement principle

A fast working non-invasive method for the in-situ characterization of the micellar state was needed for the measurement of micelles under micro fluidic conditions. Osmometry was regarded as non suited for the microfluidic application. Optical methods were preferred as non-invasive techniques. Light-scattering was excluded due to the low scatter efficiency of the expected micelles.
Fluorimetry with quencher distribution-dependent fluorescence quenching was chosen as a promising method for the in-situ characterization. Therefore, an ionic fluorescence dye and a rather hydrophobic quencher are applied. Both molecular species - dye and quencher - are molecular disperse distributed inside the solution in case of absence of surfactants or at low surfactant concentration, when no micelles are formed. If the surfactant concentration increases and micelles appear, they show a higher affinity with response to the quencher. As result, quencher molecules go into the hydrophobic centre of the micelles and the local quencher concentration outside micelles is lowered. The consequence is an increase of fluorescence quantum yield of the dye. Here, the tris(bipyridyl)ruthenium cation was used as fluorophore and methylanthracene was applied as hydrophobic quencher [40]. The introduction of quencher in the liquid system was realized by dissolution in acetonitril (Neolab-Migge) and subsequent mixing of the acetonitrilic quencher solution with the aqueous solution. The volume ratio of aqueous solution to acetonitril amounted to 500: corresponding to 0.2 Vol% organic solvent.
The optical measurements were carried out using a blue light emitting diode (DS 25, royal blue, Luxeon, USA) with a wavelength of 455 nm. An optical filter (blue, BG 7, Linos, Germany) was applied in order to suppress a small red light emission from the light source. A photo counting module (MP 900, Perkin-Elmer) was applied in connection with two optical filters (R. 630, Linos) for the fluorescence intensity measurement. The optical measurements occur directly inside the PTFE tube (inner diameter 0.8 mm, 0.4 mm wall thickness) with the streaming liquid.

3. Results and discussion

3.1. Reference measurements at static conditions

At first, the chemical system was tested inside the PTFE tube without fluidic motion after premixing the components by hand outside the microfluidic system. Therefore a series of solutions with the same quencher (10^{-4} mol/L) and constant dye concentration was characterized in dependence on the concentration of surfactant (SDS). The premixed solution was introduced into a syringe pump. The fluorescence measurements were carried out after the complete substitution of water inside the PTFE tube by the process mixture.

The SDS concentration was varied between 20 and 50 mmol/L. As expected, a considerable increase in fluorescence intensity was found with increasing SDS concentration (Fig. 2). In the log I/c plot (I_o/I vs concentration of SDS) the linear relation was confirmed. The effect of micelle formation and extraction of quencher from the liquid into the micelles is strong. It results in an increase of fluorescence intensity of a little less than 2 orders of magnitude by a concentration increase of the surfactant by a factor of 2.5, only. The aggregation number at flow-free conditions can be estimated to about 55 molecules per micelle.

3.2. Characterization of micro fluidic arrangement

The experimental arrangement was fluidically characterized by pumping for the fluorescence dye solution into a carrier stream of pure water. At a flow rate of 250 :l/min, the fluorescence intensity increases quickly within the first 60 s after starting the fluorescence dye stream. After about 250 s the fluorescence signal came into saturation indicating a stationary state in the fluidic system. The time to saturation is independent on the concentration of fluorescence dye.

The applicability of fluorescence measurements inside PTFE tubing by the above mentioned fiberoptical system as a quantitative method at applied flow was tested by varying the dye concentration at a certain flow rate. Therefore, solutions of tris(bipyridyl)ruthenium with concentrations between 2 and 21 :mol/L were pumped through the system. The fluorescence intensities were determined 250 s after switching on the syringe pump, and the optical signals were averaged over 10 s. In result, a good linear relation between the dye concentration and the measured fluorescence intensity was found (Fig. 3).

3.3. Fluorescence intensity and micelle formation under micro fluidic conditions

Fluorescence measurements in dependence on flow rate showed a slight increase of fluorescence intensity in absence of any quencher if SDS is present. The enhancement of the over-all flow rate from 400 :l/min to 800 :l/min lead to an increase of the intensity of emitted light of 2.5 % in case of solutions containing 27.5 mmol/L SDS. This observed increase is small, but it is significant (Fig. 4). Obviously, the increasing shear stress with increasing flow rate improve the fluorescence quantum yield. But, the increase of fluorescence with flow rate is weaker than linear.

The function of fluorescence increase with increasing flow rate is dependent on SDS concentration. The enhancement of fluorescence intensity with increasing flow rate increases with increasing SDS concentration. This fact supports the assumption, that the molecular self organization, the structuring inside the liquid and the influence of the shear stress at laminar flow conditions is responsible for the fluorescence increase. It is to assume, that the formation of micellar structures lowers slightly the molecular mobility of the dye and the laminar flow causes an certain enhancement of regularity inside the liquid.

3.4. Fluorescence behaviour in presence of quencher and determination of aggregation number at micro fluidic conditions

The strong effect of SDS on the fluorescence intensity in case of presence of a hydrophobic quencher was confirmed at micro fluidic conditions. The enhancement of SDS concentration from 20 to 30 mmol/L in experiments with a total flow rate of 600 :L/min causes an increase of fluorescence intensity up to about 500 %. A linear correlation was found in the dependence of 1/log(I/Io) versus the surfactant concentration (Fig. 5). The slope of the plot allows the estimation of the aggregation number in the dynamic system. With 62+/-3, this number deviates only slightly from the value under static conditions. The investigations over a flow rate range from 300 up to 800 :L/min confirmed this finding: the aggregation number is nearly independent on flow rate, in this flow rate range.

In contrast, a certain effect of decreasing micelle density with increasing flow rate can be concluded from the flow-rate-dependent measurements. This observation corresponds to a slight increase of the concentration of the free surfactant molecules. It seems, that the shear stress is not effecting the size of micelles, but causes a certain shift in the ratio of micelles and free surfactant molecules. Probably, the cmc is also enhanced in micro fluidic systems. This interpretation agrees well with the observation of a slight, but significant reduction of fluorescence intensity with increasing flow rate at different SDS concentration. The effect of flow rate on fluorescence increases with decreasing SDS content (Fig. 6).

4. Conclusions

Micellar solutions can be characterized in micro fluidic systems by use of fluorimetric measurements. The determination of fluorescence intensity in the presence of a hydrophobic quencher was successful applied for the determination of the dependence of micelle formation on flow rate and surfactant concentration.

It was found, that the aggregation number can be determined rather precisely with tis method.

For the applied microsystem and at medium volume flow rates the value found for the aggregation number was in the same range as at static conditions. Despite this finding, enhanced flow rates cause a decrease in fluorescence intensity. This effect is stronger for smaller surfactant concentrations. This leads to the conclusion, that not the size, but the density of micelles is reduced due to shear stress in micro fluidic systems.

The investigations show the importance of fluorimetric in-situ measurements for the monitoring of surfactant-containing solutions in micro reactors and other micro fluidic systems. In future, studies should address the behaviour of systems at higher shear stress, for example of other the influence of other surfactants, of lipophilic and macromolecular components.

Acknowledgement

We thank S. Schneider for the construction of the optical measurement arrangement, J. Albert and the clean room staff of IPHT (Jena) for the design and the preparation of the static mixer and J. Wagner for support and helpful discussions. The financial support of the German Environmental Foundation (DBU) is gratefully acknowledged.

References

[1] J. R. Burns and C. Ramshaw, C., *Lab on a Chip* 2001, **1**, 10

[2] T. Thorsen, R.W. Roberts, F.H. Arnold and S.R. Quake, *Phys. Rev. Lett.* 2001, **86,** 4263

[3] T. Nisisako, T. Torii and T. Higuchi, *Lab on a Chip,*2002, **2**, 19

[4] T. Taniguchi, T. Torii, and T. Higuchi, *Lab on a Chip,* 2002, **2**, 24

[5] D.R. Link, S.L. Anna, D.A. Weitz, H.A. Stone, *Phys. Rev. Lett.,* 2004, **92**, 54503

[6] T. Nisisako, S. Okushima, T. Toru Torii, *Soft Matter,* 2005, **1**, 23

[7] J.M. Köhler, Th. Kirner, *Sensors and Actuators A,* 2005, **119**, 19-27

[8] A.C. Balasz, R. Verberg, Ch.M. Pooley, O. Kusenok, *Soft Matter,* 2005, **1**, 44

[9] H. Song, H., J.D. Tice, R.F. Ismagilov, *Angew. Chem.* 2003, **115**, 792

[10] Ismagilov R F (2003) Integrierte Mikrofluidsysteme. Angew. Chem. 115: 4262-4264

[11] I. Shestopalov, J.D. Tice, R.F. Ismagilov, *Lab Chip,* 2004. **4**, 316

[12] B. Zheng, J. D. Tice and R.F. Ismagilov, *Anal. Chem*., 2004. **76**, 4977

[13] J.M. Köhler, Th. Henkel, A. Grodrian, Th. Kirner, M. Roth, K. Martin and J. Metze, *2004, Chem. Engn. J.* **101**, 201

[14] T. Henkel, T. Bermig, M. Kielpinski, A. Grodrian, J. Metze and J.M. Köhler, *Chemical Engineering Journal,* 2004, **101**, 439-445.

[15] P.M. Günther, F. Möller, T. Henkel, J.M. Köhler, G.A. Groß, *Chem. Eng. Technol.*, 2005, **28**, 520

[16] K. Martin, T. Henkel, V. Baier, A. Grodrian, T. Schön, M. Roth, J.M. Köhler and J. Metze, *Lab on a Chip* 2003, **3**, 202

[17] A. Grodrian, J. Metze, T. Henkel, K. Martin, M. Roth and J.M. Köhler, *Biosensors & Bioelectronics* 2004, **19**, 1421

[18] A. Brösing, J.M. Köhler, *Proc. Kolloquium Heiligenstadt,* 2004, **12**, 381

[19] V. Hessel and H. Löwe, *Chemie Ingenieur Technik*, 2002, **74**, 17

[20] V. Hessel and H. Löwe,. *Chemie Ingenieur Technik*, 2002, **74**, 185

[21] N. Jongen et al., Chem. Eng. Technol. 26 (2003), 303

[22] R. Schenk et al., Chem. Eng. Trans. 2002, 909

[23] Chan E M, Mathies R A, Alivisatos A P (2003) Nano Letters 3: 199-201

[24] Fritzsche, W.: Rev. Mol. Biotechnol. 82 (2001), 37-46

[25] J. Wagner, T. Kirner, G. Mayer, J. Albert, J.M. Köhler, Chem. Eng. J. **2004**, *101*, 251

[26] J.Wagner, J. M. Köhler, Nano Letters **2005,** *5,* 685

[27] J.M. Köhler, J. Wagner, J. Albert, J. Mater. Chem. **2005**, *15,* 1924

[28] Chang Z, Liu G, Tian Y, Zhang Z (2004), Materials Letters 58: 522-524

[29] P.M. Günther, J. Wagner, G.A. Groß, J.M. Köhler: 9th Internatl. Conf. On Miniaturized Systems μ-TAS (Boston, October 9-13, 2005), 918-920

[30] K. Noorgard, M.J. Weygand, K. Kjaer, M. Brust, T. Bjornholm: Faraday Discussions 125 (2004), 221

[31] R.C. Doty, T.R. Tsikhudo, M. Brust, D. Fernig: Chemistry of Materials 17 (2005), 4630

[32] K. Sönnichsen: 3rd Ilmenauer Workshop on Micro Laboratory Techniques (2006), oral presentation

[33] Dörfler, H.-D.: Grenzflächen- und Kolloidchemie (Weinheim 1994)

[34] Y. Moroi: Micelles: theorectical and applied aspects (New York 1992)

[35] N.Turro, A. Yekta: J. Am. Chem. Soc. 11,6 (1978), 5951;

[36] F.H. Quina et al.: J. Phys. Chem. 99 (1995), 17028

[37] Schwesinger, N.; Frank, T.; Wurmus, H. 1996. J. Micromech. Microeng. 6, 99

[38] Hardt S, Drese K S, Hessel V, Schönfeld F (2005) Passive micromixers for applications in the microreactor and mTAS fields. Microfluid Nanofluid 1: 108:118

[39] Kirner T, Albert J, Günther M, Mayer G, Reinhäckel K, Köhler J M (2004), Chemical Engineering Journal 101: 65-74

[40] Ch. Draeger: Mesoskopische Mizellen aus amphiphilen Bipyridiyliumkomplexen mit Ruthenmium (II), Osmium (II) und Palladium (II), Diss. Techn. Univ. Berlin (2001)

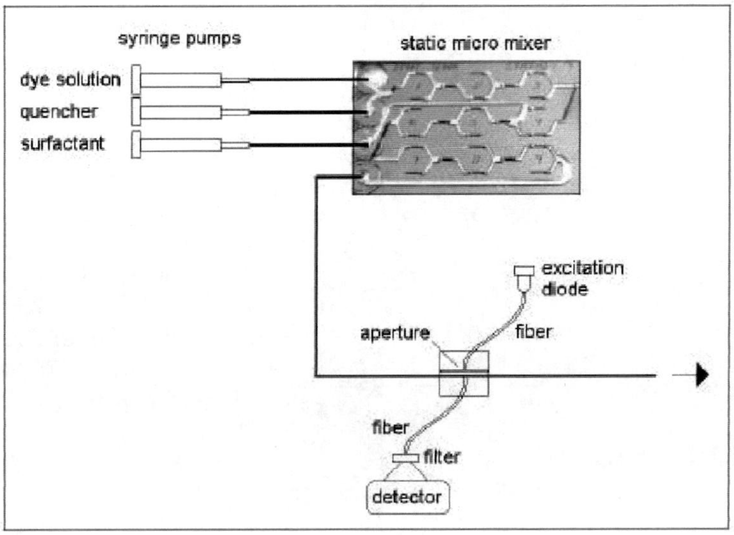

Fig. 1
Experimental arrangement for formation and characterization of micelles in a microfluidic system

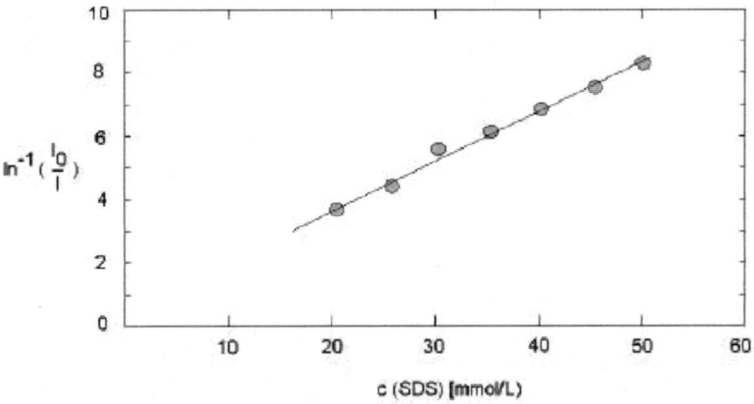

Fig. 2
Dependence of fluorescence intensity ($Ru(bipy)_3^{2+}$) in solutions with constant quencher concentration (99 :mol/L) on surfactant concentration without flow inside PTFE tube

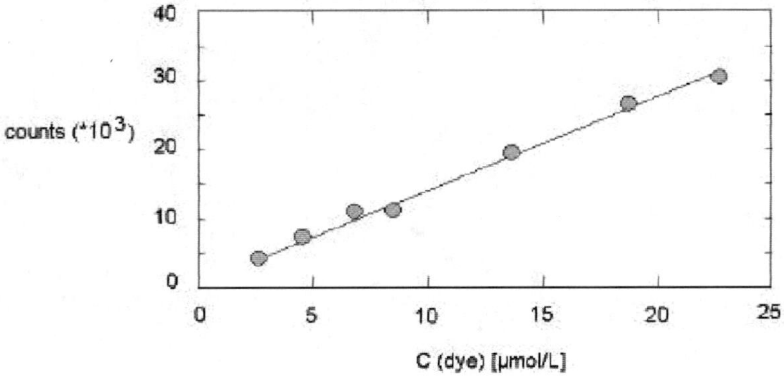

Fig. 3
Fluorescence intensity measured inside the PTFE tube in dependence on dye concentration under stationary micro fluidic conditions

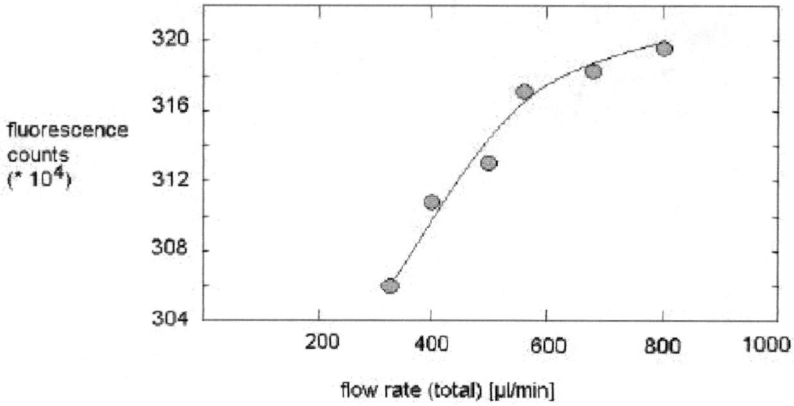

Fig. 4
Increase of fluorescence intensity in dependence on flow rate in dye solutions containing 27.5 mmol/L SDS but no quencher (constant dye concentration)

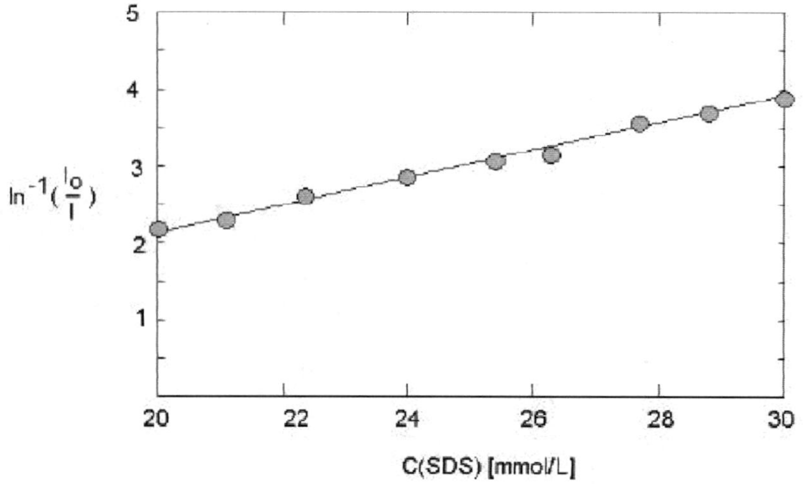

Fig. 5
Dependence of fluorescence intensity on the surfactant concentration at micro fluidic conditions (flow rate: 600 :L/min; quencher concentration: 10^{-4} mol/L; dye concentration: 10^{-4} mol/L) (Abb. 4.1.1)

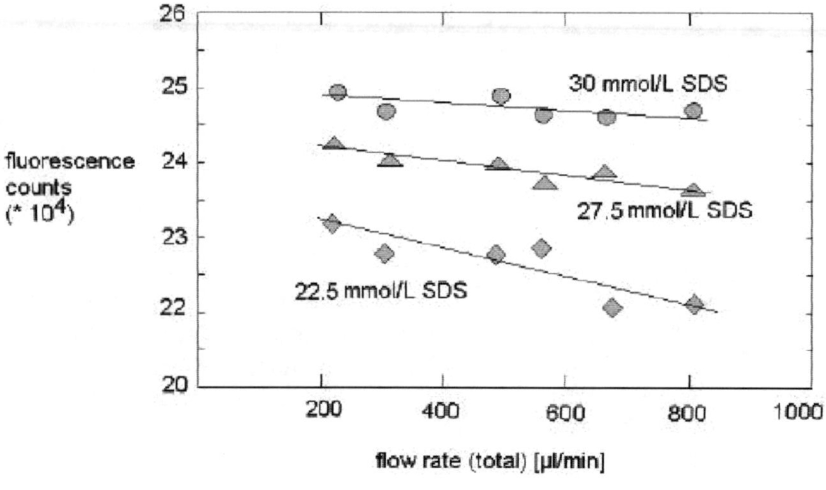

Fig. 6
Flow-induced reduction of fluorescence intensity for different SDS concentrations (Abb. 4.2.4.)

Influence of gravity on a two-layer laminar flow in a microchannel

Yoshiko Yamaguchi,[a] Takeshi Honda,[a] Maria Portia Briones,[a] Kenichi Yamashita,[a] Masaya Miyazaki,[ab] Hiroyuki Nakamura,[a] and Hideaki Maeda[ab]*

[a] Nanotechnology Research Institute, National Institute of Advanced Industrial Science and Technology (AIST), Tosu, Saga 841-0052, Japan
[b] Department of Molecular and Material Sciences, Interdisciplinary Graduate School of Engineering Sciences, Kyushu University, Kasuga, Fukuoka 816-8580, Japan

It is well known that microchannel forms laminar flow. However, the gravity influence on laminar flow has not been much investigated because its influence has been considered to be very small due to largeness of surface effects. Therefore, misperception that in a microchannel the effect of gravity is considered close to weightlessness sometimes occurs. In this work, quantitative investigation on gravity influence on laminar flow was carried out using computational fluid dynamics (CFD) simulation. As a result, it was revealed that gravity field affects laminar flow behavior. Relationship between its behavior and density, viscosity, shape of the microchannel cross-section, inner-wall characteristic, interfacial tension and diffusion were evaluated quantitatively. It was shown that the gravity influence is proportional to density difference, average residence time and channel size, and inversely proportional to viscosity, in the condition that the microchannel cross-section is square and interfacial tension and diffusion are small. This simple relationship enables a rough estimation of the laminar flow behavior. The results in this work are useful for microdevice design for chemical analysis, mixing, and chemical reaction.

Introduction

Microchannel forms laminar flow because of its low Reynolds number. This characteristic has been used for many applications such as biochemical analysis and chemical synthesis. Particularly, side-by-side laminar flow is applied for several applications. Examples are micromixers[1-2], intracellular compartments dyeing using laminar flow[3], particle separation by size in a pinched flow[4], and DNA diagnosis which utilizes fluidic behavior difference between single and double strand DNAs in a microchannel[5]. All of these applications require stable side-by-side laminar flow. However, it has not been discussed whether stable side-by-side laminar flow is formed in any conditions. In most cases, there has been a trend to neglect gravity effect because surface effects are stronger than that. This is true in the case of strong interfacial tension. However, there are not a few microsystems which require miscible two-layer laminar flow which does not have interfacial tension. It has been revealed that inertial force, which is a kind of body force similar to gravity force, is exerted in microchannel[6-8]. Therefore, it is important to investigate the effects of gravity in a microchannel. Recently, gravity effects in vertical microchannel[9] and migration of biological objects by gravitational force[10] have been examined. The purpose of this work is to investigate the gravity effects on the behavior of side-by-side two-layer laminar flow. CFD simulation was carried out in various densities, viscosities, cross-section of microchannel, interfacial tension, and diffusion coefficients. The results enabled a rough estimation of laminar flow behavior by simple equation.

Experimental

Three-dimensional simulations were carried out using FLUENT 6.1 (Fluent Inc.) based on the finite-volume scheme. The microchannel width and depth were basically set to 200 μm, and length was 10 mm. Mesh sizes were 8 μm in width direction, 8 μm in depth direction, and 200 μm in length direction. It was confirmed that this mesh size is sufficiently small by performing other calculation in half-size mesh and comparing their results. Iterative calculation for solving Navier-Stokes equations was carried out with second order accuracy for pressure; the SIMPLE algorithm was used for pressure-velocity coupling. The microchannel inlet was divided into two flow regions formed horizontally. Two kinds of liquids were assumed to be injected from the two inlets. Velocity profile of the inlet was adopted as the developed profile, that is, the Hagen-Poiseuille type profile (velocity is zero at the wall surface and is fastest at the center). With one exception, uniform velocity was assumed in the section below where fluid slip on the wall was considered. Euler mixing model was adopted in the case of miscible fluids. On the other hand, volume of fluid (VOF) model was applied on the condition that surface tension was taken into account.

Results and discussion

Density and channel size

First, calculation was performed under the condition where channel width and depth were 200 μm. Diffusion coefficient was assumed to be almost zero (1×10^{-40} m^2/s), meaning, fluids were assumed to be almost immiscible. Fluid densities of the two liquids were 1.0 g/cm^3 and 1.1 g/cm^3, respectively. Viscosity and average fluid velocity for each liquid were 1 cP (= 0.001 Pa•s) and 10 mm/s, respectively. Acceleration of gravity was 9.8 m/s^2 in the direction of minus z in Fig 1a. The result of the calculation is shown in Fig. 1a. The interface which has been vertical became almost horizontal after 1 second of the average residence time. In order to evaluate the interface tilting quantitatively, θ was defined as the

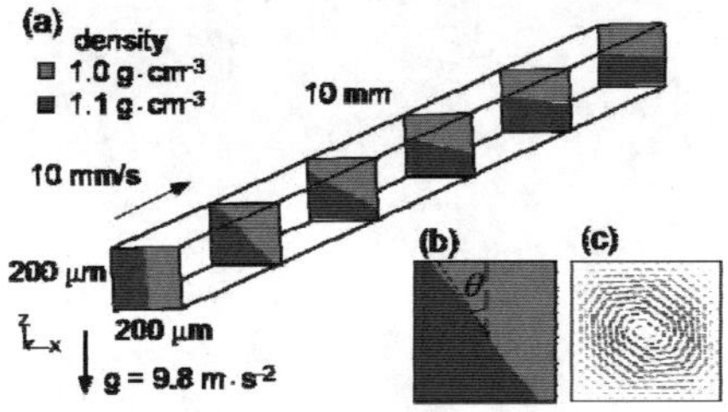

Fig. 1 Gravity effect on two liquids side-by-side laminar flow with different densities.

angle of the interface line at the center of the microchannel cross-section as shown in Fig. 1b. $\theta = 0°$ refers to the vertical interface and $\theta = 90°$ refers to the horizontal interface. Fig 1c shows velocity vectors in a cross-sectional plane. It revolves around the center in a spheroidal way in accordance with gravity.

Fig. 2 shows the change of θ under varying average residence time and density. This figure indicates that the product of average residence time and density difference affects interface tilting. Values of θ were fitted well on the second order curve (R > 0.99). The curve is shown as a dashed line in the figure.

Second, surface effect was investigated by changing the microchannel size. Calculation results for varying channel width and depth from 10 μm to 200 μm are shown in Fig. 3. Density difference is 0.1 g/cm^3. This result indicates that interface tilting becomes slower as microchannel size decreases. Gravity effect is depressed by surface effect. The values of θ were fitted well again on the second order curve ($\theta = -pt^2+qt$; R > 0.99), where t is the average residence time. Values of p and q are shown in Fig. 4. They were fitted on the second order curve and linear line versus the channel size (R > 0.99), respectively. Using this relationship, interface tilting can be expressed in the following equation,

$$\theta \approx -2.3\times10^{-1}(aLt)^2 + 9.0aLt \quad \text{(deg)} \qquad 1)$$

where a is the density difference (g/cm^3), L is the microchannel width and depth (μm), and *t* is the average residence time (s). From this equation, it is possible to roughly estimate the residence time till θ becomes 90° as follows:

$$t \approx 2.0 \times 10^{1} \frac{1}{aL} \qquad \text{(s)} \qquad 2).$$

It is estimated that interface becomes almost horizontal at the point where average residence time is 20 seconds on the condition that microchannel width and depth = 100 μm and density difference = 0.01 g/cm^3.

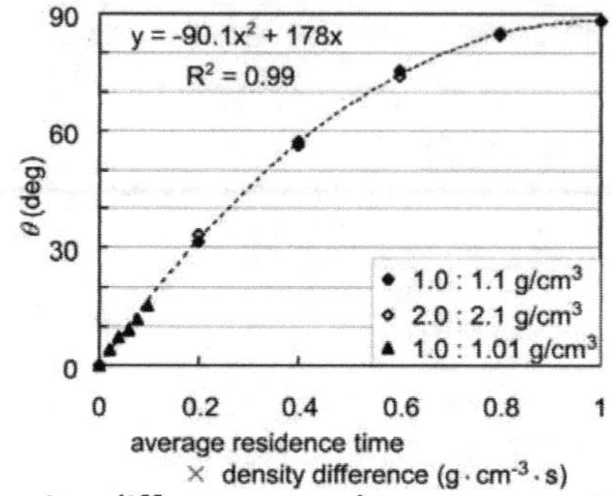

Fig.2 Effects of the density difference and average residence time on interface tilting.

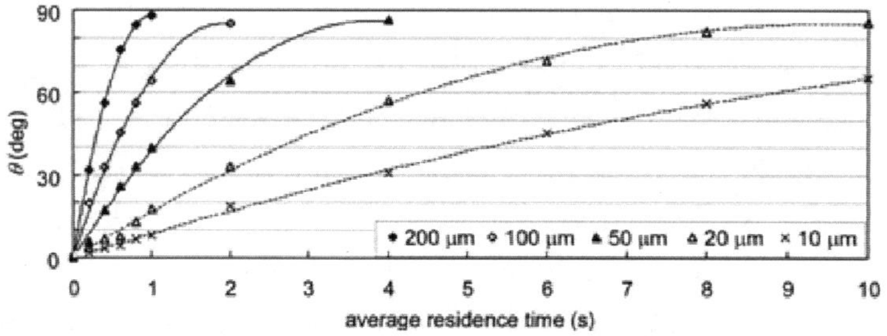

Fig. 3 Effects of the microchannel size and average residence time on interface tilting.

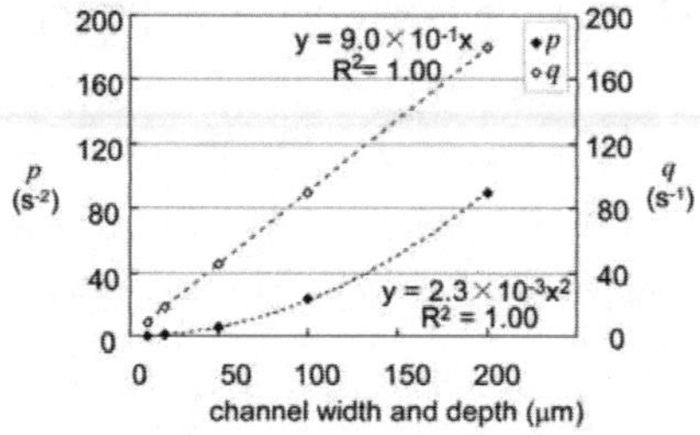

Fig. 4 Relation between parameters of the second order curve ($\cdot = -pt^2+qt$) and microchannel size.

Viscosity

The effect of fluid viscosity on gravity effects mentioned above was investigated.

Viscosities for the two liquids were changed from 1 to 5 cP (= 0.001 to 0.005 Pa•s), simultaneously. As a result, interface tilting was depressed as fluid viscosity became high. Transition of the interface tilting over residence time was fitted well on the second order curve (R > 0.99) in the same manner as Fig. 2. The relationship between parameters (*p* and *q*) of the second order curve and viscosity is shown in Fig. 5. In this figure, microchannel width and depth were 200 μm, and density difference and average velocity was 0.1 g/cm^3 and 10 mm/s, respectively. As indicated in Fig. 5, *p* was almost inversely proportional to viscosity and *q* was almost proportional to the square of viscosity. Substitution of this relationship into eqn. (1) leads to the equation related to viscosity as follows:

$$\theta \approx -2.3\times10^{-1}\left(\frac{aLt}{\mu}\right)^2 + 9.0\left(\frac{aLt}{\mu}\right) \quad \text{(deg)} \qquad 3)$$

where unit of μ is cP. Eqn. (3) is an approximate expression which adapts to the case where viscosities of the two liquids are the same. When viscosities are different for each liquid, it is impossible to measure the interface tilting at the center of the microchannel as shown in Fig. 1, because the interface is apart from the center due to viscosity ratio[9]. Therefore, the tilting θ' was measured at the point near the microchannel center. The values of θ' were fitted again to the second order curve (R > 0.99). The values of *p* and *q* are shown in Table 1. From the equation $\theta = -pt^2+qt$, the residence time when the interface becomes horizontal can be expressed as $q/2p$ (s). Fig. 6 represents the graph for $q/2p$ versus average viscosity. This result indicates that average viscosity $\mu^* = (\mu_1+\mu_2)/2$ is appropriate for the viscosity μ in eqn. (3). For eqn. (2) considering viscosity, the residence time when the interface becomes horizontal can be roughly expressed by the following equation:

$$t \approx 2.0\times10^{1}\left(\frac{\mu^*}{aL}\right) \quad \text{(s)} \qquad 4)$$

where unit of μ^* is cP.

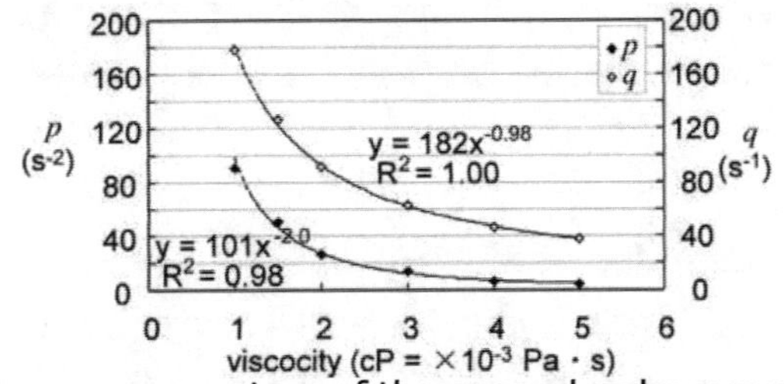

Fig. 5 Relation between parameters of the second order curve ($\theta = -pt^2+qt$) and fluid viscosity.

Table 1 The values of the parameter of the fitted second order curve ($\theta = -pt^2+qt$) in various fluid viscosity. μ_1 and μ_2 are viscosities of fluid whose densities are 1.0 g/cm^3 and 1.1 g•cm^{-3}, respectively.

viscosity μ_1:μ_2 (cP=×10^{-3} Pa•s)	1:1	1:1.5	1:2	1:3	1:4	1:5
p (s^{-2})	90	62	42	29	20	13
q (s^{-1})	178	145	120	97	79	64

viscosity μ_1:μ_2 (cP=×10^{-3} Pa•s)	1:1	1.5:1	2:1	3:1	4:1	5:1
p (s^{-2})	90	68	51	28	20	12
q (s^{-1})	178	149	128	95	80	66

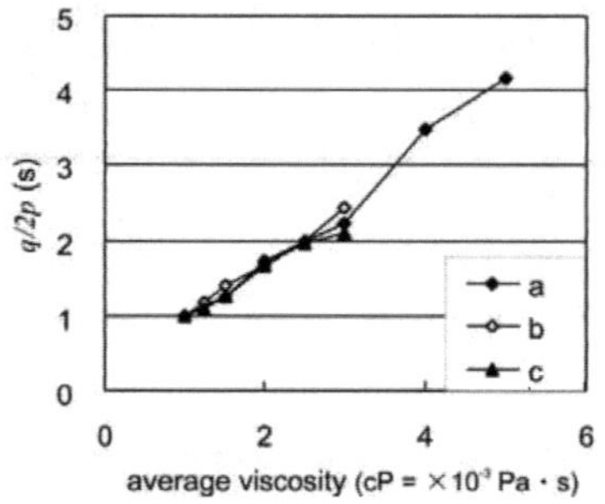

Fig. 6 Relationship between parameters of the second order curve ($\theta = -pt^2+qt$) and average viscosity. Density of each fluid is 1.0 g/cm^3 and 1.1 g/cm^3. (a) in the case where viscosity for each fluid is the same, (b) in the case where viscosity of fluid whose density is 1.0 g/cm^3 is fixed, (c) in the case where viscosity of fluid whose density is 1.1 g/cm^3 is fixed.

Interfacial tension

In this section, the effect of interfacial tension was examined. Fig. 7 shows interface tilting versus interfacial tension. Interfacial tension was varied from 1×10^{-6} to 4×10^{-5} N/m. Interface tilting decreases as the interfacial tension increases. Interfacial tension attenuates the effect of gravitational force and interface tilting. It is possible to speculate from this figure that interfacial tension greater than 10^{-3} N/m negates gravity effect. This result is in good agreement with the previous reports[11]. Since interfacial tension between water and commonly used organic liquid is about 10^{-3} to 10^{-2} N/m, the the effects of gravity could be practically neglected. However, the effects of gravity should be considered for microsystems which require multi-phase liquids with small interfacial tension.

Effects of Diffusion

The above results were for immiscible liquids. Finally, diffusion effect was considered. Diffusion coefficient varied from 1×10^{-9} to 5×10^{-9} m^2/s. Interface of the liquids was defined here as the plane where mass fraction of one liquid is 0.5. Density difference and average velocity were set to 0.1 g/cm^3 and 10 mm/s, respectively. Interface line became asymmetric with respect to the microchannel

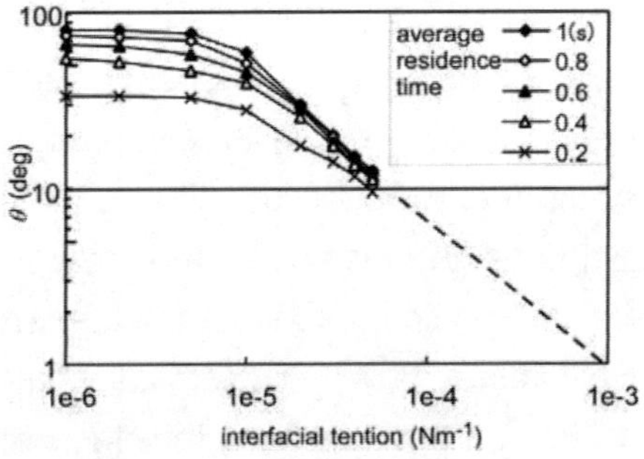

Fig. 7 Effect of interfacial tension on tilting.

center point, because gravity is exerted on diffusing liquid. Since interface was apart from the microchannel center, interface tilting θ was measured at the point near the center. Fig. 8 shows the effects of diffusion on interface tilting. This figure indicates that tilting can not be neglected when the diffusion coefficient is about less than 2×10^{-8} m^2/s. In such a case, the effects of gravity works faster than diffusion. When the density difference is 1/100 times (that is 0.001 g/cm^3), gravity effect becomes equivalent to that when the residence time is 100 times by eqn. (4). Considering the diffusion equation $< x^2 >= D\cdot t$ where x is diffusion length, D is diffusion coefficient, and t is residence time, gravity effect should be considered when D is about less than 2×10^{-10} m^2/s. The results indicate that gravity effect should be examined in macromolecules or biological solutions wherein diffusion coefficient approximates less than 10^{-10} m^2/s.

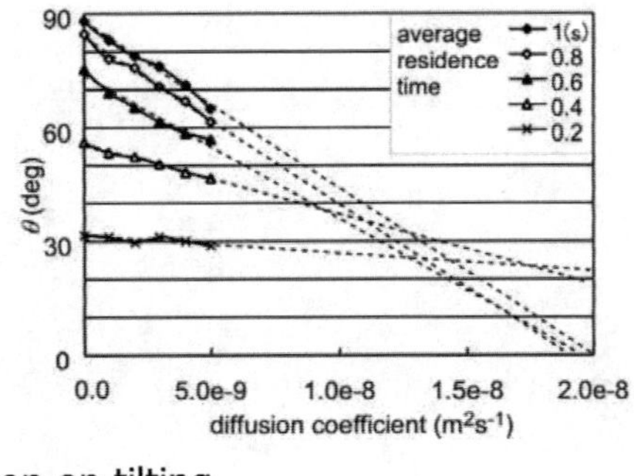

Fig. 8 Effect of diffusion on tilting.

Conclusion

The influence of gravitational force on the behavior of two-layer laminar flow in a microchannel was investigated by CFD calculation. In the case where cross-section of the microchannel was square and interfacial tension and diffusion between two liquids are small, it was possible to express the interface tilting by a simple equation with fluid characteristic parameters. This equation led to the conclusion that gravity effect is proportional to density difference, average residence time and channel size, and inversely proportional to viscosity. In addition, effects of microchannel aspect ratio, slip on the wall, interfacial tension and diffusion were investigated. It was suggested that gravity effects should be considered when interfacial tension or diffusion are small. There are many applications which utilize laminar flow for chemical analysis, micromixing, and chemical reaction. In a laminar flow system, interface tilting must have much influence on its performance. The results shown here can be useful for microdevice design in the light of prediction of the interface configuration.

References

1 V. Hessel, S. Hardt, H. Löwe and F. Schönfeld, *AIChE J.*, 2003, **49**, 566.
2 H. Chen and J.-C. Meiners, *Appl. Phys. Lett.*, 2004, **84**, 2193.
3 S. Takayama, et al., *Nature*, 2001, **411**, 1016.
4 M. Yamada, M. Nakashima and M. Seki, *Anal. Chem.*, 2004, **76,** 5465
5 K. Yamashita, et al., *Lab Chip*, 2004, **4**, 1.
6 Y. Yamaguchi,et al., *AIChE J.*, 2004, **299**, 200.
7 S. Ookawara, et al., *Chem. Eng. J.*, 2004, **101**, 171.
8 T. Brenner, T, Glatzel, R. Zengerle and J. Ducrée, *Lab Chip*, 2005, **5**, 146.
9 P. J. Stiles and D. F. Fletcher, *Lab Chip*, 2004, **4**, 121.
10 Y. -C. Lin and C. -P. Jen, *Lab Chip*, 2002, **2**, 164.
11 A. Hibara, et al., *Anal. Sci.*, 2001, **17**, 89.

71l Laminar Flow Based Electrochemical Microreactor

Paul J.A. Kenis and Seong Kee Yoon. Chemical & Biomolecular Engineering, University of Illinois at Urbana-Champaign, 600 S. Mathews Ave, Urbana, IL 61801
The concept of a laminar flow-based electrochemical microreactor that utilizes the occurrence of laminar flow in microfluidic channels to achieve the necessary compartmentalization of the reductant and oxidant streams in a single channel without a physical barrier such as a membrane has been introduced. Two specific examples were explored using laminar flow-based electrochemical microreactors: (i) Membraneless microfluidic fuel cells in which the fuel and oxidant streams are kept separate by flowing laminarly in parallel rather than by a polymer electrolyte membrane [1]; (ii) Microreactors for regeneration of nicotinamide cofactors for biocatalysis that can only be pursued at the microscale due to the ability to focus a stream of reactants closely to the electrode utilizing multistream laminar flow [2]. Diffusional mixing of the two or more streams in these electrochemical microreactors still can be detrimental to their performance. To circumvent these issues, we recently integrated a porous separator into a laminar flow-based electrochemical microreactor. This separator minimizes the inter-diffusion width while the system still takes advantage of the characteristics of the microscale. A second challenge for laminar flow-based electrochemical microreactors is the formation of depletion boundary layers on the electrodes. Through the integrated of chaotic micro mixers (see Stroock et al., Science 2002) we have been able to continuously replenish of the boundary layers on the electrode of the microreactor, thereby increasing the conversion rates in both applications.

[1] See for example: (a) Micro-Fluidic Fuel Cell Based on Laminar Flow. E.R. Choban, L.J. Markoski, A. Wieckowski, P.J.A. Kenis, J. Power Sources, 2004,128, 54-60. (b) Air-Breathing Laminar Flow Based Microfluidic Fuel Cell, R.S. Jayashree, L. Gancs, E.R. Choban, A. Primak, D. Natarajan, L.J. Markoski, P.J.A. Kenis, J.Am.Chem.Soc., 2005, in press. [2]Laminar Flow-Based Electrochemical Microreactor for Efficient Regeneration of Nicotineamide Cofactors for Biocatalysis, S.K. Yoon, C. Kane, E.R. Choban, T. Tzedakis, P.J.A. Kenis, J.Am.Chem.Soc., 2005, 127(30), 10466-10467.

A Microstructure Heat Exchanger for quenching the metabolism of mammalian cells

C. Wiendahl[1], J.J. Brandner[2], C. Küppers[1], B. Luo[1], T. Noll[3], M. Oldiges[1], U. Schygulla[2], C. Wandrey[1]

[1] Institute of Biotechnology 2, Research Center Juelich GmbH, Leo-Brandt-Str. 20, 52425 Juelich, Germany
[2] Institute for Micro Process Engineering IMVT, Research Center Karlsruhe, Hermann-von-Helmholtz-Platz 1, 76344 Eggenstein-Leopoldshafen, Germany
[3] Institute of Cell Culture Technology, University of Bielefeld, Universitätsstr. 25, 33615 Bielefeld, Germany

ABSTRACT

In this publication, the possibility to cool down a mammalian cell fermentation broth from physiological 37°C to 0°C or even less within a few seconds by using a metallic microstructure heat exchanger has been determined. As it is known from literature, a very rapid shut-down of the metabolism of cells is mandatory to obtain metabolome data. This can be performed by changing the environmental temperature very quickly by, e.g., use of liquid nitrogen. This is possible with bacteria cells, but not with the more fragile mammalian cells which might be destroyed to a very high percentage. The approach presented here is the use of a extremely efficient microstructure heat exchanger to cool the mammalian cell suspension down to around 0°C within seconds or faster. This will provide a kind of quasi-static state for the metabolism, which can be used for obtaining the desired data.

Keywords: metabolome analysis, quenching, microstructure heat exchanger, heat transfer

NOMENCLATURE

Latin symbols		
A	*effective heat transfer area*	m^2
C	*microchannel circumference*	*m*
F	*microchannel face area*	m^2
Nu	*Nusselt number*	
P	*pressure*	*Pa*
Pr	*Prandtl number*	
$\dot{Q}$	*thermal power*	*W*
Re	*Reynolds number*	
T	*temperature*	*K, °C*
c_{mean}	*mean specific heat capacity*	$J \cdot kg^{-1} \cdot K^{-1}$
c_p	*specific heat capacity*	$J \cdot kg^{-1} \cdot K^{-1}$
d_h	*hydraulic diameter*	*m, µm*
k	*overall heat transfer coefficient*	$W \cdot m^{-2} \cdot K^{-1}$
l	*microchannel length*	*m*
$\dot{m}$	*mass flow*	$kg \cdot h^{-1}$
n_{Ch}	*number of microchannels per foil*	
n_f	*number of microstructured foils per passage*	
o.r.	*of the range*	

o.v.	of shown value	
s	path length, foil thickness	m
Greek symbols		
α	heat transfer coefficient	$W \cdot m^{-2} \cdot K^{-1}$
λ	thermal conductivity	$W \cdot m^{-1} \cdot K^{-1}$
ε	efficiency of the heat transfer	%
$\Delta\vartheta_m$	mean logarithmic temperature difference	K
ξ	form factor	
Subscripts		
cold	cold water passage	
hot	hot water passage	
in	inlet position	
lam	laminar	
out	outlet position	
turb	turbulent	

Introduction

Many modern products as pharmaceuticals, food products or chemicals are produced via biotechnological processes. Besides the fermentation of bacteria or fungi the cultivation of mammalian cells in suspension culture has become more and more important during the recent years. Especially for high value products like Interferone, Erythropoetin or tPA which are used in medical care, processes with mammalian production cell lines are established. Being high profit production processes, the optimization is of great interest. To enhance the productivity and yield, the improvement of the process could be done through optimization of the process parameters like e.g. temperature, nutrient composition or media osmolality [1-6]. The addition of chemicals as Sodium Butyrate could enhance the productivity, too [2, 7]. But a deeper understanding of cellular metabolism and physiology is necessary for further improvements in process performance of animal cells for the industrial production of useful compounds. Such knowledge will contribute to a better understanding of which are the main substrates and metabolites that need to be monitored and controlled in the culture medium to increase cell growth and product yields [8]. The first steps were done. Many examples are found in literature where the monitoring of metabolites shed light on biochemical reactions and their connection. As the desired protein comprises only about 10-20 % of the total cellular protein [9, 10], clearer understanding of the interactions between substrate consumption, product formation and energy management could enhance productivity and reduce the costs of culturing mammalian cells [11].

Bonarius [12] based the metabolic flux analysis in continuously cultured hybridoma cells on ^{13}C-lactate NMR spectroscopy. Several other experiments were done with labeled glucose or other substrates. But the main problem with this method is that only distinct pathways could be examined with one experiment. Research with labeled glucose for example can only give results about the metabolism of glucose. Another problem is the broadening of the labeled compound. When it is converted into other chemicals the labeling is broadened beyond detection. This means that from a certain biochemical product on it is not possible to say whether the isotope comes from the labeling or from natural appearance of an isotope.

In the case of bacteria many efforts have been undertaken to reveal intracellular metabolite fluxes. Until now, there is a huge amount of established methods to stop the metabolism and show a snap-shot of a distinct state of metabolism by analyzing the sample via Liquid Chromatographie Mass Spectrometry (LC-MS) or Gas Chromatographie Mass Spectrometry (GC-MS) [13, 14]. This approach overcomes the disadvantages of labeling substrates to examine the metabolism. MS is a method with high sensitivity and high-throughput in combination with the possibility to confirm the identity of the components present in the complex biological samples as well as the detection. In most cases the identification of unknown and unexpected compounds is possible. The combination of separation techniques as chromatography expands the capability of the chemical analysis of highly complex biological systems [13, 14]. Using this tool for metabolome analysis, special product strains could be developed to increase the productivity and yield. An adaptation of these methods would push forward metabolome analysis in mammalian cells but is a difficult task.

To obtain the metabolome data, a very rapid stop of metabolism is required. This can be done by fast changes in the environment of a cell e.g. with quick changes of temperature or pH. To separate the extra- from the intracellular metabolites, the quenching is followed by separation of biomass. Villas-Bôas et al. [15, 16] gave a summary of the main different quenching methods. All these methods yield a very rapid stop of the metabolism of the target cells. With bacteria, the most commonly methods used are cold methanol or Perchloric Acid, with animal or plant tissues it is liquid Nitrogen. Transferring this to mammalian cells in suspension culture, the cell suspension has to be dropped or injected into cold methanol. But as mammalian cells are much more fragile than bacterial cells they would burst, and a separation of intra- and extracellular metabolites is not possible. Even if some of the metabolites are found intracellularly only, the high dilution would prevent the detection. The high salt content after concentration of the sample might disturb the analytics as well.
Another idea is not to stop the metabolism entirely by reaching temperatures of about -20 °C, but to try to slow down the metabolism within a short time. This would give an instant picture of the metabolome which is at least near to the real state of metabolism of the cells. To reach this objective, a metallic microstructure heat exchanger was connected to a stirred vessel bioreactor and the cell suspension was pumped through this device.

Microstructure heat exchanger

Since the early 90's of the last century, microstructure devices have been in discussion for use in chemical, pharmaceutical and bio-technological processes. But unlike in chemical or pharmaceutical reaction engineering, where multi-channel microstructure devices are now considered as very powerful tools for process intensification (see e.g. [17]), most microstructure devices used in the field of bio-tech are either single channel systems or analytical systems based on the standard microtiterplate. Devices like multi-channel microstructure heat exchangers or micromixers are not commonly used for biotech applications yet. Nevertheless, microstructure devices are well known for their superior heat and mass transfer behaviour [17, 18] and for the possibility to change temperatures very rapidly [19].

Manufacturing of metallic microstructure heat exchangers

Metallic microstructure heat exchangers for bio-technology are necessarily made out of materials which are suitable to the bio-processes on one hand, but also to the demands against pressure, temperature and corrosion. Therefore, it is necessary to choose a metal which can be microstructured quite easily and can also be bonded together to form stable and more or less complex structures. Due to the corrosion resistance as well as the suitability for bio-processing, stainless steel is the first choice. Microchannels or other designs of microstructures in stainless steel foils may be obtained either by mechanical precision micromachining (e.g. milling, turning or drilling, as it is described by Schaller et al. [20, 21], or by wet chemical etching (see, e.g. [22]). The cross section of the channels differs by the manufacturing method: While mechanically cut channels will take on the shape of the microtool used to cut, wet chemically etched microchannels have a semi-elliptic cross section due to the isotropic etching effect. In Figure 2, a SEM photo of a mechanically cut microchannel foil with rectangular microchannels is shown, while Figure 3 shows a SEM picture of an etched microchannel stainless steel foil.

To form the core of a microstructure heat exchanger, a number of microstructured foils is stacked together in the desired order and arrangement. The complete stack is then brought into a furnace and bonded by a diffusion bonding process. Depending on the material and the connecting face area, a strictly defined temperature and force program is applied to run the diffusion bonding process. The forces used for diffusion bonding are in the range from some kilonewton to several ten kN, depending on the connection area, while the applyable temperature range is from 500°C to more than 1000°C.

After the diffusion bonding step, the microstructure device is similar to a monolithic structure, like it is shown in Figure 4, where the view to the microchannel system of a crossflow heat exchanger is shown. The second passage is not shown but laying across the visible channels.

Fig. 2: SEM picture of a mechanically cut microstructure foil. The rectangular microchannels are 200 µm wide and around 100 µm deep.

Fig. 3: SEM picture of a wet chemically etched microstructure foil. The semi-elliptic microchannels are about 200 µm wide and around 70 µm deep.

Fig. 4: Crossflow microstructure heat exchanger made of stainless steel after diffusion bonding. Only small hints of the single layer structure are visible.

After the diffusion bonding step, the core bodies of the microstructure devices are cleaned and welded into a housing by electron beam welding. The housing may be equipped either with conventional tube fittings or any other fitting system suitable for the desired application. In Figure 5, an example photo of a crossflow heat exchanger core with an active volume of 1 cm^3 and another device welded to standard fittings is shown.

Throughout the complete manufacturing process a quality control is done, starting with measurements of the channel dimensions and ending with a nitrogen flowthrough measurement and a leak tightness measurement done with helium as test gas. More details on the manufacturing process and the microstructure heat exchanger devices can be found in [18, 23].

Fig. 5: Core of a stainless steel crossflow microstructure heat exchanger (left) and a similar core welded to conventional tube fittings. The active heat transfer volume of both devices is 1 cm^3.

Precalculation and modeling

For most microstructure devices consisting of systems of linear channels, and for most objectives, the precalculation using standard Nusselt theory equations is suitable. Therefore, the hydraulic diameter of each microchannel can be calculated using the simple equation

$$d_h = \frac{4 \cdot F}{C} \qquad (1)$$

where d_h is the hydraulic diameter, F is the face area of the channel and C is the circumference of the microchannels.

For further calculations with non-circular channels, a correction factor has to be applied to approximate a circular cross section. Those factors may be found, e.g., in [24]. For the pressure drop, equations like the Hagen-Poiseulle equation for laminar flow with a Reynolds number lower than approx. 2.300 or an equation developed by Petukhov and Gnielinski for turbulent flow with a Reynolds number Re between 3.000 and 10^5 may be applied (see [24]).

For a microstructure heat exchanger having a defined geometry and known mass flows as well as inlet and outlet temperatures, the heat transfer therefore was calculated by the energy balance equation (2).

$$\dot{m} \cdot c_p \cdot \Delta T = k \cdot A \cdot \Delta \vartheta_m \qquad (2)$$

The effective heat transfer area A was calculated by

$$A = n_f \cdot n_{Ch} \cdot C \cdot l \qquad (3)$$

The mean logarithmic temperature difference $\Delta\vartheta_m$ was calculated from the formula given in VDI-Wärmeatlas [24].
The heat transfer coefficient was defined for both a cold and a hot passage providing the same heat transfer area like it is shown in equation (4). Then, the overall heat transfer coefficient was calculated using equation (5).

$$\alpha = \frac{Nu \cdot \lambda_{Fluid}}{d_h} \qquad (4)$$

$$\frac{1}{k} = \frac{1}{\alpha_{hot}} + \frac{1}{\alpha_{cold}} + \frac{s}{\lambda_{solid}} \qquad (5)$$

For laminar flow (Re < 2.300), the Nusselt number was calculated using equations (6) to (9) (VDI Wärmeatlas).

$$Nu_{lam} = \sqrt[3]{\left(Nu_1^3 + 0.7^3 + \left(Nu_2 - 0.7\right)^3 + Nu_3^3\right)} \qquad (6)$$

$$Nu_1 = 3.66 \qquad (7)$$

$$Nu_2 = 1.615 \cdot \sqrt[3]{\mathrm{Re} \cdot \mathrm{Pr} \frac{d_h}{l}} \qquad (8)$$

$$Nu_3 = \left(\frac{2}{1 + 22 \cdot \mathrm{Pr}}\right)^{1/6} \cdot \sqrt[3]{\mathrm{Re} \cdot \mathrm{Pr} \frac{d_h}{l}} \qquad (9)$$

For turbulent flow (Re > 2.300), the Nusselt number was defined by equations (10) to (11).

$$Nu_{tur} = \frac{\frac{\xi}{8} \cdot (\mathrm{Re} - 1000) \cdot \mathrm{Pr}}{1 + 12.7 \cdot \sqrt{\frac{\xi}{8}} \cdot \left(\mathrm{Pr}^{2/3} - 1\right)} \cdot \left(1 + \left(\frac{d_h}{l}\right)^{2/3}\right) \qquad (10)$$

$$\xi = \left(1.82 \cdot \log(\mathrm{Re}) - 1.64\right)^{-2} \qquad (11)$$

When fluids with similar specific heat capacity are applied to both passages, the heat transfer efficiency ε can be used easily for comparison of differently designed (microstructure) devices. A definition is given in equation (12) and can be found in [24, 25].

$$\varepsilon = \frac{\dot{Q}}{\dot{m} \cdot c_{mean} \cdot \left(T_{hot,in} - T_{cold,in}\right)} \qquad (12)$$

In the specific case of symmetric fluid flow and using the same fluid for both passages, the capacity flow of both fluids is almost the same. Then, equation (12) can be rearranged to calculate the heat transfer efficiency ε with equation (13) (see [24]).

$$\varepsilon = \frac{\dot{m}_{cold} \cdot c_{p,cold} \cdot (T_{cold,out} - T_{cold,in})}{\dot{m}_{hot} \cdot c_{p,hot} \cdot (T_{hot,in} - T_{cold,in})} \qquad (13)$$

For modeling of the microstructure devices, the commercially available CFD software FLUENT was used. It had to be decided whether to perform a simulation of a complete microstructure device, which means the results would have been the plenum outlet temperature at a certain mass flow, or to perform a temperature distribution for a system consisting of two microchannel foils only and therefore calculate the temperature distribution across the complete area of the foil. Both simulations have been done, as it is described in [18, 23, 26].

Microstructure heat exchanger properties

The microstructure heat exchangers pre-calculated and manufactured are very powerful devices for process intensification. Stainless steel devices are corrosion-resistant and stable up to maximum static pressures of some ten MPa. With water as test fluid, a maximum thermal power of several kW can be transferred from one fluid to another, depending on the temperature difference and the fluids used. The heat transfer efficiency is up to 90% and more, as it is reported e.g. in Schubert et al 2001. A heat-transfer-area-to-volume ratio of 30.000 m^2/m^3 and more was reached. Overall heat transfer coefficients up to 56.000 $W/(m^2 K)$ and more have been reported (see: [23, 24]). Maximum mass throughput, determined with water, is in the range of 700 kg/h, obtained with the device presented here. The pressure drop arising on each passage is then in the range of about 0.5 MPa.

Nevertheless, none of the described microstructure devices was used for heating or cooling biological material before. Thus, it was to clear whether the living cells could withstand the shear stresses arising in the microstructures. The simplest method to find out was to try.

Experimental Materials and Methods

Microstructure heat exchanger: The microstructure heat exchanger used for the experimental setup was made out of stainless steel W316L. It provided a total number of 850 channels per passage. Each channel was 200µm wide and 100µm high, the sidewalls between the channels had a width of 100µm. Channel length was 14mm. The total active device volume was $1cm^3$. All microstructures have been manufactured by mechanical precision machining, the single microchannel foils have been connected by diffusion bonding, as it was mentioned in details before. The microstructure system was connected to standard Swagelok fittings by electron beam welding.
Cell line and medium: The cell lines CHO-Muc1-IgG2a PH 3931 and the mouse-mouse Hybridoma SP2/0 were used. The CHO K1 was transfected with pCDNA3/MUC1(16TR)-IgG by electroporation. As medium ProCHO4-CDM (Cat. No. BE12-029Q Biowhittaker Europe, Verviers, Belgium) was used. Spinner flask cultures were performed at 37 °C in a humidified incubator containing an atmosphere with 5% CO_2. 5 L batch cultures were performed using Cellferm-pro® (DASGIP, Jülich, Germany), which is a modular cultivation system allowing a parallel culture in up to 4

pH- and oxygen-controlled batch-, fed-batch- or continuously operating spinner flasks or bioreactors. Control of pH was realized by variation of CO_2 in the gas-phase. Cultivations parameters were 40 % pO_2, pH 7.
The microstructure heat exchanger was connected to the vessel by a coupling (Staeubli, Bayreuth, Germany) and was autoclaved between the samplings. The cooling fluid was cooled and pumped by a cooling device (LAUDA DR.R.WOBSER GMBH & CO. KG, Lauda-Königshofen, Germany), cell suspension was pumped with approximately 150 mL/min with a flexible-tube pump (Watson-Marlow 501 U).
For measurement of cell disruption the cytotoxicity detection kit (LDH, Lactate Dehydrogenase) (Cat. No. 1 644 793 Roche Diagnostics GmbH, Penzberg, Germany) was adapted.
The taken samples were cetrifuged at 200 g for 8 min at -8 °C, the pellet was extracted with ice cold methanol (60%), centrifuged again with a following ultracentrifugation (5kDa cut-off, Cat. No. VS0611, Vivascience GmbH, Hannover, Germany) of the supernatant at 3200 g for 1 h at -5 °C (Allegra 6 KR Cetrifuge, Beckmann Coulter, Krefeld, Germany).
For LCMS measurements Applied Biosystems 4000 Q trap was used. Column: Synergi Fusion RP (Phenomenex, Aschaffenburg, Germany), flow rate 200 µL/min, injection volume 10 µL, solvent A Acetic Acid/Tributyricacid, solvent B Methanol, oven temperature 20 °C.

Results

Mechanical cell disruption caused by pumping through the microstructure heat exchanger was tested with an LDH (Lactate Dehydrogenase) measurement in the supernatant of the pumped cell suspension. LDH is an intracellular enzyme, the concentration of the enzyme in the supernatant is proportional to the amount of cell disruption. Cell numbers from 2,49 to $7{,}67 \cdot 10^6$ cells /mL were used. As it can be seen in Figure 6, even with high cell numbers the mechanical stress caused only 11 % of total cell disruption, which is an acceptable value.

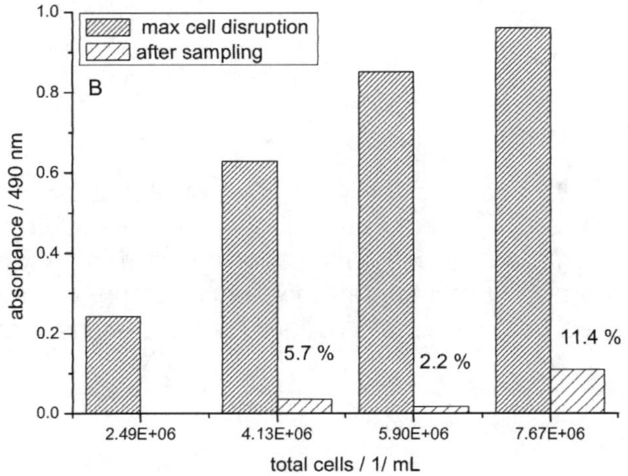

Fig. 6: Mechanical cell disruption (SP 2/0) caused by pumping cell suspension through heat exchanger determined by LDH concentration in supernatant. Max cell disruption: total cell disruption caused by TritonX100; after sampling: LDH

measurement in supernatant of pumped cell suspension. LDH concentration is given by absorbance at 490 nm.

The microstructure heat exchanger was connected to a 5 L bioreactor and three samples at three different cultivation times were taken without loss of sterile conditions. Each sample had a volume of 400 mL, which consisted of 50 mL pre run, five 50 mL aliquots taken with a cooling microstructure heat exchanger and two 50 mL aliquots taken with the not-cooled microstructure heat exchanger. The LDH concentration in all samples were measured and compared to the LDH concentration in the cultivation supernatant in the bioreactor, which was taken by regular sampling at the same time. The results show, that there is no significant difference in the LDH concentration between the different aliquots (Fig. 7). Due to different cell numbers, the temperature of the samples decreases down to – 4 °C, for this reason the temperature of the cooling fluid must be adapted. But apparently even these varying temperatures of the cooled cell suspension (Fig. 7) caused no cell disruption compared to culture supernatant and the warm aliquots.

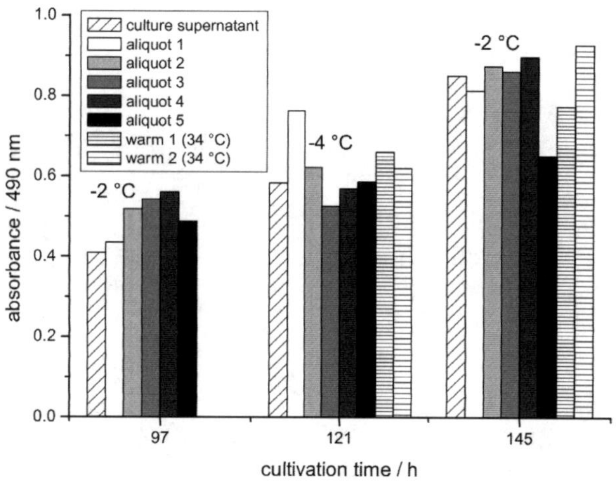

Fig. 7: Mechanical cell disruption (CHO) caused by cooling samples with the microstructure heat exchanger at three different cultivation phases. Culture supernatant: LDH concentration in supernatant of cell suspension in the bioreactor. Aliquots 1-5: LDH concentration in supernatant of samples 1-5, each sample volume 500 mL. Warm 1,2: Two 50 mL samples taken without cooling the heat exchanger. LDH concentration is given by absorbance at 490 nm over cultivation time. Temperatures given in the graph show temperature of the aliquots 1-5 directly after the heat exchanger.

After extraction the samples were stored at -80 °C until measurement via LCMS/MS. The LCMS/MS measurements showed that with this sampling method intracellular metabolites could be detected (Fig. 8). The chromatogram shows metabolites out of the glycolysis (Glucose-6-Phosphate/Fructose-6-Phosphate and Pyruvate) and the TCA cycle (Citrate, Succinate and α-Ketoglutarate).

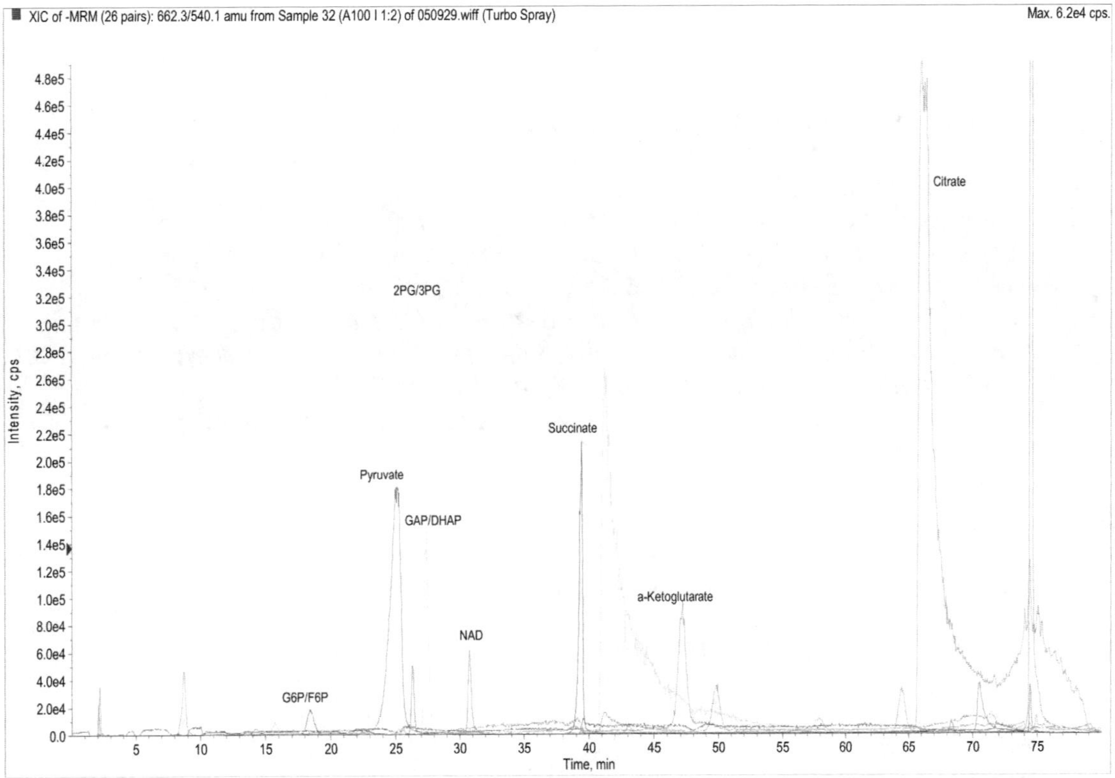

Fig. 8 Chromatogram of LCMS/MS measurements of a sample at 97 h cultivation time, taken with cooled heat exchanger. G6P/F6P: Glucose-6-Phosphate/Fructose-6-Phosphate, 2PG/3PG: 2-Phosphoglycerate/3-Phosphogylcerate, GAP/DHAP: Glyceraldehyde-3-Phosphate/Dihydroxyacetonephosphate, NAD: Nicotine-Adenine-Dinucleotide.

Summary

For investigation of a cells metabolome it requires a rapid quenching of the metabolism (at least within seconds), quantitative extraction and highly sensitive analysis of the metabolites. While in microbial fermentations the positive impact of this technology has already been shown in process optimizations and strain development, the transfer to mammalian cell culture is still hampered because of the cells' sensitivity and complexity. Due to the mechanical sensitivity of mammalian cells (because of their lack of a cell wall) quenching methods used for microorganisms like filtration or transfer of the cells into cold methanol (-40°C) are not applicable. We have investigated the applicability of a metallic cross-flow microstructure heat exchanger to cool mammalian cell fermentation broth from physiological 37°C to 0°C or even less within a few seconds. The microstructure heat exchanger was connected to a stirred vessel bioreactor, samples were taken through the microstructure heat exchanger. Only little cell damage induced by the passage could be detected. The sterility of the bioreactor was maintained throughout the whole cultivation (150 h). After centrifugation in the cold the cells intracellular metabolites were extracted and analyzed by LC-MS/MS. In summary, the microstructure heat exchanger used in this study seems to be a suitable technology to overcome one of the major obstacles in metabolomics in mammalian cell culture.

References

[1] Altamirano, C., et al., *Analysis of CHO cells metabolic redistribution in a glutamate-based defined medium in continuous culture.* Biotechnology Progress, 2001. **17**(6): p. 1032-1041.

[2] Chun, B.H., et al., *Enhanced production of recombinant B-domain deleted factor VIII from Chinese hamster ovary cells by propionic and butyric acids.* Biotechnology Letters, 2003. **25**(4): p. 315-319.

[3] Fox, S.R., et al., *Maximizing interferon-gamma production by Chinese hamster ovary cells through temperature shift optimization: Experimental and modeling.* Biotechnology and Bioengineering, 2004. **85**(2): p. 177-184.

[4] Kaufmann, H., et al., *Influence of low temperature on productivity, proteome and protein phosphorylation of CHO cells.* Biotechnology and Bioengineering, 1999. **63**(5): p. 573-582.

[5] Kim, M.S., et al., *Biphasic culture strategy based on hyperosmotic pressure for improved humanized antibody production in Chinese hamster ovary cell culture.* In Vitro Cellular & Developmental Biology-Animal, 2002. **38**(6): p. 314-319.

[6] Lee, M.S. and G.M. Lee, *Hyperosmotic pressure enhances immunoglobulin transcription rates and secretion rates of KR12H-2 transfectoma.* Biotechnology and Bioengineering, 2000. **68**(3): p. 260-268.

[7] Kruh, J., *Effects of Sodium-Butyrate, a New Pharmacological Agent, on Cells in Culture.* Molecular and Cellular Biochemistry, 1982. **42**(2): p. 65-82.

[8] Cruz, H.J., J.L. Moreira, and M.J.T. Carrondo, *Metabolic shifts by nutrient manipulation in continuous cultures of BHK cells.* Biotechnology and Bioengineering, 1999. **66**(2): p. 104-113.

[9] Bonarius, H.P.J., et al., *Metabolite-balancing techniques vs. C-13 tracer experiments to determine metabolic fluxes in hybridoma cells.* Biotechnology and Bioengineering, 1998. **58**(2-3): p. 258-262.

[10] Bonarius, H.P., et al., *Metabolic flux analysis of hybridoma cells in different culture media using mass balances.* Biotechn Bioeng, 1996. **50**: p. 299-318.

[11] Forbes, N.S., D.S. Clark, and H.W. Blanch, *Analysis of metabolic flux in mammalian cells*, in *Bioreaction engineering*, K. Schügerl and K.H. Bellgardt, Editors. 2000, Springer: Heidelberg. p. 556-94.

[12] Bonarius, H.P.J., et al., *Metabolic-flux analysis of continuously cultured hybridoma cells using (CO2)-C-13 mass spectrometry in combination with C-13-lactate nuclear magnetic resonance spectroscopy and metabolite balancing.* Biotechnology and Bioengineering, 2001. **74**(6): p. 528-538.

[13] Villas-Boas, S.G., et al., *Mass spectrometry in metabolome analysis.* Mass Spectrometry Reviews, 2005. **24**(5): p. 613-646.

[14] Villas-Boas, S.G., et al., *Mass spectrometry in metabolome analysis.* Mass Spectrom Rev, 2004.

[15] Villas-Boas, S.G., et al., *Global metabolite analysis of yeast: evaluation of sample preparation methods.* Yeast, 2005. **22**(14): p. 1155-1169.

[16] Villas-Boas, S.G., S. Rasmussen, and G.A. Lane, *Metabolomics or metabolite profiles?* Trends in Biotechnology, 2005. **23**(8): p. 385-386.

[17] Hessel, V., et al., *Chemical Micro Process Engineering*. 2005: Wiley-VCH.

[18] Brandner, J.J., et al. *Comparison of crossflow micro heat exchangers with different microstructure designs*. in *3rd Int. Conf. on Micro- and Minichannels*. 2005. Toronto, Canada.

[19] Brandner, J.J., et al., *Fast Temperature Cycling in Microstructure Devices*., Chem. Eng. J. 2004, 101 (1-3): p. 217-224

[20] Schaller, T., et al., *Mechanische Mikrostrukturierung metallischer Oberflächen*. F&M Feinwerk und Mechanik, 1994. **102**(5-6).

[21] Schaller, T., et al., *Microstructure grooves with a width of less than 50 mu m cut with ground hard metal micro end mills*. Precision Engineering-Journal of the American Society for Precision Engineering, 1999. **23**(4): p. 229-235.

[22] Madou, M., *Fundamentals of Microfibration*. Vol. 2nd ed. 2002: CRS Press LLC, Boca Raton, FL, USA.

[23] Brandner, J.J. and K. Schubert. *Fabrication and Testing of Microstructure Heat Exchangers for Thermal Applications*. in *3rd Int. Conf. on Micro- and Minichannels*. 2005. Toronto, Canada.

[24] *VDI Wärmeatlas*. 7. ed. ed. 1994: VDI-Verlag, Würzburg, Germany.

[25] Wagner, W., *Wärmeaustauscher*. 2nd ed. 1999: Vogel-Verlag, Würzburg, Germany.

[26] Schubert, K., et al., *Microstructure devices for applications in thermal and chemical process engineering*. Microscale Thermophysical Engineering, 2001. **5**(1): p. 17-39.

71n Micro Technology for Large Scale Application: Direct Synthesis of Hydrogen Peroxide

Lin Li and Kurt VandenBussche. UOP, 25 E Algonquin Rd., Des Plaines, IL 60016
The direct synthesis of hydrogen peroxide is an attractive alternative for the traditional alkylanthroquinone process. To make this process economically feasible it is highly desirable to achieve total conversion of H2 and O2 without recycle, which means the H2/O2 in the feed should be at stoichiometric ratio. On the other hand, stoichiometric H2/O2 ratio is also a requirement of maintaining high catalyst selectivity. In addition, no dilutant should be used in order to maintain high H2 and O2 partial pressure and therefore to enhance reactor productivity. However, handling of hydrogen/oxygen mixture at 1:1 ratio without any dilution presents a significant safety concern, and achieving full conversion without compromising selectivity poses a big challenge for reactor design.

In this paper, we will present our experiences of using micro technology to meet such challenges. A mixer using micro technology was used to pre-mix hydrogen and oxygen before entering the catalyst bed. Both the micromixer and catalyst bed were designed to make sure the reactor can be operated safely for H2/O2 mixture in the explosion range as required to achieve optimal performance. Such concept was first tested in a small lab setup, and then a fully automated pilot plant was built to further demonstrate the reactor performance and to study the effect of various operation conditions.

The pilot plant data demonstrated the advantages of using micro technology for direct synthesis of hydrogen peroxide. Very high productivity with ~80% conversion and 75% selectivity has been demonstrated when using methanol as solvent. Based on pilot plant data, the scale up of this technology was proposed and the design of commercial reactor will be presented. Our experiences confirmed that when implemented properly the novel micro technology can be applied at large scale and become an enabler for a breakthrough new technology.

MICROFLUIDING ACTIVITIES IN ARTIFICIAL AND BIO-MEMBRANE

Nitosh Kumar Brahma
Department of chemical Engineering,
Indian Institute of technology, Kharagpur-721302, India.

EXTENDED ABSTRACT

Along with development of nano-particles and bio-nano-technology and the support of genetic engineering, it is today we have developed additional concept for continuations of drug release, monitoring of different units, starting from chemical to biological system including the organ specific disease control with the help of nono-divices. To realize the process the relevant question was, how do you introduce the process in continuous mode. To describe the process and the concept of microfluiding activities in artificial and bio-membrane, the concept of fluid; mass and heat transfer through capillary system must be studied, where the Reynolds should be less than < 30. In general micro fluid transport obeys the low Reynolds number $\boldsymbol{Re = L.V.\rho / \mu}$, where L denotes the length of tube, V the velocity, ρ as density and μ as viscosity of fluid. Investigations of low-Reynolds number hydrodynamics naturally fall into the fluid dynamics realm of long interest to the chemical engineering and nano-biotechnology community. Gad-el-Hak, M (1999); Kim and Karrila, (1991) [1,2]. The fluid transport in microtubules is conceptualized by means of dragging, that the molecules maintain two fundamental criteria namely surface tensions of fluids and thermo dynamical equilibrium. However surface force play a more important role than the standard continuum of transport processes. So the molecular organizations of polymers and their charges on the surface of the molecules will take the major role to handle the movements and driving force of small fluid volumes through mico-nano-capillaries at small length scales. To compare the properties of micro fluiding activities changing from the range of macro-to micro- and nano-scale the following conditions are to be taken care;

1.The fluids must possess less inertia force compared to electrostatic/electrodynamic viscous or capillary effects.

2. The fluid must show the responsible molecular layering to control the local density fluctuations, to develop the slip length with non-linear shear rates at the junctions of capillary pores to convert non-Newtonian fluids to Newtonian fluids.

3. It develops a fluid nature by increasing diffusion velocities at the entrance of capillaries avoiding the clogging caused by O_2 or some other means.

4. Smooth surface with less frictions and higher surface tensions could support the movements of fluidic activities to nono-microtubules.

5. Rheology of the fluid pays a significant role in manipulations and controlling of fluid flow.

The liquid motions is maintained basically by the (a) van-der Waals molecular forces, (b) molecular interactions, (c) the electric charge effects, and (d) the surface tensions of liquids characterized by following criteria;

1. Applied Pressure difference,
2. Applications of Electric field for electro osmotic flow rate (τ), which is proportional to ϖ. η, Where (ϖ) is the width of tube and at given potential drop, that converts to ϖ^3. η, in case of volumetric flow.
3. Small cross section of micro channels presents high electric resistance and which is more or less than > 100 V/cm for maintaining low current.
4. Thermal energy is essential to weekend the viscous damping.
5. Electro osmotic plug flow will support the hydrodynamic dispersions.

Length scale matching is important for efficient momentum and energy transfer for controlling the motions of desired fluids and molecules. Several other important physico-chemical properties are to be considered to maintain the transport of micro fluids through nano- or micro channels in following conditions.

1. Electro thermal flow,
2. Electro osmosis, AC electro osmosis, Dielectrophoresis,
3. Capillary Driving Forces Owing to Wetting of Surfaces by the fluid,
4. Free surface flows-drives caused by the gradients of interfacial tensions (Marangoni flow).
5. Biological fluid (polymer) transport.

References;

1. Gad-el-Hak, M(1999) The fluid mechanics of Micro devices-The freeman Scholar Lecture, " J fluids Engineering., 121: 5 .
2. Kim, S and Karrila SJ (1991) Micro hydrodynamics, Butter worth-Heinemann.
3. Jensen KF (1999) Micro chemical System, Status Challenges and Opportunities, AIChE J 45: 2051.
4. Ho, C.M(2001) Fluidics, The link between Micro and Nano-Sciences and Technologies, 14th IEEE Intl. Conf. On MEMS, 2001 Interlaken, Switzerland., 375.
5. Quake SR and A Scherer, 2000 From micro to Nano-fabrication with Soft materials, Science, 290: 1536.
6. Brahma NK (2005) Escherichia Coli K-12 can... (Manuscript ID 1903807617835264, communicated to Nao-biotechnoloy, UK
7. Stellwagen NC, Gelfi, C, and Righetti PG, (2001) The free solution mobility of DNA, Biopolymer, 42: 687.

Nitosh Kumar(Dipl.-Ing) Brahma(Prof), Chemical Engineering, Indian Institute of technology, Kharagpur-721302,, Kharagpur-721302,, kharagpur-721302, Midnapore, 721302, India

Text Not Available

Abstract ID#: 39832
Password: 384517
Title: Microfluiding Activities in Artificial and Bio-Mmbrane
Session Selection: T1005 Micro Process Engineering and Intensified Process Systems
Preferred Presentation Format: Poster
Lead Presenter's Email: nkbrahma@yahoo.com
Comments to Organizers: If posible to place it in other oral session, according to you descetion.

First Author

Presenting Author
Nitosh Kumar(Dipl.-Ing) Brahma(Prof)
lecturer/Scientific Officer
Chemical Engineering
Indian Institute of technology, Kharagpur-721302,
Kharagpur-721302,
kharagpur-721302
Midnapore, 721302
India
Phone Number: 913222-28-1918
Fax Number: 91-3222-25-5305
Email: nkbrahma@yahoo.com

At this point the next step is normally to add another author or submit the abstract text.
Click the button below to add another author. Or click the "Submit Text" link in the Abstract Control Panel to submit the text. NOTE: At least one presenter must be identified before the abstract text can be submitted

71o Modular Microreaction System - a Powerful Tool for Process Development and Production

Shaibal Roy, Process Technology Group, Bayer Technology Services, Americas, 8500 West Bay Rd; MS #52, Seabrook, TX 77586 and Olaf Stange, Ehrfeld Mikrotechnik BTS GmbH, Mikroforum Ring 1, 55234 Wendelsheim, Germany.

Abstract

Microprocess technology opens new pathways for the development of economical and innovative processes. In the chemical and pharmaceutical industry, this technology plays an increasingly important role in research, development and production because of its decisive advantages over conventional technology. These include:

- µ very fast mixing
- µ efficient heat exchange
- µ narrow residence time distribution
- µ fast system response for efficient process control
- µ easy process control and automation
- µ high level of safety due to low hold-up
- µ easy scale-up

With the Modular Microreaction System, this work demonstrates a powerful development and production tool combining the advantages of microprocess technology with great flexibility and a wide scope of application for process development and intensification. The system comprises of individual (microstructure-equipped) modules that can be freely combined. These modules perform basic process operations such as mixing, dispersing, heat transfer, etc (Figure 1). The system can also be used for a wide range of process conditions including reactions in liquid phase, gas phase or heterogeneous gas-liquid reactions. Operating parameters include a temperature range of –20 to 200 °C (–100 °C for the CryoReactor module); operating pressures up to 100 bar; and rated throughputs of up to 25 L/h for low-viscosity media.

In the presented work, the viability of the system for investigating various multiphase reactions will be displayed using specific examples. For example, a transesterification (Scheme 1) has been performed in the Modular Microreaction System and in a conventional laboratory equipment. Transesterifications are usually associated with moderate to slow reaction rate constants (Otera, 1993). The aim of this study was to elucidate whether microreaction technology offers advantages for these types of reactions.

Thomas.daszkowski.b@bayer.com; Ph: 281-383-7941; Fax: 281-383-5403

Fig. 2 shows a comparison of the yield obtained using conventional glassware reactor and the Modular Microreactor System. With conventional laboratory glass ware, the reaction temperature of 75 °C was limited by the boiling point of the reaction mixture. Under these conditions one observed a conversion of ~90% after ~ 9 hours (Fig. 2a). However, the Modular Microreaction System allows the measurement of an entire parameter space in that time. Due to the ability to work under elevated pressures and reaction temperature, the reaction rate can be increased. For example, at 150 °C and 10 bar, the conversion of 90% is achieved within 15 min (Fig. 2b). Due to the continuous operation mode and the small scale, the heat transfer from the reaction mixture become more efficient than in a corresponding batch process and possible thermal decomposition of the product or other unwanted losses can be avoided.

Reference

J. Otera, Transesterification, Chem. Rev. 93 (1993) 1449-1471

Figures

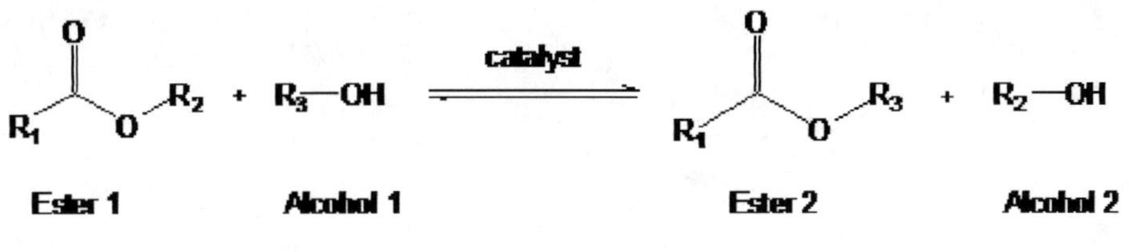

Scheme 1: Transesterification reaction scheme

Figure 1: Example configuration of a Modular Microreaction System for heterogeneous reactions

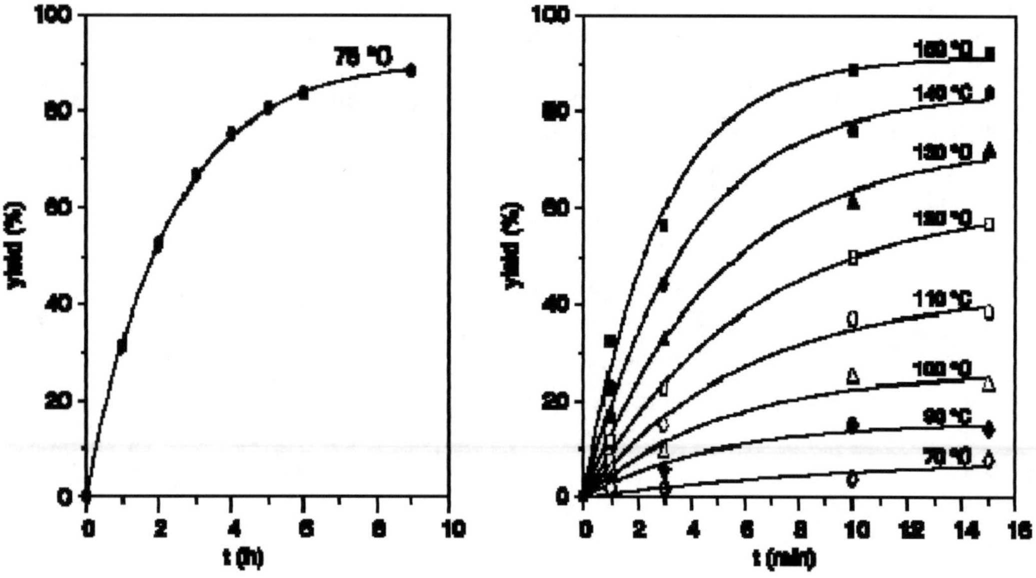

(a) (b)

Figure 2: Comparison of yield measurements for a trans-esterification using (a) standard batch laboratory equipment at 1 bara and (b) the continuous flow Modular Microreaction System at 10 bara.

71h Parallel Peptide Synthesis on Microfluidic Arrays for Immunological Applications

Suparna Mandal and Erdogan Gulari. Chemical Engineering, University of Michigan, Ann Arbor, 2300 Hayward, 3074 H.H. Dow, Ann Arbor, MI 48109
Since most natural ligands of proteins are peptides, a study of the highly specific interactions between proteins, peptides and other biomolecules on microarrays is a promising technique for high throughput identification and consequent modification of these molecular interactions. Peptides being shorter and more stable than proteins are popular probes for screening interactions at single amino acid resolution. With advances in combinatorial chemistry a number of techniques have emerged to immobilize a multitude of peptides on a solid surface either by chemical ligation by a covalent bond or by in situ synthesis. In situ synthesis eliminates the need for individual synthesis and purification of a large number of peptides, thereby enabling routine synthesis. In this work we present an integrated system for routine and flexible in situ synthesis of peptides using BOC chemistry. This system was developed using a combination of solid phase peptide synthesis, photogenerated reagent chemistry and fluorescence imaging techniques to generate spatially addressable peptide microarrays with up to 9mer sequences. The synthesis substrate was a microfluidic array of 300 reaction cells connected by channels, converging into a solution inlet and outlet, etched on a flat silicon surface sealed by a glass cover. Using HATU/DIEA chemistry a coupling time of 5 minutes for Boc-amino acids was achieved. A constant synthesis efficiency of >98% was observed for sequences of up to 9 amino acids. The bioactivity of synthesized peptides was tested by performing on-chip ELISA for mapping of antibody-peptide interactions. To further elucidate the biological applications of the synthetic peptides, the on-chip binding specificity of a 8mer peptide to a murine B lymphoma cell line was demonstrated.

Potential of batch microreactors for continuous processing of particles, slurry and polymer reactants, and particle forming products at high temperatures and high pressures

R. L. Smith, Jr., H. Inomata and K. Arai, Tohoku University, Sendai, Japan

Introduction

In this work, we review some of the high temperature (>400 C) and high pressure (20 to 2000 MPa) applications that we have studied with batch type microreactors (ca. 50 nL, 10,000 m2/m3) using water as the reaction solvent. We also discuss methods to process materials on a continuous basis using batch microreactors.

Applications

The applications include:

(i) Boehmite nano-particle formation through reaction of aluminum nitrate (Wang et al, 2002; Li et al., 2002)

(ii) Polymer particle reactions (Fang, et al., 1999; Smith et al., 2000)

(iii) Dechlorination and conversion of chlorinated particles (Fang et al., 2004; Fang et al., 2005; Nagai, et al, 2005)

(iii) Microcrystalline cellulose mixing and conversion (Ogihara, et al. 2005)

(iv) Biomass gel formulation, mixing and reaction (Ogihara, et al., 2005)

In-Situ Observations

In-situ observations, which have been recorded through optical windows made of diamond or moissanite, will be presented.

Particle formation during rapid heating of microreactor contents

In particle formation steps, it can be observed that during heating of reacting mixtures that particles tend to form at the gas bubble - liquid interfaces. This is probably due to the concentration gradients that develop and is deemed to be important for handling particle forming systems.

Heat transfer characteristics at high temperatures and high pressures in the vicinity of the critical point of water and its mixtures

In the heat transfer to the mixture at the critical point, it was observed that violent cycling of fluid phases occurred. In the heating of slurries of microcrystalline cellulose, intense convection and mixing of the microreactor contents was observed when gas was added to the system.

Mass transfer during oxidization of particles

In the oxidation of particles, for the case of deca-chlorobiphenyl, it was observed that oxidation of the particle surface occurred extremely rapidly in the initial formation process of oxygen obtained from breakdown of H2O2. Superfine myriads of O2 bubbles were observed that diffused through the liquid phase and attacked the particle surface. However, as the reaction proceeded, these O2 bubbles tended to coalesce and thus probably reduced the reaction rate.

Oil, particle and gas formation during reactions

Oil, particle and gas formation during reactions occur as a result of product liquefaction, hydrolysis or decomposition. Examples include dissolution characteristics of polyethylene in supercritical water, dissolution of nylon in supercritical water, and dehlorination of polyvinylchloride in supercritical water.

Mixing characteristics for slurry systems

It is expected that for practical application of batch microreactors on a continuous basis that product feeds will be in the form of gels. Therefore, images of biomass gels undergoing reaction will also be shown. From results thus far, mixtures tend to show strong convection when heated at low densities (< 300 kg/m3).

Heating

Heating of the microreactors is an important aspect for successful operation on a continuous basis. Previous applications have mainly used electrical or resistive heating. As a point of originality of this work, heating by chemical methods will be considered and some examples will be shown including some thermographic images where possible. A conceptual design for continuous processing with batch microreactors will be presented.

Conclusions

Batch microreactors can be used in a continuous manner once chemical or other heating methods are developed. One aspect of this research is to present some conceptual designs for achieving the handling of high solids, high concentration mixtures of solids, liquids, and gases in microreactors. Direct visual images of the above applications will be given at the conference.

References

1. Wang, T., Smith Jr., R.L., Inomata, H., Arai, K., Reactive phase behavior of aluminum nitrate in high temperature and supercritical water, Hydrometallurgy, 65 (2002) 159-175.

2. Li, G., Smith Jr., R.L., Inomata, H., Arai, K., Synthesis and thermal decomposition of nitrate-free boehmite nanocrystals by supercritical hydrothermal conditions, Mater. Lett. 53 (2002) 175-179.

3. Fang, Z., Smith Jr., R.L., Inomata, H., Arai, K., Phase behavior and reaction of polyethylene terephthalate-water systems at pressures up to 173 MPa and temperatures up to 490 C, J. Supercrit. Fluids 15 (1999) 229-243.

4. Fang, Z., Smith R.L., Jr., Inomata, H., Arai, K., Phase behavior and reaction of polyethylene in supercritical water at pressures up to 2.6 GPa and temperatures up to 670 C, J. Supercrit. Fluids 16 (2000) 207-216.

5. Smith Jr., R.L., Fang, Z., Inomata, H., Arai, K., Phase Behavior and Reaction of Nylon 6/6 in Water at High Temperatures and Pressures, J. Appl. Polym. Sci. 76 (2000) 1062-1073.

6. Fang, Z., Xu, S., Butler, I.S., Smith Jr., R.L., Kozinski, J.A., Destruction of decachlorobiphenyl using supercritical water oxidation, Energy & Fuels 18 (2004) 1257-1265.

7. Fang, Z., Xu, S.K., Smith Jr., R.L., Arai, K., Kozinski, J.A., Destruction of deca-chlorobiphenyl in supercritical water under oxidizing conditions with and without Na2CO3, J. Supercrit. Fluids 33 (2005) 247-258.

8. Nagai, Y., R. L. Smith, Jr., K. Arai, Proc. SCEJ 37th Autumn Meeting (9.15-17), Okayama, Japan 2005 (in Japanese).

9. Ogihara, Y., R. L. Smith, Jr., H. Inomata, K. Arai, Proc of the 7th Intl. Sympos. Supercrit. Fluids (ISSF2005), Orlando, U.S.A.

10. Ogihara, Y., R. L. Smith Jr., H. Inomata, K. Arai, Cellulose, published online, 14 October, 2005.

11. Fang, Z., Minowa, T., Smith Jr., R.L., Ogi, T., Kozinski, J.A., Liquefaction and Gasification of Cellulose with Na2CO3 and Ni in Subcritical Water at 350 C, IEC Res 43 (2004) 2454-2463.

Process Control of Micro Chemical System for Steam Reforming Reaction

W. C. Shin, and R. S. Besser

New Jersey Center for Microchemical Systems
Chemical, Biomedical and Materials Engineering, Stevens Institute of Technology, Castle Point on Hudson, Hoboken, NJ 07030, USA.
Tel.: 201-216-5257; Fax: 201-216-8306;
E-mail: shinwoocheol@gmail.com and rbesser@stevens.edu

Abstract

As the attempts to reduce the size of chemical process capture chemical engineers' interest, there are significant requirements of control strategies utilizing miniaturized components. After considerable research efforts to reduce the size of chemical reactors, similar efforts are needed to integrate *control elements* into complete chemical process units. While many MEMS-based examples exist of sensors and actuators for control of pressure, temperature, and flow, there are few cases where these components have been integrated together with reaction units and control algorithms, even in very basic forms.

In order to elevate the understanding in this area, we integrated an investigation of an integrated microchemical system with microfabricated flow sensors, temperature sensors, and heating elements with a silicon microreactor and process control algorithms for autonomy. Each component is interfaced with Labview control software and communicates via specific schemes implemented in the software. In this work, catalytic steam reforming of methanol is used as a model reaction.

Introduction

By the emergence of Micro Electro Mechanical System (MEMS), traditionally bulky chemical systems have been dramatically scaled down to the size of IC chips. This miniaturized chemical plant, so called microchemical systems, consists of essential peripherals including micoreactor, microheaters, mixers, flow sensors and temperature sensors [1-3]. Miniaturization of the reactor system provides speeding-up of processes as well as improving the process safety. However, to be compatible with the faster reaction speed and the shorter response of the system, appropriate control strategies are required for the miniaturized system [4]. In this regard, the ultimate goal of this work is to optimize the process based on the findings from the integrated microreactor system in which the implemented feedback control scheme plays a major role on the overall reaction performance. A schematic diagram of the feedback control system is shown in Figure 1, which includes a heating block and a flow control block as actuating units with two sensing units for the temperature and the flow rate reading. By finding a proper control approach for our specific reactor system, it is expected that the overall reaction efficiency for the target reaction would be improved with enhanced system stability.

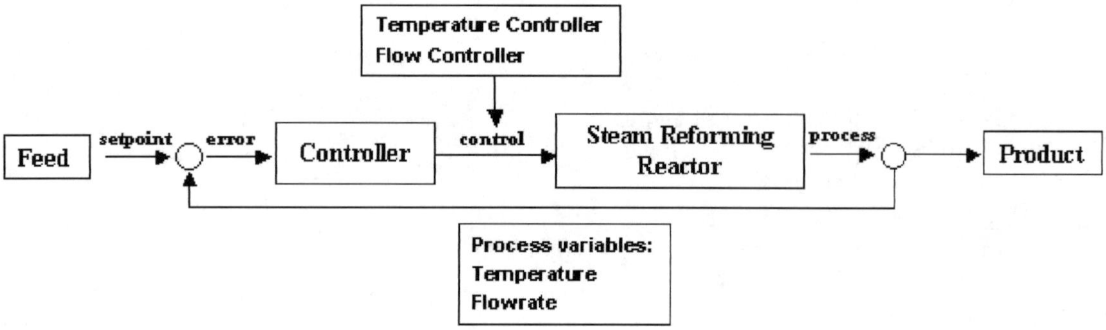

Figure 1. Block diagram of feedback control for steam reforming reaction system

Experimental

Silicon-based microreactors and thin-film heaters were fabricated by micromachining technology. A dry etching process was used to produce a reaction zone in the reactor for the catalyst loading [5]. Microfabrication of the thin-film flow sensor, heater and temperature sensor were straightforward. The thin metal film was deposited on the Pyrex wafer surface by physical vapor deposition followed by film patterning by wet etching [6]. Patterned heater and sensor resistors use DC power for heating and sensing purpose. Finally the integration of the Pyrex chip and the silicon microreactor was completed by anodic bonding technique and the bonded chips were incorporated by epoxy bonding. Completed reactor block possesses its heating unit on the top of the reaction zone and the reactor temperature was continuously monitored from the temperature sensor locating on the back of the reactor. Figure 2 shows the completed reactor block with individual heater and microreactor chips.

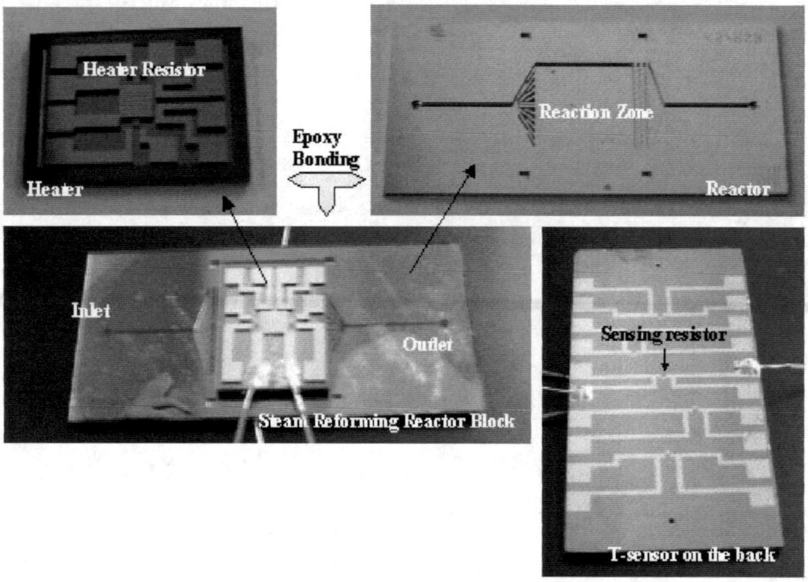

Figure 2. Steam reforming reactor block with heater and temperature sensor

Because of its effective and straightforward solution to implementation, a proportional integral derivative (PID) algorithm using LabView control software was deployed to our system. To control the heating and sensing units, LabView control software communicates with each sensing units and receives DC voltage signals from the sensors. Ultimately, an algorithm derived from the LabView program would be implemented on a microcontroller chip integrated within the microchemical system.

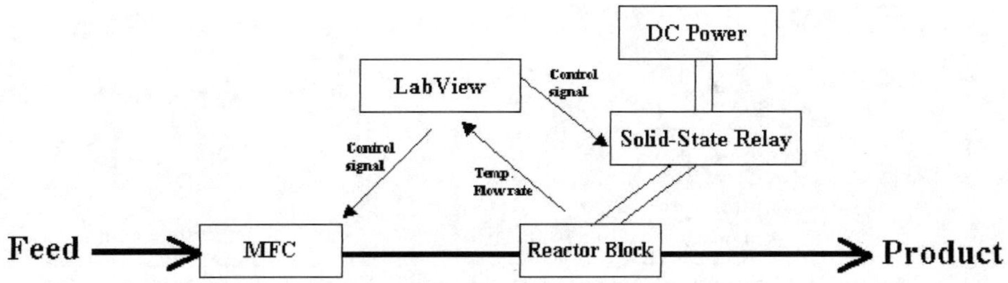

Figure 3. Experimental schematic

Figure 3 describes the experimental setup. The signals from the sensors were imported to the PID control toolset in the LabView program and according to the error ranges between process variables and set points, PID controller sends output signals to the controlling units. In this work, the thin-film heater was controlled by utilizing a solid-state relay being triggered by the PID algorithm to power the heater for a desired temperature of the reactor. A calorimetric flow sensor was used to sense the flow rate in the reactor and the flow control loop was built up by supplementing a MFC (mass flow controller), which communicates to the software and controls the flow rate of the feed stream. The MFC is a temporary flow control tool, which would eventually be replaced by a micro valve.

Measurement with the controller

The realized PID controller was tested in the laboratory with a PID toolkit in the LabView for temperature and flow control. In the following measurements the temperature sensor was first tested with changing PID tuning parameters. In Figure 4, preliminary results from the temperature control test show that as the proportional gain increases the system becomes faster (shown in Figure 4(b)) but the system tends to be more stable with lower proportional gains (shown in Figure 4 (a)). Changing the integral and derivative terms also affects the system response and stability as increased integral gain reduces the steady state error and elevating derivative gains decreases overshoot and higher gain stabilities.

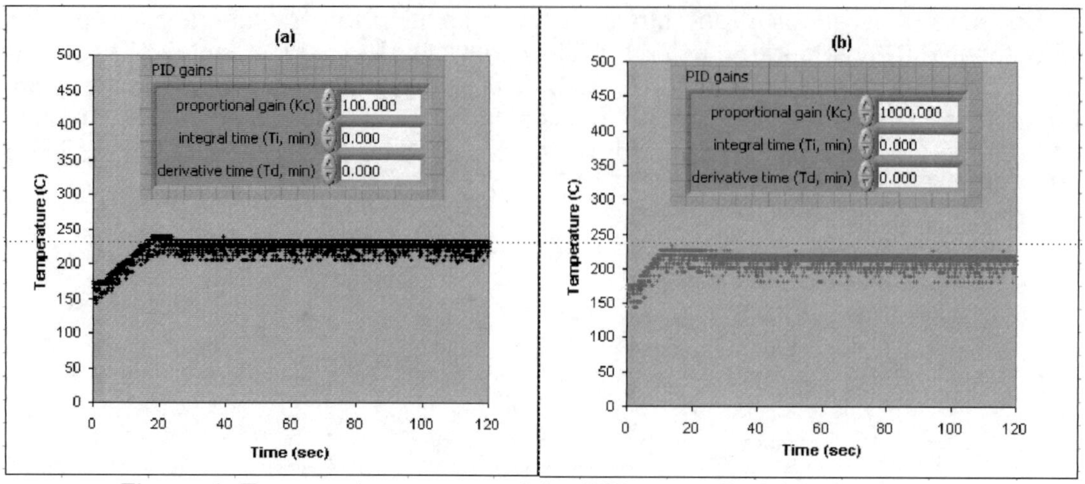

Figure 4. Temperature course of the PID controlled thin-film heater

The appropriate control settings will be decided by adjusting the optimal gains for P, I and D to get an ideal response from a control system. Once the most fitting control scheme is chosen for temperature and flow control, the target reaction (steam reforming reaction) will be carried out to understand correlations between control tuning parameters and the reaction efficiency.

Numerous research efforts have been accomplished in conjunction with the traditional chemical process optimization by well-established control theories, but there are not many attempts to apply the control theories to the emerging microreactor study. In this regard, new control strategies must be developed for the realization of independent, autonomous microchemical processing platforms.

References

1. R.S. Besser, "New Directions in Reactor Design Through Miniaturization," *3rd Annual Tulane University Engineering Forum*, New Orleans, LA, September 13, (2002).
2. K.F. Jensen, "Microchemical systems: Status, challenges, and opportunities," *AIChE Journal,* **45** (10) (1999) 2051-2054.
3. R. Srinivasan, I.-M. Hsing, P.E. Berger, M.P. Harold, J.F. Ryley, J.J. Lerou, K.F. Jensen, and M.A. Schmidt, "Micromachined chemical reactors for heterogeneously catalyzed partial oxidation reaction," *AIChE Journal*, **43** (11) (1997) 3059-3069.
4. C.M. Caruana, "Micromixing creates a stir," *Chem. Eng. Prog.,* **96** (5) (2000) 9-10.
5. R. S. Besser, W. C. Shin, "Deep reactive ion etching characteristics of a micomachined chemical reactor," *J. Vac. Sci. Technol.*, B **21(2)** 912-915.
6. W. C. Shin, R. S. Besser, "A micromachined thin-film gas flow sensor for microchemical reactors," *J. Micromech. Microeng.,* **16** 731-741.

The Influence of Pulsatile Injection Pressure on Cavity Pressure in Dynamical Injection Molding

*Weihua Zhou, Jinping Qu**

National Engineering Research Center of Novel Equipment for Polymer Processing,
The Key *Laboratory of Polymer Processing Engineering, Ministry of Education,*
South China University of Technology, Guangzhou, China

Abstract

Cavity pressure is an important issue in injection molding technology, as it determinates the quality of final parts. Compared to traditional injection molding, Cavity pressure during the filling and post-filling stages is pulsatile in dynamical injection molding. Frequency and amplitude of vibration are important factors to cavity pressure. Low-Density Polyethylene (LDPE) disc-samples were injection molded by way of traditional injection molding and dynamic injection molding. Meanwhile, by using the self-made pressure sensor, cavity pressure change with real-time is acquired. The test results show that vibration frequency of cavity pressure is equal to vibration frequency of the ram. In addition, it is show that vibration amplitude of cavity pressure, average cavity pressure and the maximum cavity pressure increase with increasing vibration amplitude and/or frequency of the ram.

Introduction

The history profile of cavity pressure is one of the important factors which affect the quality of the final part in injection molding (1). However, there are many factors which affect the cavity pressure, such as runner system, injection pressure, injection speed, packing pressure, melt temperature, mold temperature and thickness and shape of parts. Thus, cavity pressure will be different with conditions above-mentioned when the same parts are injection molded.

The application of vibration in molding of polymer is one of the hot research subjects. Early of 1980s, M. Bevis and P. Allan (2), et al. firstly introduced vibration in packing stage via hydraulic device. Results got by them and later generations show that vibration packing can enhance the quality of injection molded part (3, 4).

Jinping Qu (5) introduced vibration field generated by electromagnetic field in the process of plasticization, injection and packing, implemented dynamical plasticization metering, injection and packing. Experiments' results show that the filling pressure may reduce when the vibration is introduced. Jinping Qu, et al. (6) investigated the cavity pressure during the vibration-assisted metal powder injection molding.

In this work, the influence of vibration on cavity pressure during the dynamic injection molding will be presented. The present experiments have been carried out with LDPE

*To whom correspondence should be addressed. Email: jpqu@scut.edu.cn.

injected into a center-gated disc mold which has the facility for pressure transducer fitment at one site along its radial direction using the traditional injection molding and dynamic injection molding. The cavity pressures are real-time acquired by the pressure transducer and the data acquisition system.

Experimental

Instruments

An experimental mold was made by HUAXINKE Company and was mounted on a vertical hydraulic vibration ram-fed injection machine with a clamp force of 250 kN, as shown in Figure 1. The geometry of cavity was disc and its diameter was 150 mm with a depth of 2 mm, as shown in Figure 2. A pressure transducer was set in the mold. The transducer was located at the place of r=10 mm. The data acquisition system connected with the pressure transducer and monitored the cavity pressure.

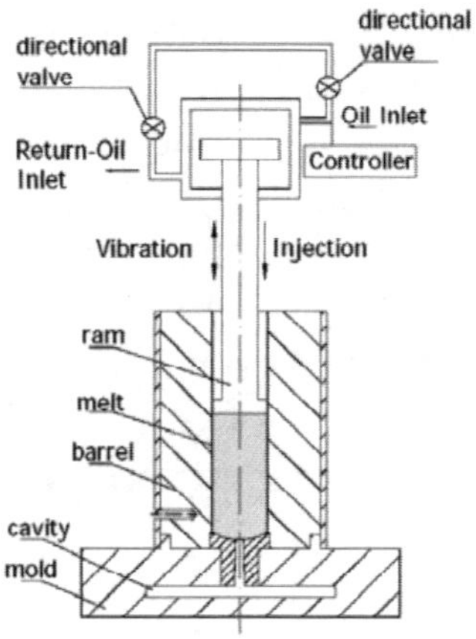

Figure 1 Schematic illustration of pulsatiled injection

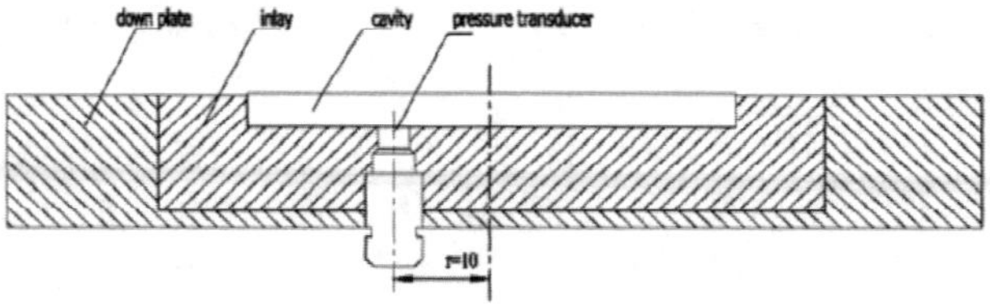

Figure 2 Schematic of collocation of the cavity pressure transducer

Materials and technical parameters

The sample material used in this work was low-density polyethylene (LDPE, 050), which made in China Petroleum & Chemical Corporation.

In experiments, melt temperature was 190℃, mold coolant temperature 40℃. Compared to traditional injection, other two parameters were vibration amplitude (A) and vibration frequency (f) of the ram in pulsatiled injection. The experiments were carried within a range of vibration frequency and vibration amplitude.

Results and discussion

The cavity pressure during filling stage

When the vibration amplitude was 0.1 mm and the vibration frequency altered, Figure 3 shows the cavity pressure profile as a function of time. When the vibration frequency was 6 Hz and the vibration amplitude altered, the profile of the cavity pressure against time was shown on Figure 4. Viewing from Figure 3 and Figure 4, the changing trend of the cavity pressure against time was the same between the traditional injection and the dynamic injection.

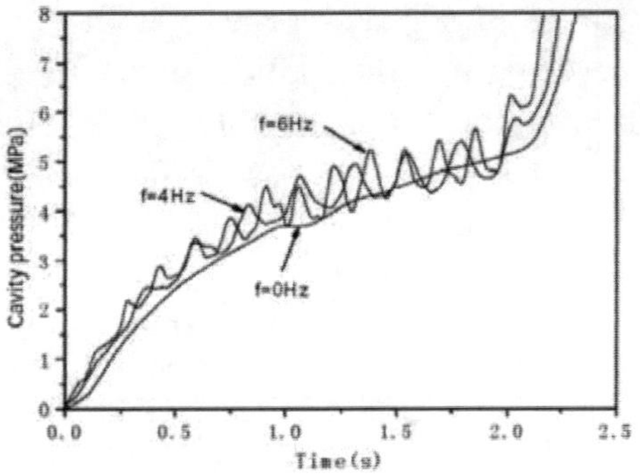

Figure 3 Cavity pressure profile during filling stage (A=0.10mm)

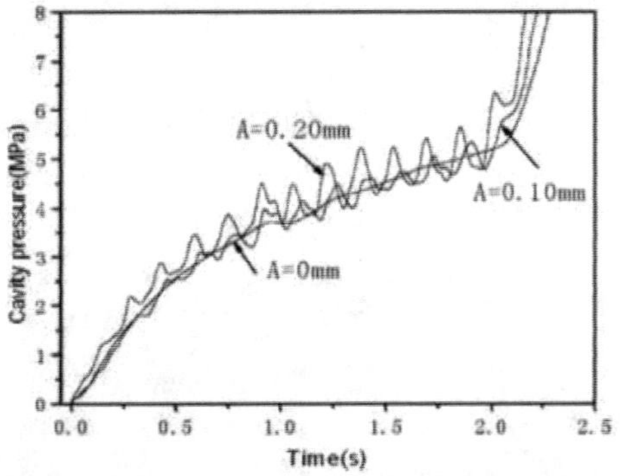

Figure 4 cavity pressure profile during filling stage (f=6Hz)

When frequency was 0 Hz (i.e. traditional injection), the variations of cavity pressure were little and irregular. The variations may be resulted from the pressure variation of hydraulic pump in the injection molding machine during injection molding, or from environment influence on the data acquisition system. When vibration frequency was 4 Hz, the fluctuation frequency of cavity pressure equaled 4 Hz. When vibration frequency was 6 Hz, the fluctuation frequency of cavity pressure equaled to 6 Hz. It is obvious that the fluctuation frequency of cavity pressure and the vibration frequency of the ram were the same.

As seen in Figure 3, the average cavity pressure at a frequency of 4 Hz was bigger than that at a vibration frequency of 0 Hz. Moreover, the average cavity pressure at a vibration frequency of 6 Hz was bigger than that at a vibration frequency of 4 Hz. It is obvious that the average cavity pressure increases with vibration frequency. The maximum cavity pressure at a vibration frequency of 4 Hz was bigger than that at a vibration frequency of 0 Hz, and it at a vibration frequency of 6 Hz was bigger than that at a vibration frequency of 4 Hz, as shown in Figure 3. So, the maximum cavity pressure increased with frequency in a certain range.

Figure 4 illustrated that the pressure fluctuations, the mean pressure and the maximum pressure in cavity increased with the vibration amplitude.

The cavity pressure during packing stage

Figure 5 showed the cavity pressure profile at amplitude was 0.2 mm, while frequency varied. Figure 6 showed the cavity pressure profile at a vibration frequency of 6 Hz, while vibration amplitude varied. As seen in Figure 5 and Figure 6, it could be found that the changing trend of the cavity pressure against time was the same between the traditional injection and the dynamic injection, the fluctuation of cavity pressure and the maximum cavity pressure increased with vibration frequency and vibration amplitude.

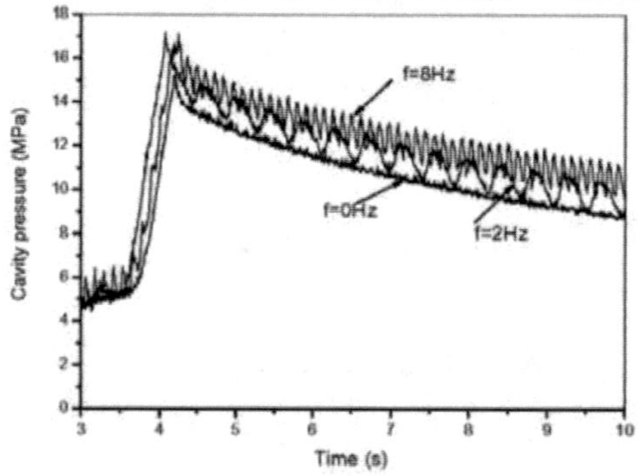

Figure 5 cavity pressure profile during packing stage (A=0.2mm)

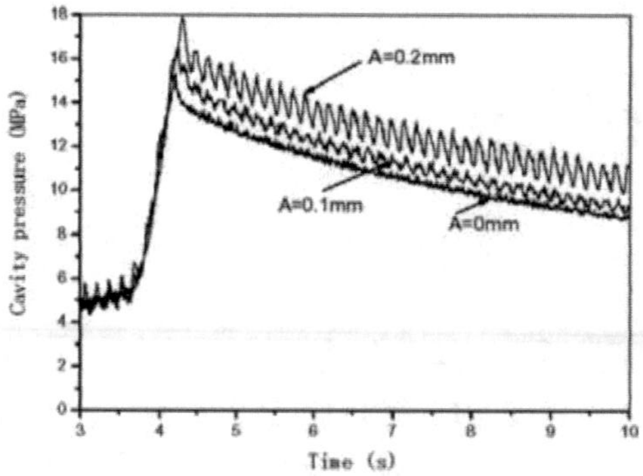

Figure 6 cavity pressure profile during packing stage (f=6Hz)

The maximum cavity pressure

According to the analysis above-mentioned, the maximum cavity pressure increased with the vibration frequency and the vibration amplitude. Figure 7 showed the relationships between the maximum cavity pressure and the vibration frequency at different vibration amplitude. Figure 8 showed the relationships between the maximum cavity pressure and the vibration amplitude at different frequency. As can be seen from Figure 7 and Figure 8, when the vibration force was superimposed, the maximum cavity pressure decrease with the increase of the vibration frequency or the vibration amplitude within a certain range of frequency or amplitude. This might be caused by the reduction of melt viscosity when vibration energy was introduced during the injection molding.

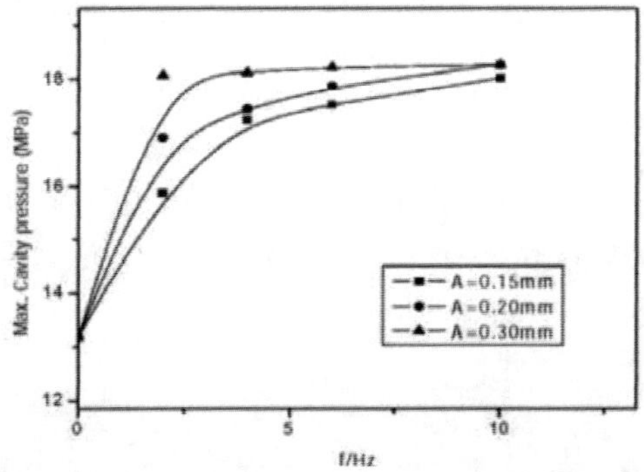

Figure 7 Relationships between the maximum cavity pressure and vibration frequency

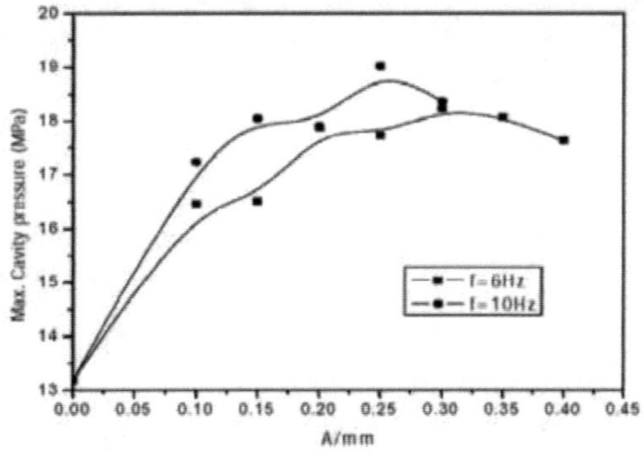

Figure 8 Relationships between the maximum cavity pressure and vibration amplitude

Conclusions

As has been revealed in this study, some conclusions as follows could be drawn:

1) The changing trend of the cavity pressure against time is the same between the traditional injection and the dynamic injection.

2) The fluctuation frequency of cavity pressure and the vibration frequency of the ram are the same.

3) The fluctuation of cavity pressure and the maximum cavity pressure increase with vibration frequency and vibration amplitude in a certain range.

Acknowledgements

The authors wish to acknowledge the National Nature Science Foundation of China (Grant 10472034) for the financial support.

References

1. M. R. Kamal, W. I. Patterson, et al., (1987), "Dynamics and control of pressure in the injection molding

of thermoplastics", *Polym. Eng. Sci.*, October, pp. 1403-1410.
2. Allan P S, Bevis M J, (1985), *US Patent*, No.4,925,161.
3. Ibar J P, (1997), "Melt viscovity of plastics reduction by vibration during filling in injection molding", *Society of Plastics Engineers*, ANTEC'97
4. Ibar J P, (1998), "Control of polymer properties by melt vibration technology: A review", *Polym. Engi, Sci.*, January, pp. 1-12.
5. Jinping Qu, Wu Hong-wu, (2000), "New technology of dynamic vibration field enhancing polymer injection molding", *Plastics*, June, pp. 9-12.
6. Jinping Qu, Yin Xiaochun, et al., (2005), "Experimental study of the cavity pressure during the vibration-assisted MIM", *ANTEC 2005 Plastics: Annual Technical Conference,* Volume 1: Processing

High Throughput Discovery of Families of High Activity Water-Gas-Shift Catalysts

Karin Yaccato, Ray Carhart, Alfred Hagemeyer, Michael Herrmann,

Andreas Lesik, Peter Strasser, Howard Turner, Anthony Volpe, Henry Weinberg

Symyx Technologies Inc., 3100 Central Expressway, Santa Clara, CA 95051, USA, www.symyx.com

Christopher J. Brooks

Honda Research Institute, USA, 1381 Kinnear Road, Columbus, OH 43212, USA

Abstract

Catalyst libraries were synthesized on wafers in 16x16 array format and screened with a high throughput scanning mass spectrometer in the temperature range of 200 °C to 350 °C. More than 200 wafers were screened under various conditions and more than 250,000 experiments conducted to comprehensively examine catalyst performance for various higher-ordered compositions. Promising compositions have been scaled up and optimized in a Celero™ 8x1 multi channel fixed bed reactor. Examples of discovered WGS catalysts that operate in the temperature regime of ~200°C-350°C are the binaries Pt-CeReCoVMoNaK, the ternaries Pt-CeReCoMoVFe-Na and Pt-CoMoVFe-K, supported on zirconia, ceria and titania carriers.

Keywords: Water gas shift, combinatorial, high throughput, catalyst for fuel cell.

Introduction

Current state-of-the-art water gas shift catalysts [1, 2] (FeCr for high temperature shift and CuZn for low temperature shift) are not active enough for use in fuel processors for the production of hydrogen from gasoline for fuel cells for power generation in vehicles or for stationary applications [3]. The need to substantially lower catalyst volumes has triggered a search for novel WGS catalysts that are an order of magnitude more active than current systems; and moreover, are resistant to air and moisture during shut down. Novel catalytic materials for the high, medium and low temperature water gas shift reactions have been discovered by application of combinatorial methodologies.

Experimental

Libraries were synthesized on round 3" diameter quartz wafers in an 11x11 element matrix or 4" quartz wafers in a 16x16 element format. The libraries are designed using Symyx' Library Studio® software. Once the library design is complete, Library Studio® generates a text file containing instructions that the Impressionist® software uses to control liquid dispensing robots, to physically prepare the library. The project makes use of Symyx' proprietary robotic slurry dispensing technology for carrier powders to prepare wafer formatted carrier pre-coated arrays that can subsequently be impregnated with metal solutions [4-6]. For secondary scale-up, the libraries are designed using Library Studio® and prepared in a similar manner as the primary screening libraries. Secondary screening synthesis involves the use of a Powdernium™ solid handling robot for support dispensing. The masses of catalyst prepared were always in excess of one gram, and designed in a 6x 4 element format. All secondary catalysts were prepared using incipient wetness techniques.

Scanning mass spectroscopy is a primary screening tool that uses flat wafer catalyst surfaces, a scanning/sniffing nozzle and a mass spectrometer to compare relative catalytic activities (figure 1). The sniffing/scanning nozzle was designed to 'touch down', surround each independent catalyst element, deliver the feed gas, and transmit the product gas stream to the quadrupole mass spectrometer. Each element is heated locally using a CO_2 laser, allowing for an accessible temperature range of approximately 200 – 600°C. The product signals from each element are analyzed for a given time period (generally 1 – 2 minutes) and an average over the last 6 data points is calculated and reported for each of the catalyst signals. This results in a screening time of approximately 3.5 hours per 11x11 catalyst grid. Secondary screening involved using an 8-channel Celero plug flow reactor. Conversion and selectivity were determined by measuring CO, CO_2, and H_2 concentration by gas chromatography using TCD and FID detection.

Results and Discussion

Pt is a well-known catalyst for both hydrocarbon steam reforming and water gas shift reactions. However, Pt alone, is both too expensive and not active enough. Metals such as Co, Mo, Ru and Ni have also been used as WGS catalysts but are normally too active for the selective WGS reaction and cause methanation of CO to CH_4 under typical reaction conditions. In other words, the hydrogen produced by the water gas shift reaction is consumed as it reacts with the CO present in the presence of such catalysts to yield methane. This methanation reaction activity has limited the utility of metals such as Co, Ru, Mo, and Ni as water gas shift catalysts. In this work, we have discovered WGS catalysts from combinations of Pt with Co, Mo, and V.

As a representative example for a discovery wafer, figure 2a-c shows library design and synthesis as well as screening data at 250C for the 5-point ternaries Pt-Co_Ru_Mo-14M supported on ZrO2 carrier ('-' signifies 'and' whereas '_' signifies 'or'). Each ternary Pt-Co-X, Pt-Ru-X and Pt-Mo-X is mapped out by different 5 compositions. The 16x16 wafer accommodates 45 ternaries (14 metals as dopants plus the 3 undoped binaries as reference for each of Pt-Co, Pt-Ru, Pt-Mo). It is seen that many hits are more active than the Pt/ZrO2 standards in row 1 and column 16. Cobalt from Na3Co(NO2)6 precursor is the most active and results in high CO2 yields. Many highly active hits in the Mo block are noteworthy: Mo ~ MoCe ~ MoLa ~ MoZr > MoRu ~ MoRe. Furthermore, Ru and CoCe are also very active. Little difference is observed for Ce(III) and Ce(IV) precursors and between zirconyl nitrate and zirconyl acetate precursors, respectively. Mo is more selective than Ru or Co.

Figure 3a-b shows the design for a focus library consisting of 56-point ternaries Pt_Ru_Rh_Re_Pd-Mo on ZrO2 carrier. The low temperature shift activity ranking is given by PtRu > PtRe > PtRh ~ Pt > PtPd and the high temperature shift selectivity ranking is given by PtRe > PtRu > PtRh. It is observed that PtRe forms a synergistic binary that is very selective over the full temperature range. PtReMo is slightly less active but more selective than PtRuMo.

For best results, the catalysts also contain alkali metals (Na and K). Figure 4 shows the primary screening results of PtNa (CoMoV) on zirconia at 200 °C. As the CO conversion increases, the points will move up the WGS diagonal trajectory. If a catalyst starts to methanate, the point will move up off the trajectory, indicating CO conversion, not CO_2 production. For reference, the Pt and Na concentrations remained constant throughout the library. At the same time, the Co, Mo and V atomic concentration varied, but their sum also remained constant. The standard for this library was Pt/ZrO_2 with a Pt loading of 2.5 times that of the rest of the library. Several facts became quite evident. First, at lower temperatures, the V and Mo rich materials showed high activity, similar to the standards (red data points in Figure 4), with much less Pt. Second, as the temperature was increased, the Co rich catalysts became more active. Finally, there was little synergistic advantages of multiple base metals. In other words, Mo or V was usually better than Mo+V. The $PtVNa/ZrO_2$

series was then scaled up for secondary screening using a Celero™ plug flow reactor system. Figure 5 shows the CO concentration of the secondary screen from 175 to 325 °C. If one focuses at 225 °C, it is easy to see a large performance difference between some catalysts, while others have similar activities. Through further screening, it was easy to learn many compositional combinations had similar catalytic behavior, both in regards to activity and selectivity. Although not presented here, further studies were conducted that led to a better understanding of the effects of metal precursor, order of metal addition, and pretreatment conditions.

Conclusion

The addition of selected base metals as well as alkali metals to a Pt WGS catalyst was shown to increase the activity dramatically, enabling a reduction in the amount of Pt. This work also validates combinatorial techniques as a valuable technique for new catalyst discovery.

Figure 1

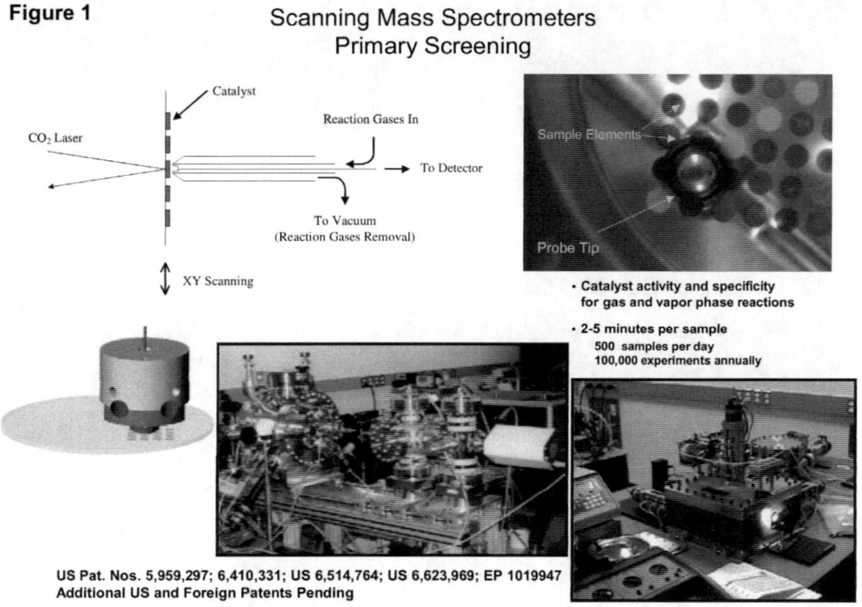

Figure 2a

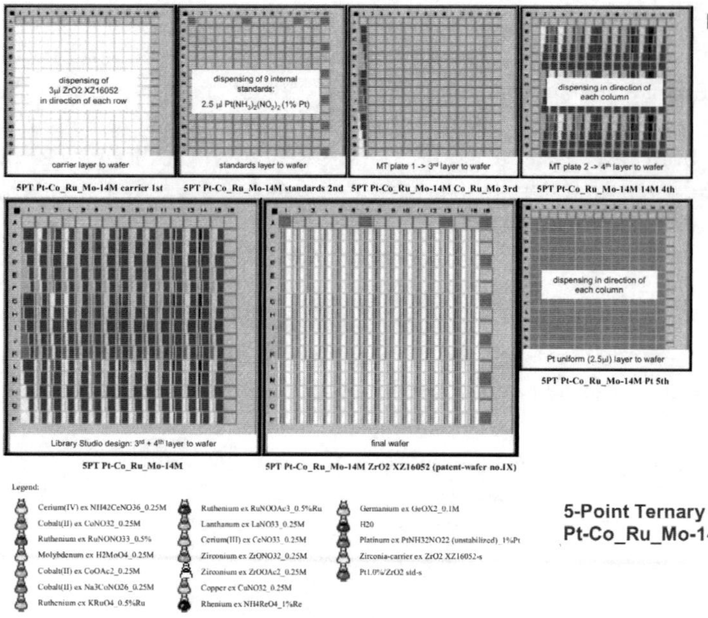

5-Point Ternary
Pt-Co_Ru_Mo-14M/ZrO_2

Figure 2b

5PT Pt-Co_Ru_Mo-14M/ZrO_2

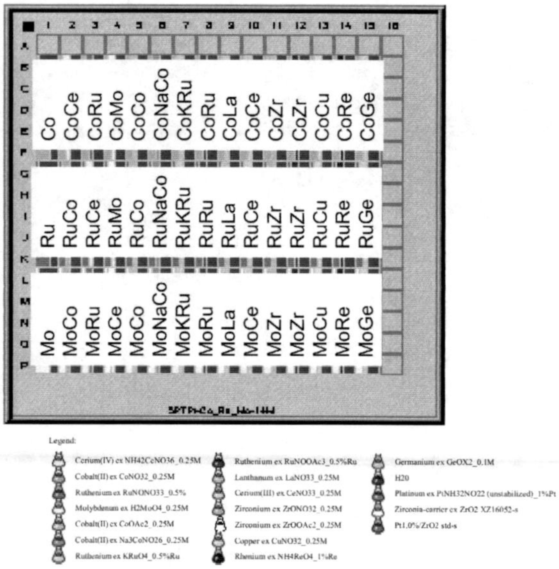

Figure 2c

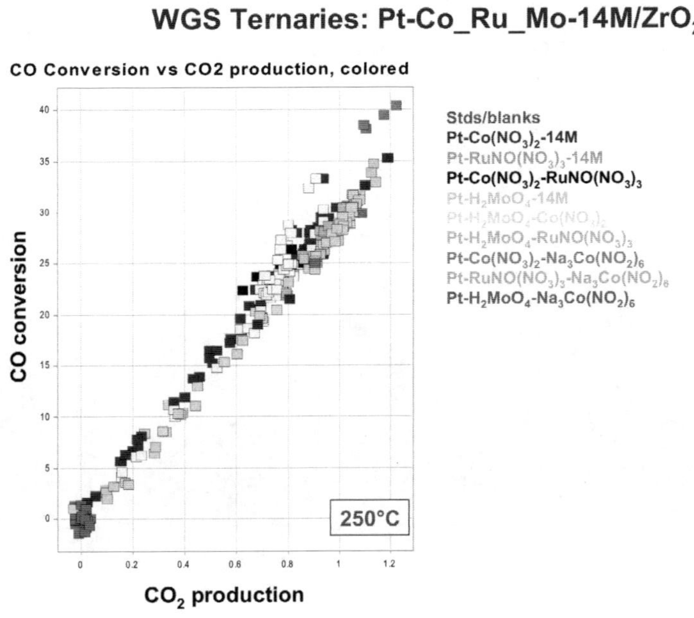

Figure 3a

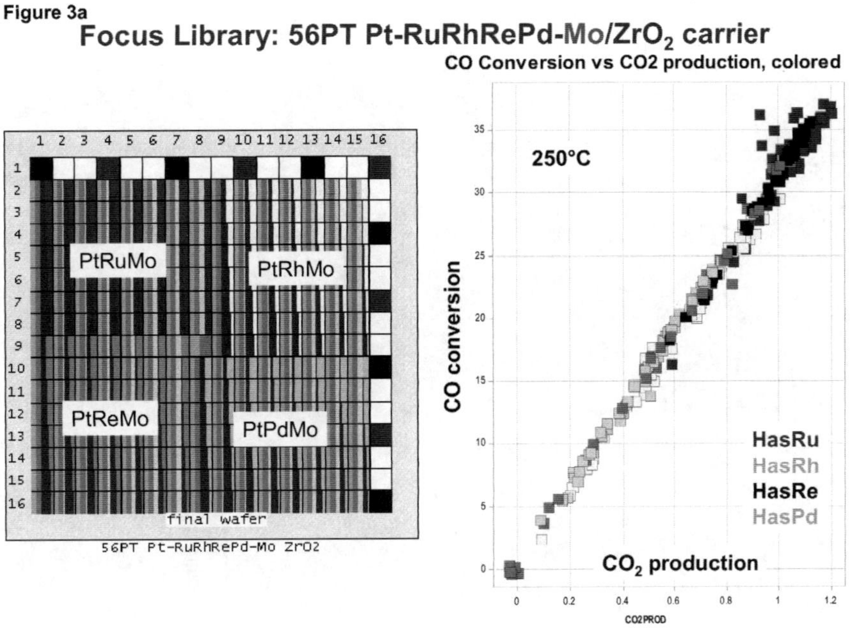

Figure 3b

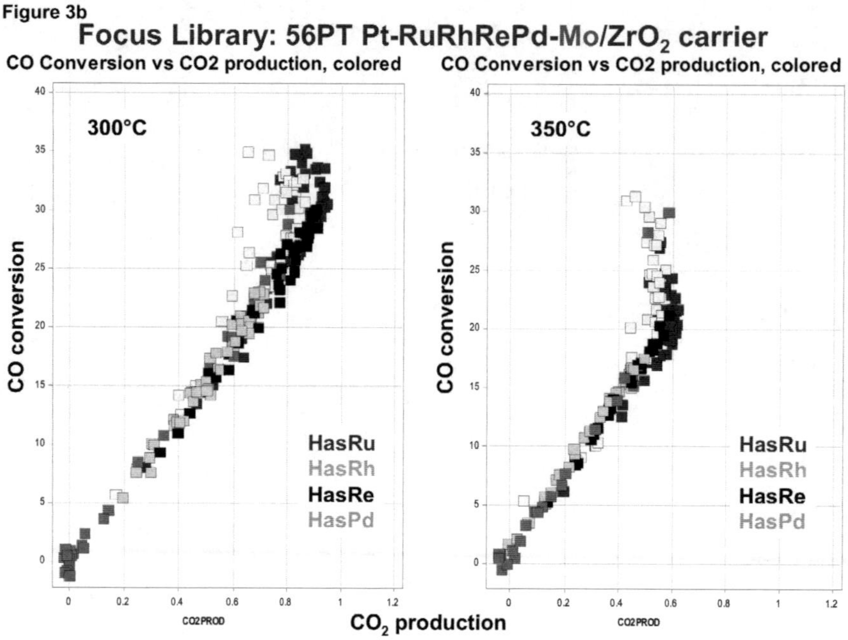

Figure 4 PtNa-(Co, Mo, V) on zirconia at 200C

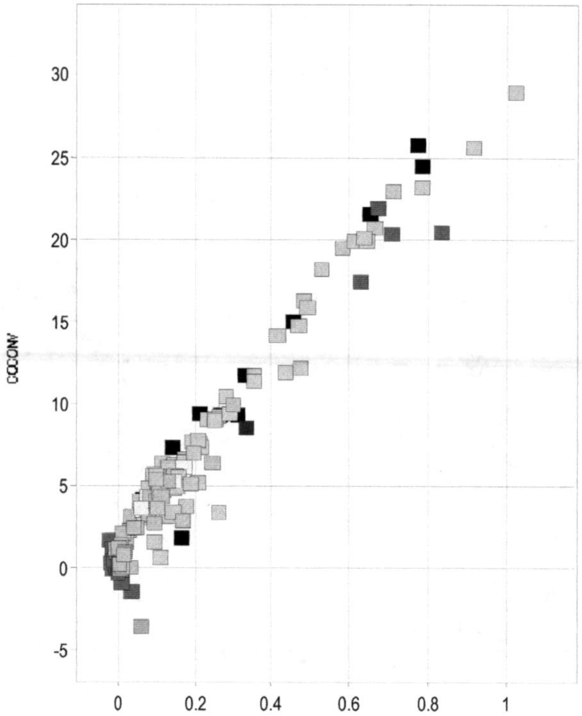

Figure 5 Secondary screening results in 8 channel Celero^{TM} plug flow reactor

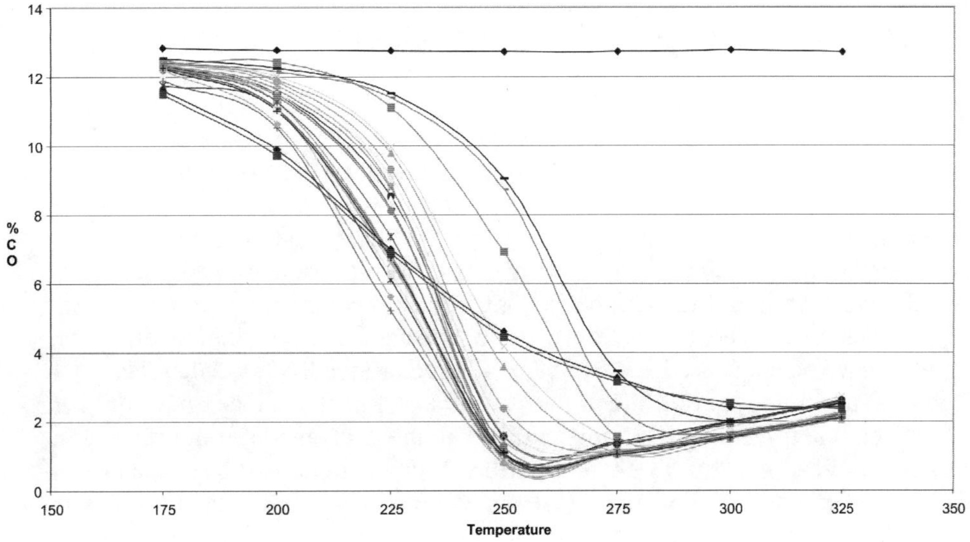

References

[1] K. Weissermel, H.-J. Arpe, "Industrial Organic Chemistry", Wiley-VCH (1993).

[2] H. F. Rase, "Handbook of Commercial Catalysts: Heterogeneous Catalysts", CRC Press (2000).

[3] D.L. Trimm, Z.I. Onsan, Catalysis Reviews 43(1&2) (2001) 31-84.

[4] A. Hagemeyer, P. Strasser, A. Volpe, Eds. "High Throughput Screening in Chemical Catalysis", Wiley-VCH (2004).

[5] A. Guram, A. Hagemeyer, C.G. Lugmair, H.W. Turner, A.F. Volpe Jr, W.H. Weinberg, K. Yaccato, Adv. Synth. Catal. 346 (2004) 215-230.

[6] K. Yaccato, R. Carhart, A. Hagemeyer, A. Lesik, P. Strasser, A.F. Volpe, H. Turner, H. Weinberg, R.K. Grasselli, C. Brooks, Appl. Catal. A 296 (2005) 30-48.

71s Microfluidic Modules for Optimized Mixing and Heat Transfer

Thomas R. Dietrich[1], Andreas Freitag[2], and Scholz Ralf[2]. (1) mikroglas chemtech GmbH, Galileo-Galilei-Str. 28, Mainz, D-55129, Germany, (2) Microreaction Technology, mikroglas chemtech GmbH, Galileo-Galilei-Str. 28, Mainz, D-55129, Germany
Since more than 10 years micromixers, micro heatexchangers and microreactors have been developed and tested for different applications. Several principles for mixers (e.g. interdigital mixers, split-and-recombine, chaotic mixers, cyclone mixers, falling films, etc.) and for heat exchangers (cross flow, counter current flow, pre-heating, electrical heating, etc.) have been realized and can be purchased as standard modules. The structurization technology for different materials like stainless steel, ceramics or glass is available. Even though all this is commercially available worldwide from different companies and research institutes, there are still special chemical issues which need optimized reaction modules. This paper will show new results about reactions in two-phase systems. Micromixers have been developed, which limit the size of plugs in a two-phase system to accelerate reactions by shortening diffusion times. This is supported by geometric channel structures which induce a chaotic behaviour of the flow. A variable dwell time module allows the adjustment of residence time according to the used chemical reaction. The system ends with a separation module which allows to stop the reaction at a certain point and to seperate the products from remaining educts and by-products. It will be shown that with such a system critical reactions, e.g. in highly viscous fluids or polymerizations, can be handled and controlled.

97a Single and Multiphase Mixing and Contacting in Microreactors

Asterios Gavriilidis, Dept of Chemical Engineering, University College London, Torrington Place, London, WC1E 7JE, United Kingdom
Contacting between reactants is a prerequisite in reactor operation. In most cases rapid mixing is required (for single-phase systems) or efficient interdispersion (for multiphase systems). In single phase systems, passive mixing has until recently relied primarily on molecular diffusion. In these cases, mixing structures have been designed so that the lamellae thickness of fluids is minimised. Many successful mixers have been developed on this basis. Recently other highly efficient mixing devices have been developed utilising chaotic advection, where mixing length is not linearly dependent on fluid velocity. In multiphase systems, microreactors open up opportunities not only for good interdispersion between phases by achieving high surface areas and short mass transfer distances but also for efficient non-dispersive contact between phases, due to the thin fluid layers that can be achieved. Relevant studies will be presented, while selected applications demonstrating the improved performance of microscale mixing and contacting will be given.

EXPERIMENTAL INVESTIGATION OF FLOW BOILING HEAT TRANSFER IN A MICROCHANNEL BY USING INFRARED THERMOGRAPHY

M. Cortina Díaz, J. Schmidt

Institute of Fluid Dynamics and Thermodynamics
Otto-von-Guericke University of Magdeburg
Universitätsplatz 2, D-39106 Magdeburg, Germany
Tel: +49-3916712558, Fax: +49-3916712762
E-mail: marta.cortina@vst.uni-magdeburg.de

1 INTRODUCTION

Flow boiling heat transfer in small channels has received growing attention in recent years due to their high heat transfer rates and to their high potential applications in many fields. Compact heat exchangers can be employed for example in mobile energetic systems like fuel cells and also in chemical processes, where many reactions take place in the gas phase and the vaporization of the usually liquid reactants is necessary. Although the number of investigations has increased rapidly in recent years, some aspects still remain unclear. For example, no generally accepted correlations are available for calculation of two-phase heat transfer coefficients and pressure drop. In conventional channels, it is assumed that both nucleate and convective boiling appear. A strong dependence of the heat transfer coefficient on the applied heat flux indicates the dominance of nucleate boiling. On the other hand, a strong dependence on both mass flux and vapor quality implies that convective boiling dominates the two-phase flow heat transfer. Nevertheless, no agreement could be found up to now in the case of small channels. Nucleate boiling seems to play an important role, either in form of elongated bubbles, which develop into slugs and fill the entire channel, or bubbles which nucleate in the thin liquid film near the wall. The time dependence of flow patterns has already been reported in literature. Time-averaged local heat transfer coefficients of water at atmospheric pressure in a vertical channel of rectangular cross-section 2×1 mm were presented in Wen *et al.* [1]. The heat transfer coefficients were found to increase nearly with the square root of the heat flux. Only a weak dependence on the mass flux was reported for vapor qualities below 0.3. That would indicate the dominance of nucleate boiling but this inference was shown to be false by observations. At any given location nucleate boiling occurred only for part of the time and the heat transfer coefficient did not change greatly between periods of nucleate boiling and convection. The authors concluded that the conventional distinction between nucleate and convective boiling mechanisms is unreliable when flow boiling in narrow channels is considered. An experimental study including high resolution visualizations was presented by Kandlikar and Balasubramanian [2]. Water flowing through six parallel minichannels, each 1054 µm wide and 197 deep, was investigated. Three gravitational orientations were compared under the same operating conditions. Bubbly flow, thin film nucleation, slug flow and churn flow were observed in all three orientations. Flow patterns were found to be time dependent. Decreasing trends in heat transfer coefficients with increasing quality were reported. These trends were in concordance with observations, where nucleating bubbles were detected at the wall during liquid flow and also in the liquid film near the wall in plug or annular flow. It was concluded that nucleate boiling was the dominant boiling mechanism. Yen *et al.* [3] carried out visualization of boiling flow modes in a 0.3 mm diameter microtube. HCFC123, FC72 and ethanol were employed as working fluids. The local heat transfer coefficients were calculated and a time dependent model on heat transfer in small tubes was proposed. A semi-periodic variation of the flow patterns was observed. Relatively good predictions were obtained by applying time fraction characteristics to current empirical correlations. By these means, a time fraction factor collected the percentage of different flow patterns in one period. A recent study of the unsteady behavior of boiling in small channels can be found in Xu *et al.* [4]. The test section consisted of ten triangular microchannels with the hydraulic diameter of 155.4 μm. Acetone was used as the working fluid. Outer wall temperatures on the heating side of the silicon wafer were measured by an infrared camera. The front side of the silicon wafer was a pyrex glass cover for visualization using a high speed camera. The authors proposed the boiling number, Bo, as the key for establishing the relative dominance of boiling mechanisms. When different working fluids are considered, the use a non-dimensional group K1 derived by Kandlikar [5] was suggested. The boiling cycle was divided in three substages, viz. liquid refilling, bubble nucleation, growth and coalescence and transient liquid film evaporation stage. The heat transfer coefficient was found to increase with quality for low Bo numbers, be independent of mass flux and quality for medium Bo numbers and decrease with quality for high Bo numbers. Lin *et al.* [6] investigated two-phase heat transfer to R141b in a 1 mm internal diameter vertical tube. The test section was electrically heated. For low heat fluxes, the heat transfer coefficients increased first with quality, indicating the onset of subcooled nucleate boiling, decreased then at low qualities, where nucleate boiling dominated and increased finally at higher qualities, indicating convective boiling. It was also observed that, after a certain value of heat flux, the heat transfer coefficients decreased with quality. This decrease was attributed to the overprediction of the heat transfer coefficient for high heat fluxes due to the linear approach of the pressure drop in the data reduction. Finally, some overviews of the progress made in the field of boiling in small channels during the past decades can be found in [7-9]. Reported discrepancies can be due to the use of different test conditions or measurement methods, as well as due to measurement accuracy. Furthermore, flow boiling patterns in narrow channels seem to be time dependent and that may also be one of the reasons for the reported discrepancies among experimental results of researchers. Therefore, further research is needed to develop reliable design tools, which will allow for the use of microchannels in the industry.

The present study focuses on the experimental investigation of boiling heat transfer in a single 0.3 × 12.7 mm^2 rectangular microchannel. Experimental results for three test fluids, viz. water, ethanol and n-hexane will be presented and discussed. The examined parameter ranges are: mass fluxes between 50 and 500 kg/(m^2s) and heat fluxes up to 400 kW/m^2. In contrast to many of the available experimental studies, where thermocouples were used to measure the wall temperature, infrared thermography is employed here to measure the outer wall temperatures of the channel. Infrared images of the test section are recorded at a frequency of 150 frames per second using the half image mode. Data are collected over 25 s and the behavior of the obtained wall temperatures is analyzed. Through these means, a continuous and contactless measurement of the wall temperature distribution is possible without time delay and the unsteady behavior of boiling can be analyzed. This transient analysis aims to contribute to a better understanding of boiling mechanisms in mini and microchannels, as well as to obtain reliable heat transfer coefficients by using time-averaged values of the wall temperature. The experimental heat transfer coefficients will be shown and compared with predictions of available correlations.

2 EXPERIMENTAL SETUP

Figure 1 shows the experimental rig. The working fluid flows from a container through a dosing pump, where the desired mass flow is adjusted. Afterwards, it passes through a heat exchanger, where the inlet temperature is set and through a mass flow meter, which operates according to the Coriolis principle. The test section is a rectangular channel made of the nickel alloy Inconel 600 and has the following dimensions: 0.3 mm height, 12.7 mm width and 200 mm length. The fluid flows upwards, and the test section is electrically heated. Two DC power supplies are connected in parallel to provide the necessary heat flux. Temperatures are measured at both inlet and outlet of the channel. Inlet pressure and pressure drop are also registered. Temperature and pressure are also controlled after the container. After leaving the test section, the fluid condenses and returns to the container. The experimental setup is open to the atmosphere after the condenser. Once both mass and heat fluxes are set and after the desired operating conditions are reached, the temperature distribution of the channel is measured by an infrared camera. The data recording is performed by a ThermaCam®SC3000, manufactured by FLIR Systems. The emissivity of the scanned surface is a very important parameter when this measurement method is used. A high value of emissivity permits to reduce the uncertainty of the temperature measurement and also the influence of the ambient. A value of approximately 0.95 could be achieved by coating the channel with a thin black lacquer. The emissivity is a strong function of both structure and temperature of the surface and, since its knowledge is essential to obtain reliable results, an intensive calibration is necessary. The calibration procedure as well as the experimental accuracy of the measured values can be found in detail in Cortina Díaz and Schmidt [10].

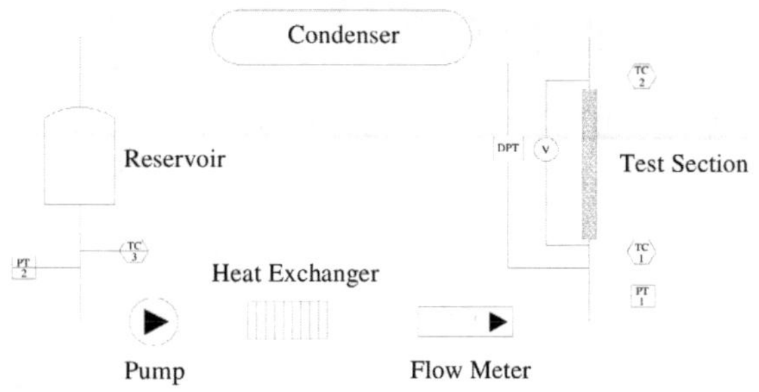

Figure 1: Experimental setup

3 DATA REDUCTION

The first step in the data reduction is the analysis of the thermographic images. The whole spatio-temporal temperature distribution of the channel surface is available. For the present study, only a middle area of the channel, of approximately 3 mm width, and the entire channel length are considered for the data reduction. The temperature for each axial position is set to be the average value of the temperatures corresponding to each pixel in the analysis area, where the influence of the radial heat conduction can be neglected, Fig. 2. The entire channel can be registered instantaneously with a spatial resolution in the axial direction of 0.7 mm/pixel. Since significant oscillations of the temperature are observed, the recorded values are analyzed for each test fluid to determine the minimum period of time over which the average wall temperatures remain constant. These time-averaged temperatures will be used later to calculate the heat transfer coefficients as follows, Eq. (1):

$$\alpha = \frac{\dot{q}}{(T_{iw} - \overline{T}_{Fl})} \qquad (1)$$

The local heat flux, $\dot{q}$, is calculated from the applied electrical power and accounting for the heat losses to the environment through natural convection and radiation. These heat losses are obtained from calibration tests, since they depend on the temperature of the outer wall and, therefore, also on the axial position. They represent between a 0.5 and 2 % of the total input power. Both voltage and current are measured and calibration tests are also carried out to determine the electrical resistance of the channel. In the single-phase flow region, the fluid temperature is computed from an energy balance between inlet and the corresponding axial position. In the two-phase flow region it is set to be the saturation temperature at the corresponding pressure. As far as the pressure is concerned, single-phase measurements are carried out to determine the friction factor for all liquid phase. These coefficients are compared with the well-known prediction of Kakaç *et al.* [11] for rectangular channels, and a good agreement is found. The pressure at the axial position where boiling begins is calculated from the measured inlet pressure and the calculated single-phase pressure drop. From there and for the two-phase flow region, the pressure drop is determined from the difference pressure measurement and accounting for the pressure drop corresponding to the single-phase region. A linear pressure drop is assumed from the onset of boiling and during the two-phase flow region.

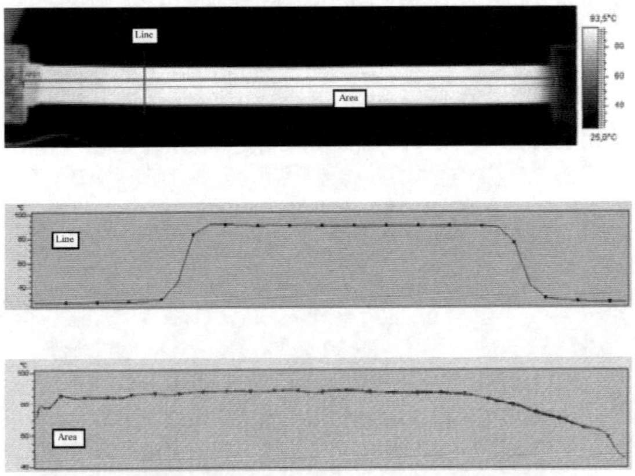

Figure 2: Infrared image with analysis area and line

The inner wall temperature is calculated approximately from the measured values of the outer wall temperature using the one-dimensional steady state heat conduction equation with heat generation. The effect of the unsteady behavior of the wall temperature will be discussed later. The vapor quality is computed from an energy balance between inlet and the corresponding axial position. The analysis of experimental uncertainties leads to an average error in the heat transfer coefficient of 15%.

4 EXPERIMENTAL RESULTS

4.1 Measured Outer Wall Temperatures

The registered outer wall temperatures for some axial positions are shown in Figs. 3-5 for water, ethanol and n-hexane. Different behaviors can be recognized depending on the working fluid. Both amplitude and frequency of the oscillations vary depending on the operating parameters and also on the axial position, i.e. on the vapor quality. Fig. 3a shows the outer wall temperature over the measuring time for water. The maximum amplitude is found in the region of subcooled boiling, just before nucleate boiling begins. The maximum temperature difference in this region is found to be 30 K. With increasing axial position, the amplitude decreases strongly and the frequency increases. Nearly constant wall temperatures are registered after the initial region. A completely different situation is observed for ethanol, Fig. 4a. The amplitude remains approximately constant and its maximum value is lower than 5 K. The frequency of the temperature oscillations increases slightly with increasing quality and nearly periodic oscillations are registered at high vapor qualities. In the case of n-hexane, Fig. 5a, low amplitude and high frequency oscillations are observed. It can be seen that the outer wall temperature remains nearly constant within the region of flow boiling. The oscillations of the wall temperature shown in Figs. 3-5 can yield different values and trends of the heat transfer coefficient. In order to obtain reliable results, the temperature used for the calculation of the heat transfer coefficient should be time-averaged. Therefore, the registered wall temperatures are averaged over increasing time periods to find out the minimum measurement time required to obtain constant values of the mean temperature. The time-averaged wall temperatures are shown in Figs. 3b to 5b for the considered test fluids. In the case of water, the wall temperatures remain constant after approximately 5 s, especially at higher values of quality. As already seen in Fig. 3a, pronounced instability of the wall temperature is found at low vapor qualities. Since the oscillation amplitude in the case of ethanol is smaller, the wall temperature remains nearly constant when averaged over only 3 s. This averaging time interval is even lower for n-hexane.

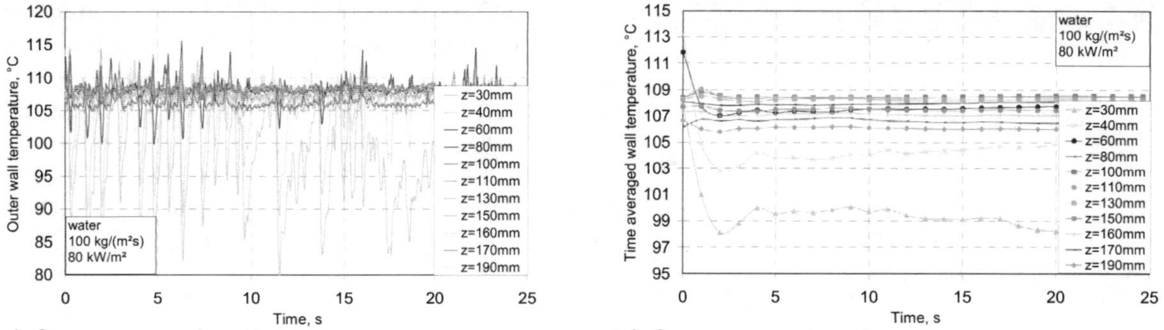

a) Over measuring time b) Over averaging time

Figure 3: Outer wall temperature for different axial positions, water, $\dot{m}$ = 100 kg/(m²s), $\dot{q}$ = 80 kW/m²

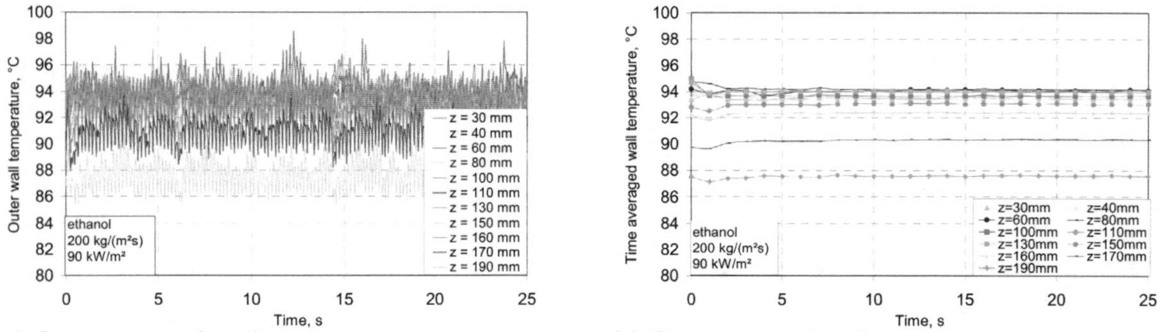

a) Over measuring time b) Over averaging time

Figure 4: Outer wall temperature for different axial positions, ethanol, $\dot{m}$ = 200 kg/(m²s), $\dot{q}$ = 90 kW/m²

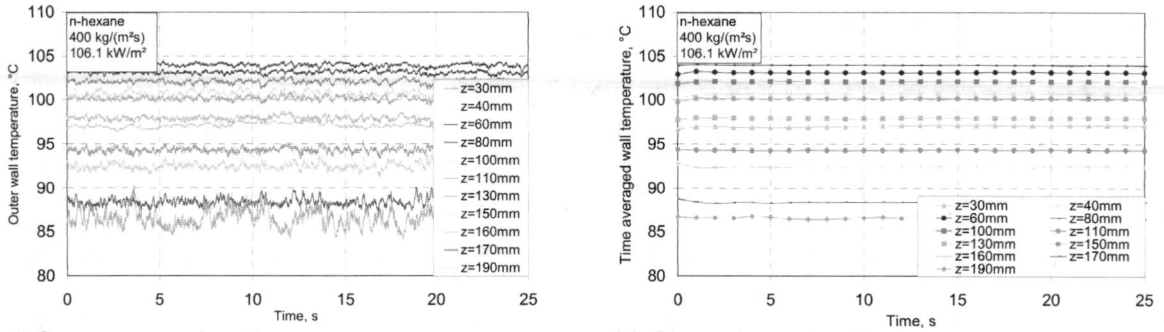

a) Over measuring time b) Over averaging time

Figure 5: Outer wall temperature for different axial positions, n-hexane, $\dot{m}$ = 400 kg/(m²s), $\dot{q}$ = 106.1 kW/m²

The oscillations are measured at the outer surface of the channel and an influence of the wall thickness, in this case 0.3 mm, on both attenuation of the amplitude and phase shift is expected. Therefore, the computer software FLUENT® is used to study this effect. Some results are shown in Figs. 6 and 7. In this case the wall temperature shown in Fig. 4a for a vapor quality of 0.328 (z = 120 mm) is obtained by approximating the inner wall temperature by a sinusoidal function, Fig. 6. Since the heat losses to the environment are low when compared to the heat generation rate, the outer wall is set to be adiabatic. One-dimensional heat conduction with heat generation is considered and periodic boundary conditions are set in the axial direction. Once the inner wall temperature is identified, the heat generation rate, i.e. also the heat flux, is varied in order to find out the influence on attenuation and phase shift, Fig. 7. It can be observed that the wall thickness has a negligible influence of the phase shift. On the other hand no influence of the heat generation rate on the attenuation is observed within the considered range. In each case an attenuation of the amplitude of 5% is found.

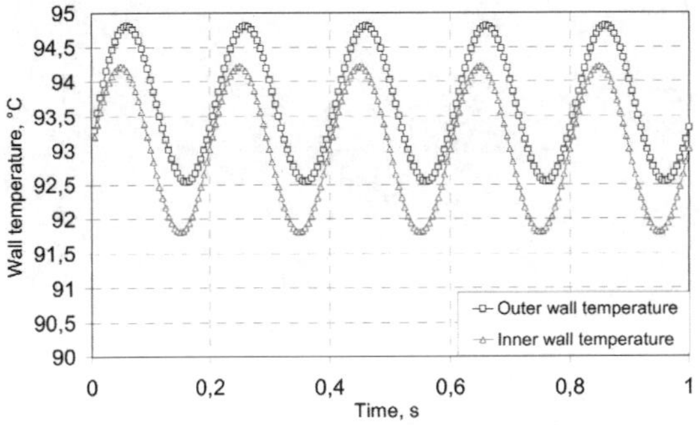

Figure 6: Inner and outer wall temperatures, ethanol, $\dot{m}$ = 200 kg/(m²s), $\dot{q}$ = 90 kW/m², x = 0.328

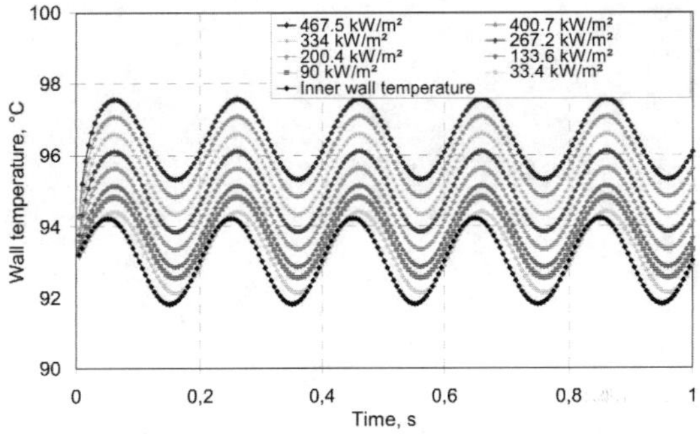

Figure 7: Outer wall temperatures at different heat fluxes

4.2 Heat Transfer Coefficients

Fig. 8 shows the calculated heat transfer coefficients over vapor quality for water at different heat fluxes and mass fluxes of 300 and 500 kg/(m²s). It can be observed that the heat transfer coefficient for water decreases with increasing quality and depends on the applied heat flux. This behavior has been already reported, for example in [2]. The decrease in the heat transfer coefficient for water can be explained by the special boiling behavior in small channels. According to [2], the presence of nucleating bubbles at the walls and in the thin liquid film during slug and annular flow is responsible for this decrease. This explanation is consistent with some visualization reported in literature. After this initial decrease and as the vapor quality increases, the heat transfer coefficients tend nearly to a common value and the lines overlap. The experimental trend for water in this study differs from that reported in previous investigations by the authors, Cortina Díaz *et al.* [12], where convective boiling dominated. On the one hand, this discrepancy can be due to the use of different channel geometries and dimensions and to the different operating parameters. On the other hand, the heat transfer coefficients are measured here only within the low quality region. It is also possible that convective boiling would appear if higher qualities were considered. In the case of ethanol, Fig. 9, the experimental trends of the heat transfer coefficient indicate that heat transfer is also influenced by convective boiling. The heat transfer coefficient increases in the region of subcooled boiling and decreases immediately after the quality reaches zero. The heat transfer coefficients in this region depend on the applied heat flux. Thereafter, two different trends can be recognized in the two-phase region. At higher qualities and low heat fluxes, the heat transfer coefficient increases with increasing quality and is independent of the applied heat flux. This behavior corresponds, according to the classical theory, to the dominance of convective boiling. For higher heat fluxes, the heat transfer coefficient decreases with quality within the investigated range. This change in the boiling behavior was already reported in [4] and [6]. Fig. 10 shows the calculated heat transfer coefficients for n-hexane. The first decrease of the heat transfer coefficient immediately after the quality reaches zero can also be recognized. Afterwards and similar to the observed trends for ethanol, the heat transfer coefficient increases with increasing vapor quality and becomes independent of the applied heat flux. It is worth to note that also in this case a trend change occurred but only at high mass and heat fluxes. A first analysis using the non dimensional group proposed by Kandlikar [5] yielded satisfying results but further research is needed to prove the applicability of this factor to the identification of this trend change.

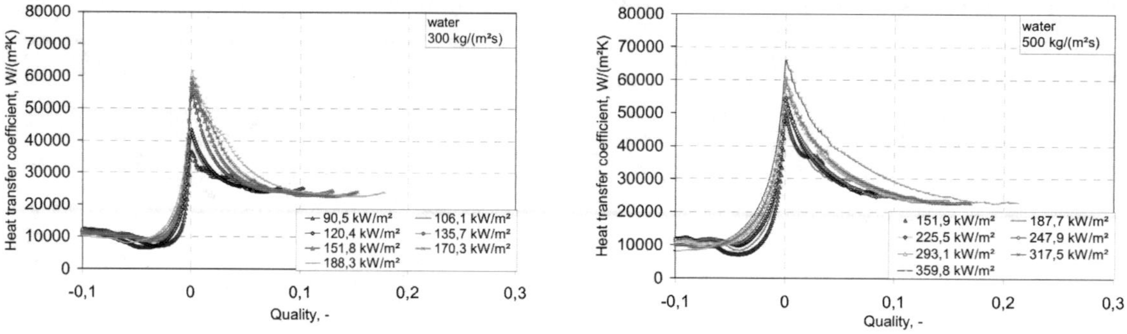

Figure 8: Heat transfer coefficient over vapor quality, water, $\dot{m}$ = 300, 500 kg/(m²s)

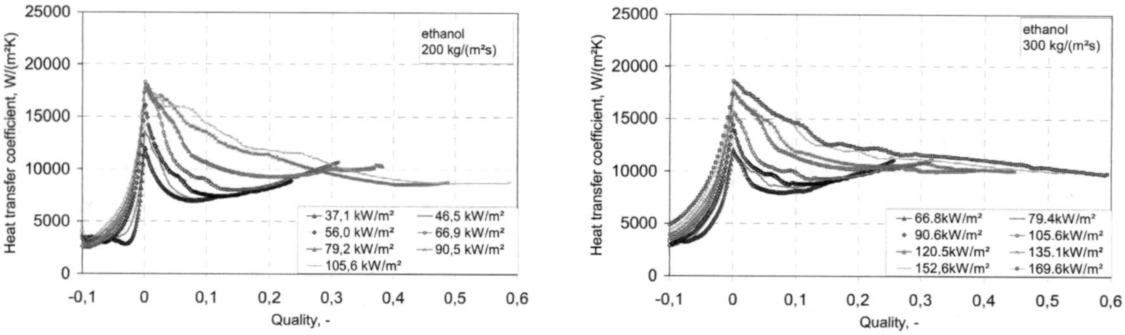

Figure 9: Heat transfer coefficient over vapor quality, ethanol, $\dot{m}$ = 200, 300 kg/(m²s)

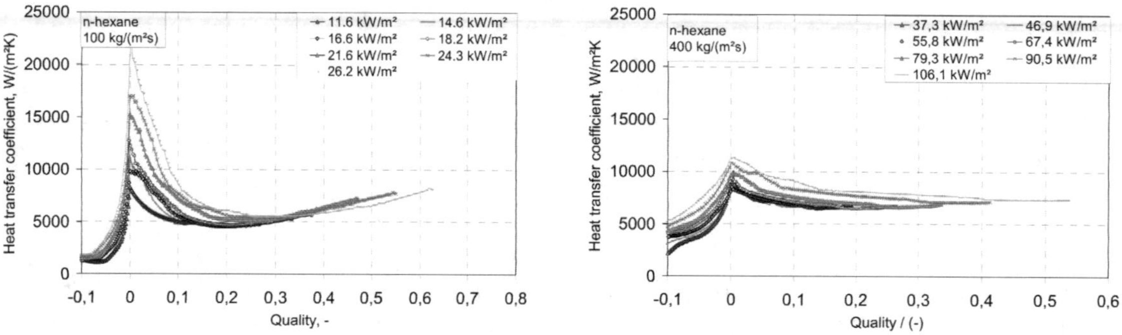

Figure 10: Heat transfer coefficient over vapor quality, n-hexane, $\dot{m}$ = 100, 400 kg/(m²s)

The experimental data are compared with the predictions of six widely used two-phase flow heat transfer coefficient correlations, viz. Kandlikar, [13], Liu and Winterton, [14], Lazarek and Black, [15], Chen, [16], and Zhang *et al.* [17]. In the case of the Kandlikar correlation, only the nucleate boiling contribution to the heat transfer coefficient is considered for the present comparison. Figs. 11 and 12 show the experimental and predicted heat transfer coefficients for all test fluids and selected mass and heat fluxes.

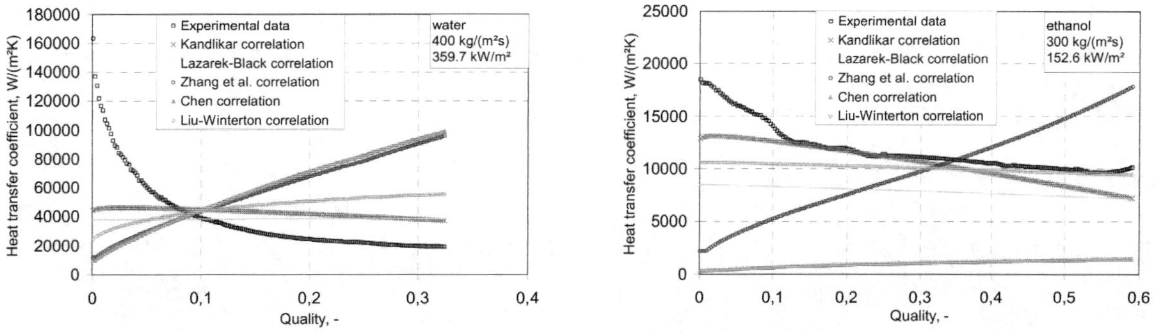

a) Water b) Ethanol

Figure 11: Comparison between experimental and predicted heat transfer coefficients

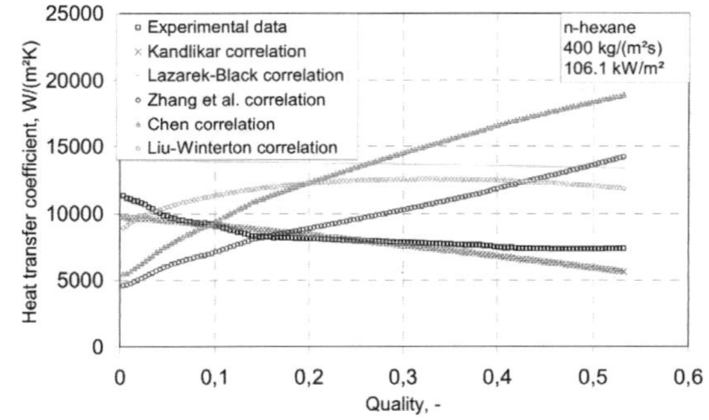

Figure 12: Comparison between experimental and predicted heat transfer coefficients, n-hexane

The experimental trend of the heat transfer coefficient for ethanol is in relatively good agreement with the predictions of the Kandlikar and Liu-Winterton correlations. In the case of water, the experimental values are overpredicted by all the correlations. An increasing trend is predicted in nearly all cases, with exception of Kandlikar and Lazarek-Black correlations. For n-hexane, a good agreement is obtained between the experimental heat transfer coefficients and the predictions of the Kandlikar correlation.

5 CONCLUSIONS

In the present study, infrared thermography was employed to obtain the wall temperature profiles. Images of the test section were recorded at a frequency of 150 frames per second. Flow instability of boiling in microchannels, which has been often reported in literature, was studied on the basis of the registered oscillations of the wall temperature. High amplitude oscillations were found for water in the region of subcooled boiling at qualities slightly below zero. However, the amplitude decreased as quality increased and no characteristic frequency was observed. Flow boiling of ethanol yielded low amplitude oscillations of the wall temperature, even within the subcooled boiling region. Increase in quality resulted in increasing oscillation amplitude. Nearly periodic oscillations were registered at high vapor qualities. Low amplitude oscillations were observed in the case of n-hexane and the wall temperature remained nearly constant within the two-phase region.

As a result of this analysis, time-averaged wall temperatures were selected for the calculation of the heat transfer coefficient. The heat transfer coefficient was found to decrease with increasing quality for water. On the other hand, different trends could be observed for ethanol after the low quality region. The heat transfer coefficient increased with increasing quality at low heat fluxes. With increasing heat flux, the heat transfer coefficient decreased with quality. In the case of n-hexane, the heat transfer coefficient decreased within the low quality region and increased thereafter. The experimental values within the high quality region were found to be independent of the applied heat flux, what could indicate the dominance of convective boiling. A trend change was also observed but only at high mass and heat fluxes. The experimental values were compared with those predicted by some correlations available in literature. The experimental values for water were overpredicted. A qualitatively good agreement was found for ethanol. A good agreement was observed between the experimental heat transfer coefficients for n-hexane and the prediction of the Kandlikar correlation.

6 NOMENCLATURE

Bo	Boiling number, $Bo = \dot{q}/(\dot{m}h_{fg})$, -
K_1	Non-dimensional group, - $K_1 = (Bo)^2\rho_L/\rho_V$
$\dot{m}$	Mass flux, kg/(m²s)
$\dot{q}$	Heat flux, W/m²
T	Temperature, °C
x	Vapor quality

Subscripts

Fl	Fluid
iw	Inner wall
L	Liquid
v	Vapor

Greek Symbols

α	Heat transfer coefficient, W/(m²K)
ρ	Density, kg/m³

REFERENCES

1. D. S. Wen, Youyou Yan and D.B.R. Kenning, Saturated Flow Boiling of Water in a Narrow Channel: Time-Averaged Heat Transfer Coefficients and Correlations, *Applied Thermal Engineering*, pp. 1207-1223, 2004
2. S. G. Kandlikar and P. Balasubramanian, Effect of Gravitational Orientation on Flow Boiling of Water in 1054×197 Parallel Minichannels, *Proc. Second International Conference on Microchannels and Minichannels*, pp. 539-550, Rochester, New York, USA, 2004
3. T. H. Yen, Y. Suzuki and N. Kasagi, Modeling on Convective Boiling Heat Transfer in a Microtube based on Flow Visualization, *6th World Conference on Experimental Heat Transfer*, Matsushima, Miyagi, Japan, 2005
4. J. Xu, S. Shen, Y. Gan, Y. Li, W. Zhang and Q. Su, Transient Flow Pattern based Microscale Boiling Heat Transfer Mechanisms, *Journal of Micromechanics and Microengineering,* 15, pp.1344-1361, 2005
5. S. G. Kandlikar, Heat Transfer Mechanisms During Flow Boiling in Microchannels, *Journal of Heat Transfer,* vol. 126, pp. 8-16, 2004
6. S. Lin, P. A. Kew and K. Cornwell, Two-Phase Heat Transfer to a Refrigerant in a 1 mm Diameter Tube, *International Journal of Refrigeration,* vol. 24, pp. 51-56, 2001
7. M. Groll and R. Mertz, Minichannel Heat Transfer: An Overview on Activities in Europe, *Proc. First International Conference on Microchannels and Minichannels*, Rochester, New York, USA, 2003
8. S. G. Kandlikar, Fundamental Issues Related to Flow Boiling in Minichannels and Microchannels, *Experimental Thermal and Fluid Science,* 26, pp. 389-407, 2002

9. J. R. Thome, Boiling in Microchannels: a Review of Experiment and Theory, *International Journal of Heat and Fluid Flow,* 25, pp. 128-139, 2004
10. M. Cortina Díaz, J. Schmidt, Experimental Investigation of transient boiling heat transfer in microchannels, *Proc. ECI International Conference on Heat Transfer and Fluid Flow in Microscale*, Castelvecchio Pascoli, Tuscany, Italy, 2005
11. S. Kakaç, R. K. Shah, W. Aung, *Handbook of Single-Phase Convective Heat Transfer,* Wiley-Interscience, USA, 1987
12. M. Cortina Díaz, H. Boye, I. Hapke, J. Schmidt, Y. Staate, Z. Zhekov, Flow Boiling in Mini and Microchannels, *Proc. Second International Conference on Microchannels and Minichannels*, pp. 445-452, Rochester, New York, USA, 2004
13. S. G. Kandlikar, P. Balasubramanian, Extending the Applicability of the Flow Boiling Correlation to Low Reynolds Number Flows in Microchannels, *Proc. First International Conference on Microchannels and Minichannels*, Rochester, New York, USA, 2003
14. Z. Liu, R. H. Winterton, A General Correlation for Saturated and Subcooled Flow Boiling in Tubes and Annuli, Based on a Nucleate Pool Boiling Equation, *International Journal of Heat and Mass Transfer,*34, pp. 2759-2766, 1991
15. G. M. Lazarek, S. H. Black, Evaporative Heat transfer, Pressure Drop and Critical Heat Flux in a Small Vertical Tube, *International Journal of Heat and Mass Transfer,* vol. 25, No. 7, pp. 945-960, 1982
16. J. C. Chen, A Correlation For Boiling Heat Transfer to Saturated Fluids in Convective Flow, *Ind. Eng. Chem. Proc. DD.* 5, pp. 322-329, 1966
17. W. Zhang, T. Hibiki, K. Mishima, Correlation for Boiling Heat Transfer in Mini-channels, *International Journal of Heat and Mass Transfer,* 47, pp.5749-5763, 2004

97c Numerical and Experimental Study of Mixing Enhancement in Micromixers Using Residence Time Distribution (Rtd)

John T. Adeosun and Adeniyi Lawal. Dept. of Chemical, Biomedical and Materials Engineering, New Jersey Center for MicroChemical Systems, Stevens Institute of Technology, Castle Point on Hudson, Hoboken, NJ 07030

The small physical dimensions of micromixers/reactors in combination with low processing flow rate, imply laminar flow, which precludes convective turbulent mixing normally relied upon for effective mixing in conventional-size processing equipment. Laminar flow poses a challenge, and mass transfer enhancement is therefore a critical issue in micro-channel mixers/reactors. The objective of this work is to utilize a combination of well-designed micromixers, a unique numerical approach, and a matching experimental set-up and procedure along with a mixing characterization measure for the study of mixing in micromixers/reactors, the development of which will greatly benefit the numerous potential applications of microchemical systems, and other related novel technologies. In this mixing enhancement study, three multilaminated/elongational flow mixers (MEFMs), uniquely designed to effect mixing using passive mixing structures by creating folding as well as local and global re-orientation of fluid interfaces, are investigated along with the standard T-junction mixer for their mixing performance. The knowledge base necessary for establishing the experimental platform for this mixing study is created first using Computational Fluid Dynamics (CFD) approach. Using the residence time distribution (RTD) and species mass fraction distribution (SMFD) as the mixing characterization measures, the results obtained from the numerical study show that the MEFMs give excellent mixing performance when compared with the standard T-junction mixer and a cross-shaped micromixer reported in the literature. The RTD is an innovative technique for experimental characterization of mixing in micromixers/reactors. The preliminary results from the experimental procedure that utilizes UV-vis absorption spectroscopy detection technique to obtain RTD in the micromixers will also be presented.

SHAPE DESIGN OF A TUBULAR MICROREACTOR WITH DESIRED PRODUCT DISTRIBUTION

Osamu Tonomura, Satoshi Nagahara, Manabu Kano, and Shinji Hasebe
Dept. of Chemical Engineering, Kyoto University, Kyoto, JAPAN
E-mail: tonomura@cheme.kyoto-u.ac.jp

Introduction

Micro chemical process technologies offer much improved controllability of reaction conditions such as residence time, temperature and concentration distributions. A large number of papers say that the product yield and selectivity in microreactors are higher than those in conventional reactors. However, the design and operation conditions for microreactors are actually determined by trial and error. In this research, the influence of a tubular microreactor dimensions on the residence time distribution (RTD) and the polymer molecular weight distribution (MWD) is investigated by numerical simulations. In addition, a systematic approach for designing tubular microreactors with the desired product distribution is proposed, and the effectiveness of the proposed approach is assessed through case studies and experiments.

Residence Time Distribution

The RTD is one of the key design specifications for microreactors, because the broad RTD reduces the yield and selectivity of reaction products. To investigate how dispersion affects the concentration profile in a tubular microreactor, we suppose the injection of a tracer pulse at time $t = 0$. The pulse starts spreading due to many things such as velocity profile and molecular diffusion, and the RTD curve becomes symmetrical and Gaussian in the downstream region of the microreactor. The narrower RTD can be achieved by reducing the channel diameter and lengthening the channel length. The validity of the above-mentioned dispersion process is confirmed by our Computational Fluid Dynamics (CFD) simulations and experiments.

A design approach for the microreactor with a desired RTD of a viscous fluid is proposed in this work. The proposed approach can systematically determine the microreactor dimensions by using Taylor dispersion model and CFD simulation. In the dispersion model, there is an axial dispersion of material, which is governed by an analogy to Fick's law of diffusion, superimposed on the flow. The design procedure is as follows:

i) Define design specifications such as flow rate (Q), average (τ) and variance (σ^2) of the desired RTD, and lower size boundaries (d_1).

ii) Determine the upper limit of channel width (d_2), which insures the gaussian RTD curve, by using Eq. (1).

$$\tau \geq \frac{d^2}{D} \tag{1}$$

where d is channel width and D is diffusion coefficient.

iii) Derive the channel width (d^*), which can achieve the desired RTD, by using Taylor equation (2).

$$\sigma^2 \geq \frac{\tau d^2}{96D} \qquad (2)$$

iv) Validate the design result as follows:

If d^* is between d_1 and d_2, d^* is feasible solution.

If d^* is smaller than d_1, recalculate σ^2 of RTD for d_1.

If d^* is larger than d_2, the proposed design method is not appropriate.

The effectiveness of the proposed method is confirmed through a case study. The proposed approach enables us to effortlessly determine the design parameters of the tubular microreactors that can achieve the desired RTD.

Polymer Molecular Weight Distribution

In addition to the control of RTD, the molecular weight distribution (MWD) for continuous radical polymerization in microreactors is investigated since the MWD may have a significant impact on the final polymer properties. We develop the model of the styrene free radical polymerization in a tubular microreactor, which is expressed by the partial differential equations resulting from the hydrodynamics, thermal and mass transfer (convection, diffusion and reaction). In the model, the polymerization kinetics consisting of initiation, propagation, chain transfer, combination termination and gel effect are considered, and reaction rate constants are calculated from a generalized Arrhenius form. And also, the method of moments is used to derive polymer MWD successfully.

After modeling, the effect of design parameter, operating conditions (imperfect feed mixing etc.), transport and kinetic parameters, and solution parameters on the polymer properties is examined. The polymer properties mean polydispersity index (DPI), number- and weight-average molecular weights (Mn and Mw), and monomer conversion. The simulation results show that the inadequate reactor design and operation generate hot-spots that give rise to a condition called thermal runaway, which is characterized by an undesired product distribution. Moreover, a design method is developed to successfully determine the microchannel dimensions, which realize the desired DPI and monomer conversion.

Conclusion

The microreactor design has the strong influence on RTD, MWD, the reaction temperature distribution, the concentration gradient of chemical species, etc. The proposed approach enables us to effortlessly determine the design parameters of the microreactors that achieve the desired RTD. In addition, the design method is also developed to successfully derive the microchannel dimensions, which realize the desired DPI and monomer conversion. The effectiveness of both approaches is confirmed through case studies.

Acknowledgment

This research was partially supported by the New Energy and Industrial Technology Development Organization (NEDO), Project of Micro-Chemical Technology for Production, Analysis and Measurement Systems.

97e Optimization of Metallic Microchannel Array Evaporators

Torsten Henning[1], Juergen J Brandner[1], Klaus Schubert[1], Marco Lorenzini[2], and Gian Luca Morini[2]. (1) Institute for Micro Process Engineering IMVT, Forschungszentrum Karlsruhe, Hermann-von-Helmholtz-Platz 1, DE-76344 Eggenstein-Leopoldshafen, Germany, (2) DIENCA, Universita degli Studi di Bologna, Viale Risorgimento 1, IT-40136 Bologna, Italy

Flow boiling in microchannels and microchannel arrays is an effective way of achieving the high heat transfer rates required in compact evaporator designs. One objective in designing compact evaporators is to generate dry vapor at a temperature not exceeding the saturation temperature more than necessary, at the lowest possible surface temperature. Water is used as a model substance in this study, in which different evaporator designs are compared primarily with the aid of high-speed videography in the visible wavelength range.

While the evaporation in single microchannels has been the subject of intensive research over the last years, relatively little effort has been expended on the situation in arrays of many (several dozen) microchannels. For a practical application, however, pressure loss over the device is often an important concern, and for a commercially relevant throughput range, there is often no alternative to designs involving such large numbers of channels. Microchannel arrays with liquids undergoing flow boiling are well suited to the removal of heat from devices such as semiconductor chips with high surface power densities. For such heat removal applications, complete dryout of the microchannels is undesirable and can be highly detrimental. For evaporator applications, however, the situation is quite the opposite since complete dryout of the microchannels towards their downstream ends is what a good design should provide. This goal can always be achieved by making the microchannels sufficiently long. The requirement of minimized pressure loss mentioned above, however, also puts restraints on the length for the individual microchannels in such arrays.

In this paper, several microchannel arrays with different channel lengths and layouts are compared. To allow for comparison between the different array designs, several parameters were held constant. The arrays always consisted of 68 channels, each 200 μm wide and 100 μm deep separated by 100 μm wide ridges, and were made by mechanical micromachining of steel or brass, respectively. Array lengths varied between 20 mm and 80 mm.

The experimental set-up consists of a steel frame with electrically powered cartridge heaters, an exchangeable microstructure foil with the microchannel arrays, and a window of glass or Makrolon™. Water was supplied by an HPLC pump at a constant flow rate of 0.5 kg h^{-1}, corresponding to a mass flux of 61 kg $m^{-2}s^{-1}$.The two phase flow phenomena can be observed through the window and microscope optics by a high-speed video camera operating at up to 1825 frames per second at a resolution of 1024×256 pixels. Illumination can be provided alternatively by halogen lamps with gooseneck light guides, or by a stroboscope lamp with a flash duration of approximately 10 μs.

Microchannel array evaporators for industrial applications will have to consist of metal, such as stainless steel. When such a device is connected to the (inlet) tubing via metallic connectors, heat conduction against the direction of water flow can cause boiling to start already in the inlet plenum. Once this happens, relatively large bubbles are formed in the inlet plenum, periodically breaking up at the microchannel array entrance. The pulsation resulting from these events is primarily responsible for the observed periodic breakdown of complete evaporation. There are two possible ways of mitigating this effect:

- Minimizing the size of the inlet plenum, and hence the residence time there, should move the boiling front downstream and into the microchannel array proper.
- Alternative multistage evaporator designs should cope better with a feed already containing some vapor.

Both methods have been implemented in a number of foils whose performances as evaporators are compared in this paper.

Understanding Critical Heat Transfer Issues in Thermal Integration of Planar Microreactor Components of A Fuel Processor

Keyur Shah and R.S. Besser
New Jersey Center for Microchemical Systems, Chemical, Biomedical and Materials Engineering Department, Stevens Institute of Technology, Hoboken, NJ 07030

The processing of high density liquid hydrocarbon fuels appears to be the most promising method of supplying a hydrogen stream for feeding portable fuel cells. Liquid hydrocarbons could quickly become a viable energy source for portable power if an efficient, lightweight, and portable fuel processor/fuel cell integrated system were made available. Considerable efforts have been made to develop an integrated micro fuel processor/fuel cell system; however there are major hurdles and several universal challenges exist with fuel processing at small scale. Miniaturization of system components and thermal management in miniature systems are perhaps the most crucial challenges for microscale fuel processors. Through this work, we are trying to understand these crucial system issues.

Microreactors seem suited to small scale power generation, as their compactness should enable integration by stacking of miniature fuel processor and fuel cell components with the fuel delivery system. [1,2]. The need for miniaturization directs the use of microchemical systems in the field of fuel processing. The demand for a greater energy per volume and weight has naturally led to the investigation of microreactors which can take the form of any of the essential fuel processor elements including desulfurizer, steam reformer (SR), preferential oxidation (PrOx) reactor, etc [3-6]. The microscale dimensions which result in extremely small heat transport resistance [7,8] make them especially attractive in applications like fuel processing, where it is desired to thermally couple highly endothermic and exothermic processes in an efficient manner.

Thermal Integration Issues in Micro Fuel Processor

The fuel processor needs effective thermal coupling to allow transfer of energy from the heat producing combustor to the endothermic SR. Coupling endothermic and exothermic components of the fuel processor and minimizing losses can achieve a high thermal efficiency. However, such coupling must be accomplished in a manner that permits the maintenance of specific temperatures in the various components as shown in Figure 1 and maintains the surface of the package near room temperature.

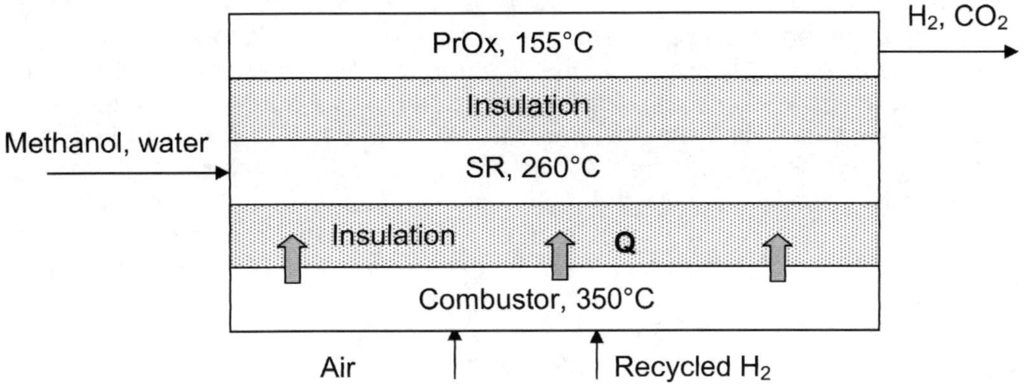

Figure 1. Integration of fuel processing components

In this study, a silicon microreactor-based catalytic methanol SR was designed and fabricated in the context of complete thermal integration to directly address the heat management issue. As shown in Figure 2, a silicon micro SR is designed and integrated with microfabricated insulation as dictated by overall system design. A thin-film microfabricated heater is used to simulate integration with a combustor by controlled heat introduction.

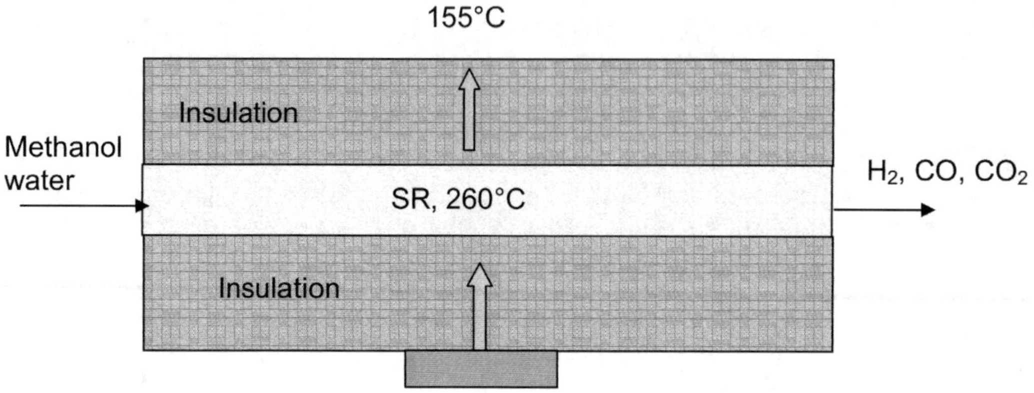

Figure 2. Demonstration of thermally integrated SR microreactor

Thermal Coupling and Insulation

Because of the short conduction paths inherent to microreactors, high performance insulators are needed between fuel processing components to isolate the high temperature components. The objective is to have insulating materials that can offer ultra low thermal conductivity to bridge these temperature differences and to minimize conductive and convective heat losses to the surroundings without adding much to the weight and volume of the total system.

The best commercial insulators have thermal conductivities in the range of 0.02 W/mK and are difficult to integrate into a microfabricated unit. However, vacuum packaging of microreactors can provide an effective means of insulation. From the kinetic theory of gases, it is known that the thermal conductivity of a gas is approximately independent of pressure for atmosphere and above, and decreases with sub-atmospheric pressures [9]. A microfabricated cavity surrounding the reactor filled with low-pressure gas offers low thermal conductivity down to approximately 0.001W/mK and can be engineered by managing pressure and cavity depth. Silicon microfabrication enables a straightforward approach to this structure as a cavity of precise depth can be etched in silicon and bonded with either silicon or glass using vacuum bonding methods [10]. Figure 3 shows a concept for microfabricated vacuum insulation.

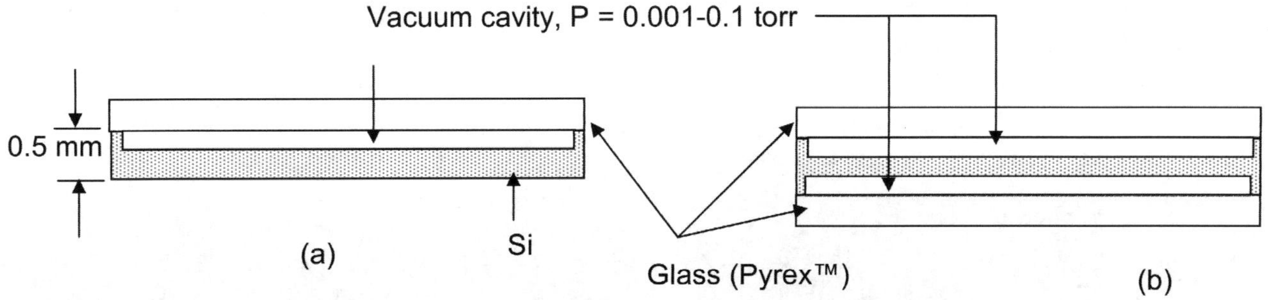

Figure 3. Vacuum insulation chip with (a) a single microfabricated cavity in Si, and (b) cavities on both sides of Si

Integrated Micro SR: Design and Fabrication

The design of an integrated SR is shown in Figure 4, where vacuum packaging chips, thin film metal heaters and temperature sensors are directly embedded with the reforming microreactor to simulate an integrated SR in an overall fuel processing scheme.

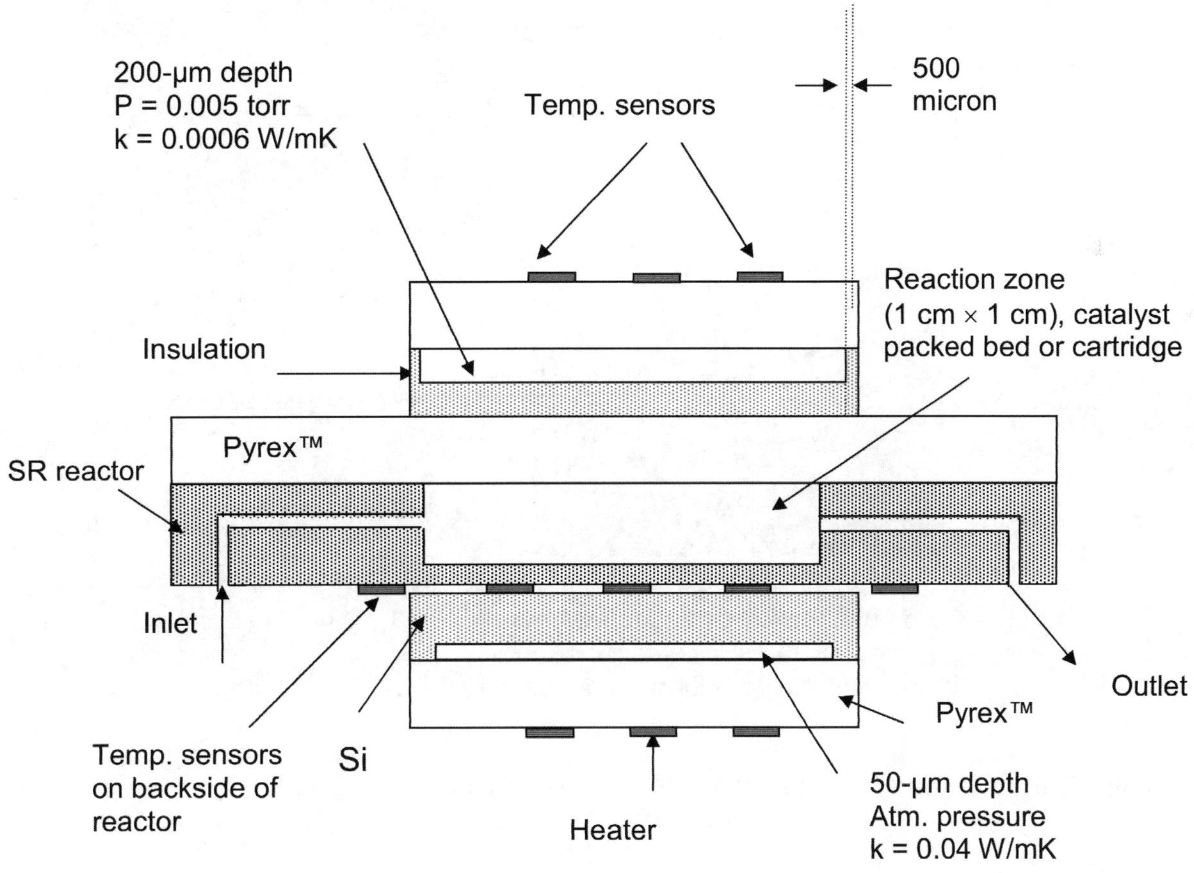

Figure 4. Design of an integrated SR

Silicon microfabrication technology is used to fabricate reforming microreactors, vacuum packaging chips, thin-film heaters and temperature sensors on silicon and glass substrates. Figure 5 shows fabricated components of this integrated device.

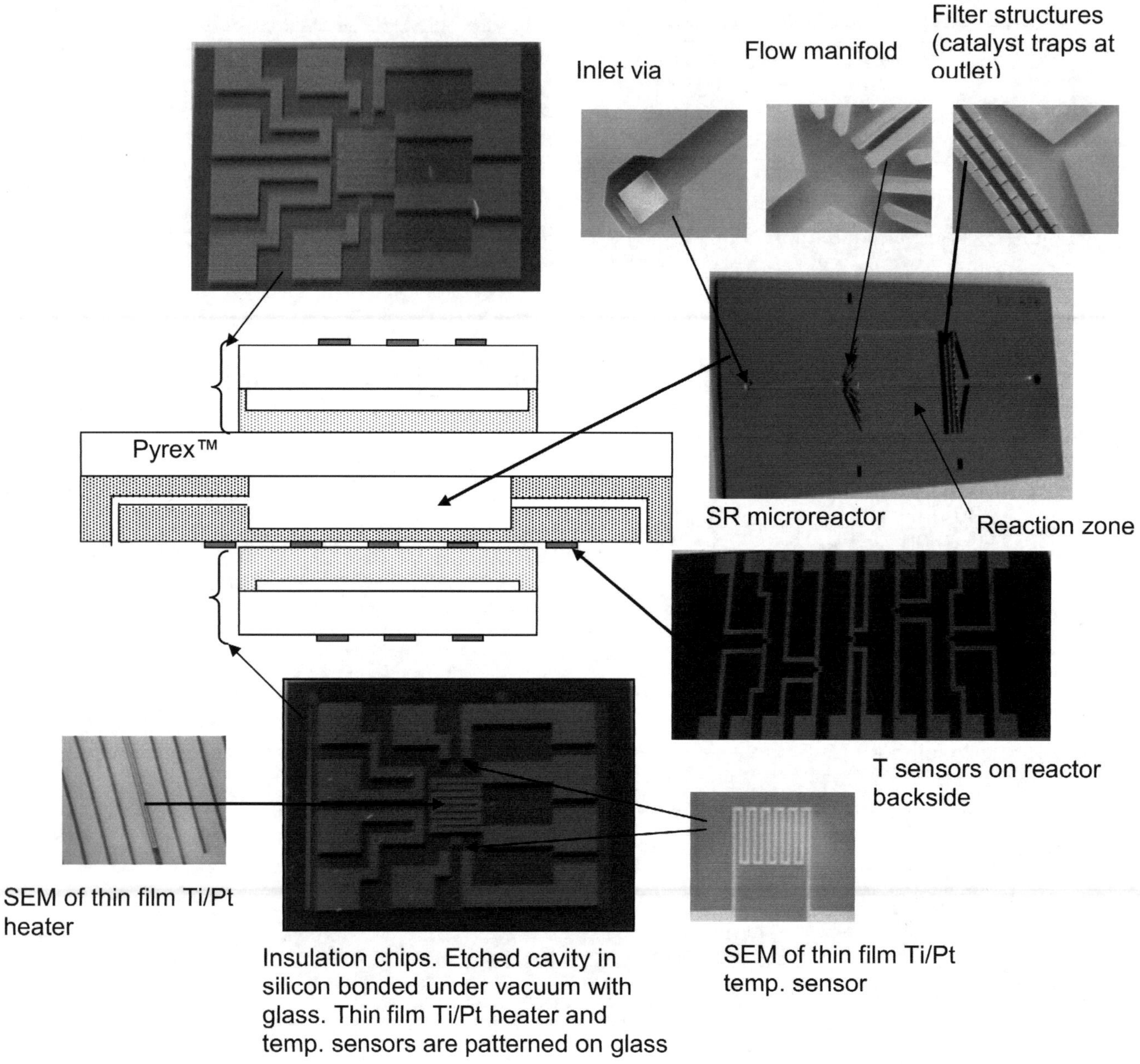

Figure 5. Fabrication of an integrated SR using silicon microfabrication techniques

The catalyst is incorporated in the form of a packed bed in the silicon microreactor chip bonded with glass. Figure 6 shows the silicon microreactor packed with catalyst particles of mean diameter 70-80 μm.

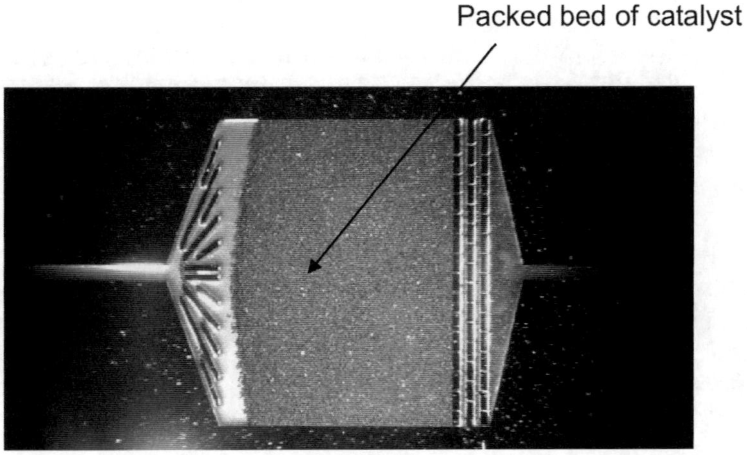

Figure 6. Incorporation of catalyst in the form of packed bed by vacuum loading
Catalyst loading achieved: 51 mg

Next, the insulation chips are bonded on the front and back sides of the catalyst-loaded microreactor using high temperature epoxy. Figure 7 shows the microreactor bonded with insulation chips.

Frontside of SR microreactor bonded with insulation chips

Backside of SR microreactor bonded with insulation chips

Assembled integrated microreformer (six-chip stack)

Figure 7

Thermal Characterization

An important aspect of integration was the thermal management at the system level, which involved heat required, transfers of heat between units, temperature profiles in different components, insulation effectiveness, and heat loss mechanisms. Prior to detailed design of individual reactors, necessary data were needed to confirm the accuracy of the assumptions used to simulate the thermal behavior. Therefore fabricated model components of an integrated device were used for the acquisition of critical thermal parameters of the overall coefficients for the transfer of heat from system fluids to one another and to the outside environment by different means of heat loss.

Thermal characterization experiments were carried out under different conditions at both atmospheric pressure and under vacuum to distinguish multiple modes of heat transport. These experiments helped quantify heat losses through various pathways from the planar microreactor structure. Well defined external heat flux is introduced to simulate the flows that result from integration with the combustor. Distributed sensors are provided to obtain the temperature profiles along the axial length.

The following sets of experiments were carried out.

1. Device suspended under ambient pressure
2. Device suspended in vacuum chamber
3. Device covered with commercial insulator under ambient pressure
4. Device covered with commercial insulator in vacuum
5. Device covered with Al foil and open to ambient
6. Device covered with Al foil in vacuum chamber

For each experiment, $Q_{combustor}$ required to raise the temperature of SR to 260°C was measured. Figure 8 and Figure 9 show thermal characterization results when tested under atmospheric pressure and vacuum respectively.

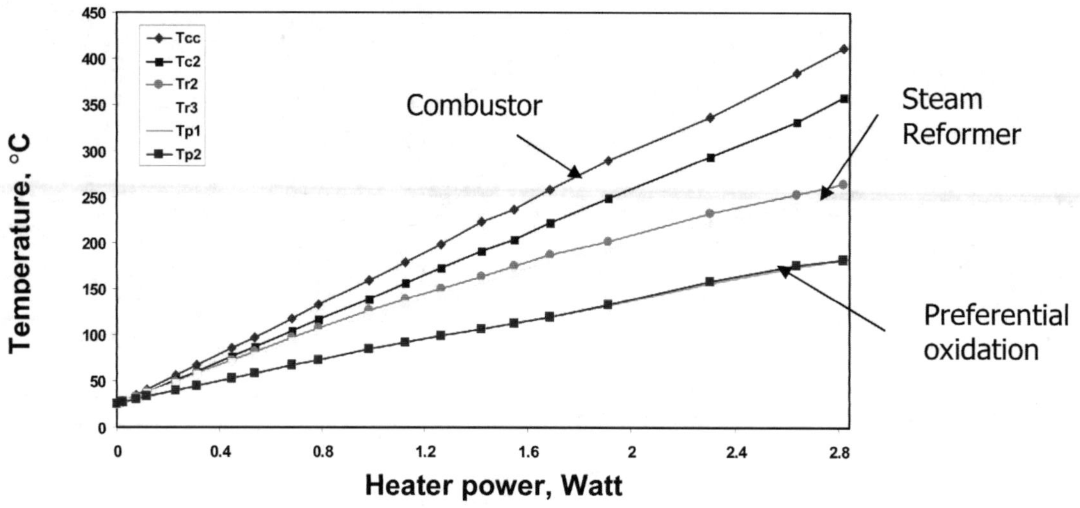

Figure 8. Thermal characterization, device covered in insulation and open to ambient. Heat required versus temperature profiles in SR, combustor, and PrOx

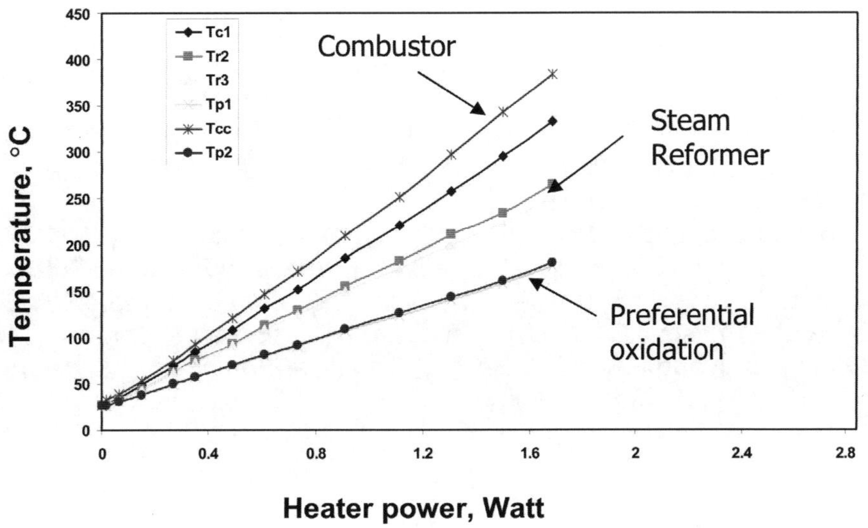

Figure 9. Thermal characterization, device covered in insulation and placed in vacuum chamber. Heat required versus temperature profiles in SR, combustor, and PrOx

Based on the results obtained from above thermal characterization experiments, modeling is under progress to quantify heat loss mechanism and to deduce free convection heat coefficients from microscale geometry. These data will then help to verify and improve thermal model. Based on understanding gained, suggestions will be made for scale up of reactor components and a packaging scheme for reduction of convective and radiative losses.

This combined experimental/modeling approach then can be used as the quantitative basis for designing an integrated portable fuel processor. The refined model can be used for energy density projections of portable fuel processor/fuel cell systems as well as for projecting energy requirements to microchemical reactors in general. The research outcome will help conceive applications and limitations of planar microchemical technology for micro fuel processing and provide impetus for subsequent development.

Acknowledgements

The authors acknowledge the financial support from the Picatinny TACOM/ARDEC and New Jersey Commission on Science and Technology (NJCST).

The silicon and glass microfabrication was performed in part at the Cornell NanoScale Facility (a member of the National Nanofabrication Users Network) which is supported by the National Science Foundation under Grant ECS-9731293, Cornell University and industrial affiliates.

References

1. J.D. Holladay, E.O. Jones, M. Phelps, J. Hu, *Journal of Power Sources* **2002**, 108, 21.
2. A.V. Pattekar, M.V. Kothare, *Journal of Microelectromechanical system* **2004**, 13(1), 7.
3. E.R. Delsman, M.H.J.M. de Croon, G.J. Kramer, P.D. Cobden, Ch. Hofmann, V. Cominos, J.C. Schouten, *Chemical Engineering Journal* **2004**, 101(1-3), 123.
4. D. Seo, W. Yoon, Y. Yoon, S. Park, G. Park, C. Kim, *Electrochim. Acta* **2004**, 50, 719.
5. L.R. Arana, S.B. Schaevitz, A.J. Franz, M.A. Schmidt, K.F. Jensen, *Journal of Microelectormechanical System* **2003**, 12(5), 1.
6. X. Oyuang, L. Bednarova, R.S. Besser, P. Ho, *AIChE Journal* **2005**, 51(6), 1758.
7. R.S. Besser, X. Oyuang, H. Surangalikar, *Chemical Engineering Science* **2003**, 58, 19.
8. H. Surangalikar, X. Ouyang, R.S. Besser, *Chemical Engineering Journal* **2003**, 93, 217.
9. P. Atkins, J.D. Paula, *Physical Chemistry*, 7th ed., W.H. Freeman & Company, NY **1997**.
10. Keyur Shah, X. Ouyang, and Ronald S. Besser, *Chemical Engineering & Technology* **2005**, 28(3), 303.

NEW SYNTHETIC METHOD OF ORGANIC PIGMENT NANO PARTICLE BY MICRO REACTOR SYSTEM

Hideki Maeta, Fuji Photo Film, Kanagawa, Japan
Tadahisa Sato, Fuji Photo Film, Kanagawa, Japan
Hideharu Nagasawa, Fuji Photo Film, Kanagawa, Japan
Kazuhiro Mae, Kyoto University, Kyoto, Japan

Abstract

A solution of an organic pigment, dissolved in an alkaline aqueous organic solvent, is mixed with a precipitation medium in a micro channel, which produce an organic pigment nano particle solution. The K-M reactor, which has been developed for both the instant mixing and the high throughput producing, can be applied efficiently on this build-up procedure. The clear dispersion, produced by this method, is extremely stable and has narrow size distribution, which is the feature difficult to realize by the conventional pulverizing method (breakdown procedure).

Introduction

In general, the coloring material is classified into two groups, pigment and dye. They are important compounds indispensable in everyday life at the present day. The pigment are now widely used in many fields such as paint, printing ink, electrophotographic toner, ink-jet ink, color filter, CCD sensor, etc.; owing to their attractive features of vivid color tone, high coloring power, light resistance and water resistance superior to those of dye-based inks. In the field of ink-jet ink so far, dyes have been used usually. But the images from dye have serious drawbacks as to resistance against water and light despite having superior wide coloration ability and diversity of structure design. However, the former images from the pigment have also some problems. One example is that the pigment could not tend to permeate through the surface of paper because of its particle size (>100nm) and wide size distribution, which has brought some disadvantages in the point of smooth surface and rubbing against eraser as a result. Therefore, the pigment having nano size (<50nm) with narrow size distribution is desired, although it is difficult to prepare on a large scale with cost effective manner.

Generally the method to produce organic fine particle is roughly classified into two categories: one is breakdown method, in which fine particles are produced from a bulk material by pulverization or the like. The other is build-up method, in which fine particles are produced by particle-growth from gas or liquid phase. The pulverizing method has been widely used hitherto as a conventional producing method because of its high practicability. But, it has various problems such as its low productivity in producing organic pigment of nano particles with narrow size distribution. This weakness is restricting the application of the pigment made by the conventional breakdown method within some field.

From these backgrounds the build-up method has been paid great concerns for organic nano particle production in recent years, and many investigations have been made so that it may be represented by the work of Kasai [1]. A micro-reactor based on a micro chemical process technique has now come into the limelight as a new process because of its many advantages of operability. However, the channel may be highly apt to cause blockage if conditions to be controlled are incorrect. So, it is unavoidable to say that there is difficulty from the viewpoints of reproducibility and continuous production especially in the case of handling particles. In spite of this anxiety, some pioneering works in producing organic nano particle have been done in micro reactor system [2,3]. In this paper, we will show our new method named "the co-precipitation method with pH shift in micro-chemical process" for producing nano particle with small size (<50nm) and narrow size distribution [4].

Materials and Experimental

Preparation of materials

In general, pigment is the compound with low solubility to water and organic solvent. However, the pigment having dissociative protons in its chemical structure is able to form homogeneous pigment solution by adding alkali with polar organic solvent. By pouring this organic pigment solution into water, precipitation of the pigment is occurred. This process is well known as recrystallization. In the presence of suitable dispersant, dispersion of nano particle pigment is easily obtained instead of its crystal. We call this process "the co-precipitation method with pH shift", which is categorized into build-up process [5]. Feeding the homogeneous solution into micro device is favorable for reducing the risk of the channel clogging. In this study, C.I. Pigment Yellow 128 was chosen as a model compound. The reaction scheme to getting clear pigment solution is shown in Scheme 1. According to this scheme, 5wt% of PY128 in DMSO solution was obtained with appropriate dispersing agent.

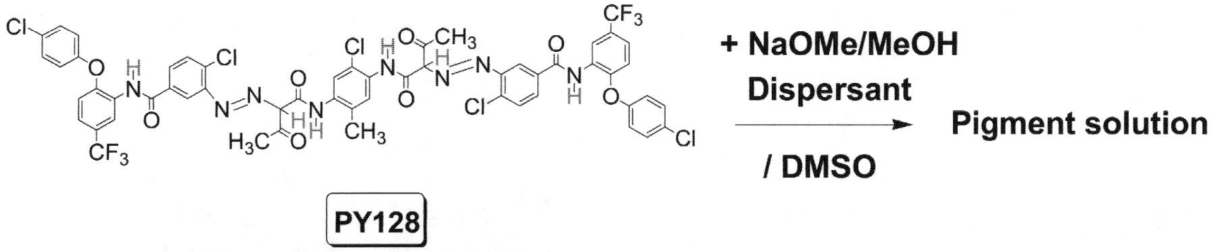

Scheme 1. Preparation of pigment solution in the case of C.I. PY128

Micromixer, which is a microreactor aimed at mixing, is an important device for micro chemical processing. There have been developed several types of micromixer on the concept of controlling molecular diffusion in laminar flow [6]. The K-M reactor, consisting of three cylindrical components, was developed as practical device by concept of using kinetic energy and molecular diffusion [7]. This new type of reactor has realized rapid mixing, high throughput, wide range of mixing ratio, prevention of clogging in micro channel and so on. We selected this K-M reactor shown in Fig.1 for producing organic pigment nano particle, because of these features is suitable for our object under nano size and narrow distribution control. And the possibility of mass production on the micro chemical plant with K-M reactor was also attractive.

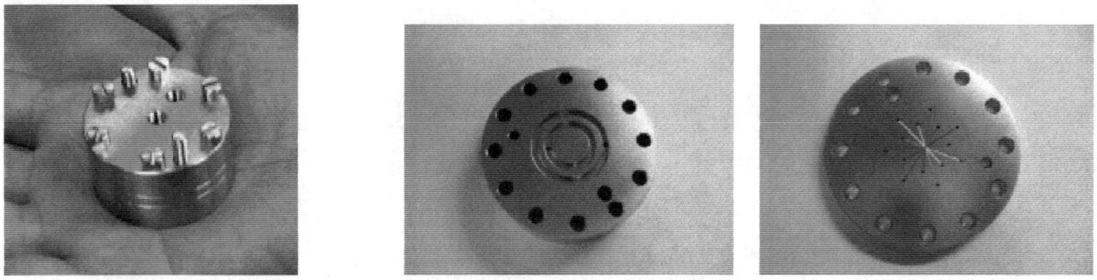

Figure 1. K-M reactor used for organic pigment nano particle (from Left: Overview, Inlet plate and Mixing plate)

The pigment solution, including dispersing agent, was contacted with water in K-M reactor as shown in Fig. 2. The two solution were fed to the inlet plate, then uniformly distributed to the plural channels on the backside of the mixing plate. The fluids were next converged at the center along with the radial channels. Finally, they met at the center of the mixing plate and discharged through the hole of the exit plate. In this experiment, we used the four radial types of mixing plate, the width are 50μm with 14 channels, 100μm with 10 channels, 150μm with 6 channels and 400μm with 6 channels, and also four sizes of the exit plate, the hole size are 100μm, 200μm, 300μm, and 800μm.

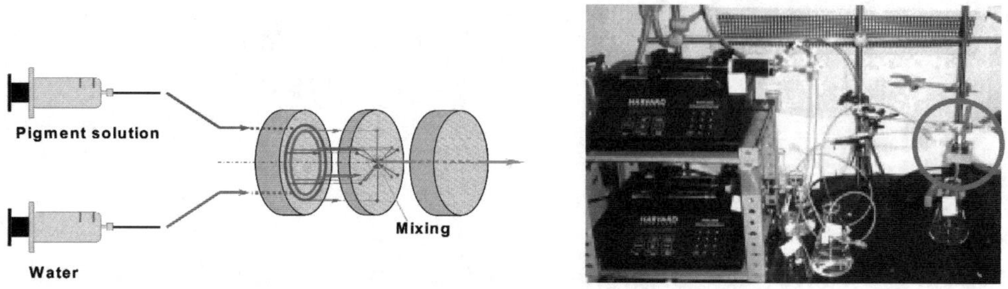

Figure 2. Micro reaction system using K-M reactor (red circle)

Procedure

The mixing ratio of 5wt% solution of PY128 and water is an important factor to produce nano particles. The ratio of pigment solution to water was set at 1/4 (v / v) in light of the size control and the stability of the dispersion. The K-M reactor combining the mixing plate of 400μm-width with the exit plate of 800μm was chosen, and the total flow rate of pigment solution and water was controlled from 10 to 250 mL/min by syringe and/or plunger type pumps. The experiment was conducted at room temperature. The particle size distribution was measured using a particle sizer (UPA 150EX, Nikkiso Co.,Ltd.).

Results and Discussion

Particle size distribution produced by micro reaction system

Figure 3 shows the size distribution of particles obtained by micro reaction system. A clear and stable yellow dispersion was obtained as shown in the photograph. The particle sizes on the volume basis, Mv, were within 15 to 25 nm, and showed extremely narrow distribution (Mv/Mn =1.3-1.4), which features were difficult to realize by the conventional pulverizing method.

Figure 3. Size, photograph and SEM image (x100k) of nano particle by K-M reactor

On the other hand, when the solution was mixed with water at the same mixing ratio (pigment solution : water = 1 : 4) in the flask with severe stirring, non-clear yellow dispersion was produced. The particle size was large and wide-spread (Mv=153nm, Mv/Mn= 2.29) as shown in Fig. 4. This was caused by a random precipitation.

Figure 4. The pigment solution performed in the flask and SEM image (x100k)

Presumed mechanism on particle formation in microreactor

At first, the logic of rapid mixing by K-M reactor is shown in Fig. 5. The mixing proceeds through 3 steps. At the step1, micro segments are regularly arranged before mixing adjusted the ratio of flow rate. At the next step2, these micro segments meet at the mixing zone (collision center) under high shear rate to deform and break them instantly. And finally at the step3, these micro segments are deformed in a restricted space again in order to make the diffusion distance shorten. This system would realize to increase the number of interface per unit time, which promotes the diffusion of molecules explosively.

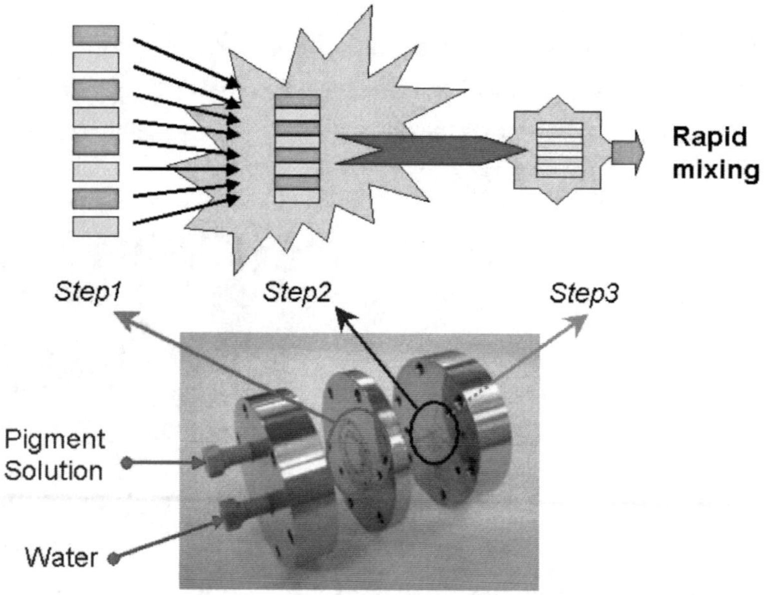

Figure 5. Logic of rapid mixing by K-M reactor

Next, we considered the mechanism for the co-precipitation method with pH shift in the channel of K-M reactor in Fig. 6. This is the magnified figure in part of micro segment of step2 to step3 in Fig. 5.

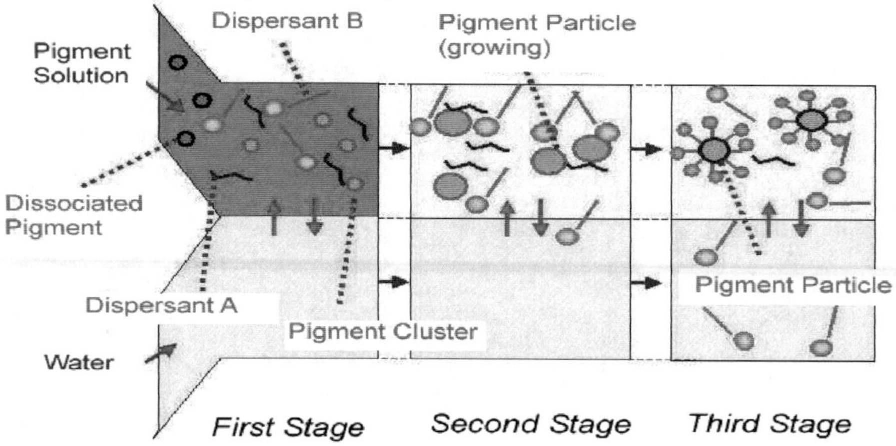

Figure 6. Mechanism for organic pigment nano particle

The mechanism would be divided into three stages. The first stage is the contact the pigment solution with water, where the pigment cluster are generated with decreasing pH value. The second is the competition process between the growth of particle and the restriction of growth by the adsorption of dispersant to the particle. And the last stage is the transformation process to solid stable nano particles protected by dispersant. The second stage is the most important process to make a fine particle. The pigment cluster formed by the uniform contact with water segment was instantly attached with dispersant, and restrict the growth of particle by collision of cluster. This precise distinction of reaction stage makes it possible to create the particle with narrow size distribution in micro reactor. In the flask, these stages would be confused and localized at random, so wide particle size distribution was formed. In addition, no clogging was observed though the experiment using the K-M reactor. This was brought about by both small particle of ca. 20 nm and the high flow rate in the reactor.

Construction of a pilot plant

Since very fine particle was successfully synthesized in laboratory scale, a pilot plant for the organic pigment nano particle process has been built up as shown in Fig. 7. We are now going to check the performance on continuous mass production at the point of operability and quality of product. We think the total volume of flow rate conducted in the pilot plant is sufficient for practical production, because it is equivalent to over 50 t/year productions theoretically by operation of the pilot plant in about 4800hr/year.

Figure 7. Pilot plant for the organic pigment nano particle with K-M reactor (red circle)

Conclusions

We proposed a new synthesis method for producing fine organic pigment by pH shift, in which the pigment solution with alkali media was instantly mixed with water in the presence of a dispersant. The method shows the effectiveness for producing nano particle of ca. 20 nm using the microreactor system with K-M reactor. Now, we are challenging a large scale production [8]. This micro production system will be also applied to the other pigment having dissociative proton such as PR122, PV19, PY74, PY93, PR254 and so on.

Acknowledgments

The research was conducted with the aid of The Research Association of Micro Chemical Process Technology (MCPT) for the Project of Micro-Chemical Technology for Production, Analysis, and Measurement and the financial support of New Energy and Industrial Development Organization (NEDO). We acknowledge MCPT and NEDO for their assistance.

References

1. H. Kasai, H. S. Nalwa, H. Oikawa, S. Okada, H. Matsuda, N. Minami, A. Kakuta, K. Ono, A. Mukoh, and H. Nakanishi, "A Novel Preparation Method of Organic Microcrystals", *Jpn. J. Appl. Phys.*, 31, L1132-L1134 (1992).
2. Patent JP-A-2002-30230.
3. Patent JP-A-2002-129050.
4. Patent JP-A-2005-307154.
5. Patent JP-A-2003-26972.
6. References of other type of micromixer are cited in ref. [7].
7. H. Nagasawa, N. Aoki and K. Mae, "Design of a New Micromixer for Instant Mixing Based on the Collision of Micro Segments," *Chem. Eng. Technol.*, 28, No. 3, pp. 324, 2005.
8. This result was partially discussed in ; H. Maeta and K. Mae, "New Synthetic Method of Organic Pigment Nano Particle by Micro Reactor System," *Proceedings of the 4th International Workshop on Micro Chemical Plants*, Kyoto, Japan, Jan. 26-27, pp.72, 2006.

98b Micro-Fluidic Strategies for Synthesis of Nanoparticles

Challa S.S.R. Kumar, Center for Advanced Microstructures and Devices, 6980 Jefferson Hwy, Baton Rouge, LA 70806

Recent research publications, including from our own laboratories, demonstrate that physical properties of semiconductor nanocrystals, palladium nanoparticles, cobalt nanoparticles, silver nanoparticles and nanocomposites can be fine tuned through control over nanocrystal growth parameters and kinetics in micro fluidic reactors. Based on these initial results, one can anticipate that micro reactors are likely to provide cost-effective and environmental friendly technologies for rational designing of nanoparticles. By utilizing several thousands of microreactors, process scale-up with continued reaction control provide several advantages in commercialization of processes for wet chemical synthesis of nanoparticles. The presentation will focus on how microfludic strategies for synthesis of nanoparticles are begining to show promise in providing control over size, size distribution and crystal structure of nanoparticles.

Formation of Au/Ag Nanoparticles in a two step micro flow through process

J. Michael Köhler[1], Marie Held[1], Uwe Hübner[2] and Jörg Wagner[1]
[1]Techn. Univ. Ilmenau, Germany,Institut für Physik, PF 10 05 65, D- 98684 Ilmenau, Germany,
[2]Inst. f. Phys. High Technol. Jena, Germany

Abstract

The optical properties of metal nanoparticles depend strongly on their composition and size. The plasmon absorption band can be shifted between the blue range and the red range of the optical spectrum, if silver and gold are mixed and aggregation of particles is induced, in addition. The formation of different nano particles and nano particle aggregates was studied in a two step flow-through process using a glass/Si/glass static micro mixer. The educt solution was pumped through the mixer chips by programmable syringe pumps. The process was monitored online by a compact spectrometer connected with the outlet tube by glass fibers. The product solution was collected and investigated in addition by an UV/VIS spectraphotometer. Selected samples were investigated by Scanning Electron Microscopy (SEM) in order to evaluate the particle morphology.
The typical Ag nanoparticle absorption was found, if an excess of Ag ions is introduced into the system. The intensity of the peak is decreased with the reduction of the flow rate of $AgNO_3$ solution in a reaction system including $HAuCl_4$ solution and polyvinylalcohol beside the reducing agent (ascorbic acid). The experiments show that educt concentration ratios as well as the total flow rates influence the shape and composition of the nanoparticles. Therefore, microreactors with fast mixing and well reproducible mixing conditions are interesting for the synthesis of nanomaterials.

Keywords

micro reactors, nanoparticles, gold, silver, fast mixing, plasmon absorption

1. Introduction

Nanoparticles attract increasing interest due to their unique properties. Nanoparticle materials are characterized by extremeous high specific surface and are accordingly used as catalysts. They possess specific size-dependent optical properties and can be applied, therefore, as pigments or as labels in biomolecular diagnostics. In addition, they are of interest as building units for nanotechnology. The composition of nanoparticles is wide spread and include dielectric inorganic, organic, polymer, semiconductor and metallic materials. Nanoparticles can be built by pure materials or mixtures, They can be represent one homogenous phase, single crystals or can be composed by several phases. Composite nanoparticles exist in form of spherical aggregated elementary particles, as core shell or onion-like multi-shell particles or can show more complex structures. The size of nanoparticles can range from well defined molecule-like clusters and bulk material covering a range of some dozen up to some million atoms per particle, These numbers correspond to a size range between a little less then one nanometer in diameter up to some hundreds of nanometres [1,2].
Particular interest is paid to metallic nanoparticles. Metal nanoparticles are used in handcrafts like glass and ceramics since antiquity. So, the deep red colour of dispersed gold nanoparticles was

applied for the coloring of glass (rubin glass). The nature of colloidal solutions containing dispersed metallic particles was correctly interpreted by Michael Faraday in the 19th century. The particular optical and the electronical properties of metal nanoparticles make them very interesting for nanotechnological applications. So, they are under investigation for nanoelectronic and nanooptoelectronic applications [3]. For handling and positioning and for the construction of well defined electronic gaps. The formation of monomolecular films at the surface of nanoparticles and specific interactions between nanoparticles and macromolecules were investigated [4-10]. DNA-nanoparticle aggregates are of interest as building blocks for molecular nanotechnology as well as for the optical and electrical read-out of biochips [11-15]. Recently, gold nanoparticles comes into focus of interest due to astonish catalytical activity [16,17].
The specific colour of metallic nanoparticles is a result of the resonant electronic states of these small objects with a high number of surface-near or surface atoms. The characteristic energies of transitions between the highest electronic states cause characteristic absorptions of light, which are found in case of gold, for example around 520 nm. The analogue transition of silver causes a transition at higher energy resulting in an absorption band in the region of about 420 nm. Because the electronic states are influenced by elementray composition, by size and shape of the particles and are additionally influenced by the surface state, the interaction of particles with the surrounding solution and interactions between particles, changes in the absorption bands indicate changes of particle parameters or properties of the colloidal system [18].
The formation of nanoparticles of special types is a time-critical process. The sensitivity of nucleationand the growth of primarely formed seeds from surrounding conditions leads to a strong dependence of nanoparticle properties from the reaction conditions. Therefore, micro reactors are tested for the generation of nanoparticles, because the conditions for convective mixing and interdiffusion can be controlled very well in micro flow-through systems. In general, the introduction of heterogenous systems into micro reactors must be regarded as a complication. Particles can cause disturbances due to wall adhesion, fouling, shifting of fluid resistances and blocking of micro channels. But, it was demonstrated that micro reactors can be applied for dielectric inorganic nanoparticle generation despite these difficulties [19, 20]. Micro mixers were also used for the generation of emulsions for micro flow-through emulsion polymerization resulting in the formation of polymer nanoparticles [21-23]. Although the high surface to volume ratios of micro channels represent a serious problem for particle generation inside micro reactors, it is possible to growth seeds of gold by metal catalyzed reduction of tetracholoroaurate with ascorbic acid at room temperature and to generate gold nanoparticles directly inside micromixers [24]. The influence of molecular effectors like macro molecules can also be studied using micro flow-through systems [25].
The metal composition is of particular importance for the optical properties of nanoparticles. Therefore here, the formation of nanoparticles composed by silver and gold was studied. Experiments in micro flow-through arrangements with and without macromolecular effectors are presented and product spectra of micro-flow-through syntheses are discussed in comparison with batch reactions.

2. Experimental

2.1. Mixing devices and experimental arrangements

Micro flow through experiments for synthesis of AuAg-nanoparticles were made by use of glass/Si/glass static micro mixers. The reactors are constructed following the split-and-recombine-scheme, wherby splitting and recombination occur in different planes and a reshaping of the cross section of fluid channels ensure a good lamination [26]. The chip reactors are built in a sandwich construction with a bottom and a top glass chip and a Si(100) chip between them. The micro channels use all three levels in order to realize the above mentioned split-and-recombination principle. Several sections of channel structures consist on completely removed Si. These channel sections are arranged in parallel to the (111) crystal planes resulting in vertical side walls of the channels. These sections of micro channels are completely transparent due to the transparent top and bottom chips [27].

In the experimental arrangement 2- and 3-step mixers were applied. Each mixing step consist on a group of split-and-recombine units. We use the 2+1 mixer (Statmix 7, IPHT Jena) and the 2+2 mixer (Statmix 9, IPHT Jena). Both type of static multi mixers possess the same chip format but are distinguished by the number of fluid ports and the number and arrangement of mixing units [28]. The mixing in the 2+1 mixer starts with four split-and-recombine-units for the both first educts followed by five split-and-recombine-units after the addition of the third educt stream. The fluid ports (3 inlets, 1 outlet) are arranged linearly (Fig. 1, top). Two separate premixing steps can be realized in the 2+2 mixer. Two pairs of educt streams are mixed in parallel, but separated from each other by 5 split-and-recombine-units in each pair. After these premixing, these both streams are conducted together and pass additional 6 split-and-recombine-units (fig. 1, below).

For some experiments, a 2+1+1 -mixing arrangement was realized by combination of the 2+1 chip mixer with a PTFE tube T-junction. In this arrangement, a PTFE tube of 30 cm length was introduced between the outlet of the chip mixer and the T-junction in order to realize a longer residence loop between the second and the third mixing step ("2+1+T"). The internal volume between the mixer and the T-junction amounted to about 60 µL.

The liquids were pumped through the micro fluid arrangement by syringe pumps. A compact spectrometer (Ocean Optics) was connected with the outlet tube by a fibre optical arrangement and used for monitoring purposes.

The product solutions were collected in Eppendorf cups. They were characterized by UV/VIS spectralphotometry (Specord 200, Analytik Jena) after dilution of the product solution in the ratio of 5:1 or 10:1. The spectra were taken between 1 hour and 24 hours after preparation of the sample. In some cases, aggregation and sedimentation of particles from product solution was observed in the following days and weeks. But, a great part of the product solutions remains stable over months. The properties of samples were investigated by SEM imaging and DCS (differential centrifugal sedimentation) measurements (CPS 20 000 Disc Centrifuge, CPS Instruments, USA).

2.2. Materials

All samples were prepared as aqueous dispersions in ultrapure water (Aqua purification G 7795, Miele Germany). The used chemicals were pA. graded. Solutions of ascorbic acid were freshly prepared due to the reactivity of the reducing agent against oxygen.

Gold ions were introduced in form of tetrachloroauric acid, silver ions were applied in form of silver nitrate solutions. The applied concentrations were chosen in the range between 0.5 and 2 mM for the metal compound educt solutions. The reducing agent ascorbic acid was applied in a concentration of 0.1 M.

3. Results and discussion

3.1. Influence of Ag/Au-ratio on spectral properties of generated nanoparticles

If silver nitrate solution is brought together with ascorbic acid at room temperature, no reaction is observed. But, small amounts of tetrachloric acid or a certain content of Au seeds starts the reduction of silver. In case of the reduction of silver ions by ascorbic acid, a typical example of a metal-catalyzed metal salt reduction is observed. Gold particles, primarely formed, catalyze the further redox reaction between ascorbic acid and silver ions. The reaction can be monitored easily by observing the appearance of color. In dependence of composition, additives and mixing conditions, a wide spectrum of colors is found. It covers the range from deep red over violett and blue (excess of gold) to green, brown and yellow (excess of silver). Examples of product solutions are shown in figure 2. Therefore, optical spectra are very well suited for the estimation of product characteristics [29-31].

The influence of metal salt ratio on the quality of product was well reflected by micro flow-through experiments. A characteristic shift of color was found in the two step synthesis of nanoparticles under flow-through conditions in dependence on composition. In these experiments, the concentration of educt solution was constant, but the educt volume flow ratio was changed by changing the flow rate ratios. At low $HAuCl_4$: $AgNO_3$ flow rate ratio, the product absorption spectrum is dominated by the typical plasmon absorption peak of silver nanoparticles in the short wavelength range (fig. 3, curve A). At invers flow rate ratios, a long wave length shoulder arises beside the silver peak at a wavelength of about 520 nm (Fig. 3, curve B), which is characteristic for gold particles. At further increased flow rate ratio ($HAuCl_4$: $AgNO_3$), this shoulder growths forming the dominating peak in the product spectrum (Fig. 3, curve C).

The influence of composition of educt solutions on the optical properties of AuAg nanoparticles generated by reduction with ascorbic acid was investigated with reference experiments in batch. Therefore, tetrachloric acid was mixed with silver nitrate in different concentration ratios. This mixtures (200 µL) were brougth together with ascorbic acid solution (100 µL, 0.1 mol/L) and mixed quickly. In result, the strong and characteristic shift of color in the formed product solutions was confirmed (Fig. 4). At low gold concentrations, the silver peak is dominating. The yellow solution is characterized by a spectrum showing a strong peak at about 400 nm. A shoulder at the long-wavelength side of this peak increases with increasing Au concentration. A further increase in the gold concentration lead to a wide peak and shifts the maximum to higher wave lengths. The color changes over brown and blue grey to a deep blue corresponding to a absorption peak at about 600 nm. This absorption represents a longer wave length than the absorption band of small spherical particles of pure gold indictaing a more complex particle formation behaviour. A further increase in the gold:silver ratio leads to an inversion of the wave length shift of the absorption maximum. The peak becomes sharper and drifts towards the 520-nm-peak of pure gold particles, if the gold concentration is further increased.

3.2. Influence of mixing order, mixing time and flow rate

The high reproducibility of kinetic conditions is one of the most important advantages of micro reactors and micro flow-through processes. No effect of flow rates on the product properties would be expected, if mixing processes are much faster than the reactions. In contrast, time-critical mixing processes indicate reaction mechanisms with characteristic times of reaction steps in the same order of magnitude as the mixing processes.

The influence of time-management on the flow-through synthesis of gold/silver nanoparticles was studied by varying the flow rates of educt solutions in a 2+1 mixing process. The flow rate ratios were kept constant, so that all educts were always applied in the same concentration ratios. With respect to the importance of gold seeds for the silver reduction, tetrachloric acid and ascorbic acid were brought together at first in order to initiate the gold particle formation. After this, silver nitrate was mixed to the reaction solution in the second mixing step. The internal volume of the mixer in the first mixing step (4 split-and-recombine units) amounts to about 2 µL. This corresponds to a theoretical residence time of about 1 s, if both first educt solutions are introduced with a flow rate of about 60 µL/min, each. During this time a certain mixing by interdiffusion, the starting of nucleation and beginning growth of gold seeds must take place.

The experiments showed a strong effect of total flow rate on product properties in a comparatively small range of absolute flow rates (Fig. 5). Therefore solutions of 1 mM $HAuCl_4$ and 1 mM $AgNO_3$ were applied. At lower flow rate (25 µL/min) the typical plasmon absorption band of small Au particles at 520 nm arises beside a large absorption between 300 and 500 nm. This band is accompanied by a longer wavelength absorption reaching the range about 700 nm. With increasing flow rates, the longer wave length absorption is reduced, the Ag-peak at about 400 nm increasingly dominates the spectrum, whereas the gold peak (at 520 nm) shifts to shorter wave lengths and becomes a shoulder of the Ag-peak at about 500 nm.

An analogue result was obtained at reduced $HAuCl_4$ concentration (0.5 mM) and enhanced flow rates (fig. 6). At lower flow rates (100 µL/min in each educt channel) an broad absorption with two peaks at (about 450 nm and 700 nm) was observed. At higher flow rates (200 µL/min in each educt channel), the long wave length absorption band disappeared.

The complex behaviour of the reaction system in dependence on flow rate was confirmed by batch experiments with different mixing order. The optical absorption of product solutions from batch processes strongly depends on the mixing order of educt solutions. If the tetrachloroauric acid was mixed with ascorbic acid, first, and the silver nitrate solution was added later, a short wave length absorption band with a shoulder was observed, similar to the flow conditions at higher flow rate (compare fig. 5, 58 ml/min, fig. 6, premix Au+ascorbic acid (AA)). If gold and silver educt solutions were mixed first, and then ascorbic acid was added, a broad absorption band with a maximum at about 550 nm was observed (fig. 6).

The measurements at flow-through conditions and the result of the batch experiments with different mixing order lead to the assumption, that the short-time characteristics of mixing process controlls the properties of the formed products. This hypothesis was checked in reference experiments in batch. Thereby, tetrachlorauric acid was quickly mixed with ascorbic acid, and silver nitrate was added after different time intervals (fig. 7). The short wave length peak at 400 nm, which dominates in case of longer time gaps between the both mixing steps (Fig. 7, A), is lowered, if the time interval is reduced (Fig. 7, B, C). Simultaneously, the characteristic gold absorption peak at 520 nm arose (Fig. 7, C, D). In addition, an significant increase in the absorption above 600 nm was observed. This increase corresponds to the large band in the optical spectra in the cases where silver and gold salt were mixed before the ascorbic acid was added (fig. 7, D, E).

The results suggest, that the composition of the outer part of the formed nanoparticles possesses a stronger influence on the spectral properties then the inner part. This interpretation would explain the domination of the short wave length peak (silver peak at about 400 nm) in case of a sufficient long time gap between the premixing of tetrachloroauric acid and the reducing agent and the follwing addition of silver nitrate, because in this case particles are expected to consist of a pure gold core and a high excess of silver in the shell. Short time intervals between both mixing steps lead to a deposition of silver at preformed gold nuclei, whereas simultaneously, more gold seeds are formed. Thus, in result at least two classes of particles with different spectral characteristics are generated, one dominated by silver shells and another with higher gold content. In dependence on their ratio, a shorter wave length absorption band with longer wave length shoulder (fig. 7, C, for example) or a longer wave length band with shorter wave length shoulder (fig. 7, D, for example) are observed. The difference in the particle distribution can be interpreted by discussing the conditions during the liquid lamination in the split-and-recombine units (fig. 8). In case of faster flow, the lamella stack of ascorbic acid and gold tetrachloroaurate is quickly further laminated and stacked together with the silver nitrate stack. Hence, silver ions can diffuse into the whole liquid before all silver ions are converted to metal particles. In result, nanoparticles with silver dominated shells are formed (Fig. 8, A). At slower flow, the lamination of tetrachloric acid-ascorbic acid mixtures with silver nitrate solution is slower. The silver ions are reduced to metallic silver at the surface of metal seeds before they can diffuse into the complete number of Au/ascorbic acid-lamella. In result, nanoparticles with silver dominated shells are formed in the outer part of the

Au/ascorbic acid-lamella, whereas gold-dominated nanoparticles remain prevailing in the center of the liquid lamella generated by ascorbic acid and tetrachloroaurate solutions (Fig. 8, B). Thus, high flow rates and, therefore, fast mixing lead to an increase in the number of nanoparticles with Ag-rich shells(high absorption in the short-wavelength range), whereas low flow rates, respectively lower mixing rates, cause the formation of a mixture of gold-rich nanoparticles (longer-wavelength absorption and) and nanoparticles with silver-rich shells.

3.3. Influence of macromolecular effectors

It was known from earlier studies, that water-soluble polymers cause a reduction of seed growth of gold nanoparticles [32]. Here, the influence of macromolecules on the Au/Ag-nanoparticle formation in 2- and 3-step mixing processes under micro flow-through conditions was studied. In first experiments, a 2+1 and a 2+1+1 fluid arrangement with a residence loop before the last mixing step was applied. A broader absorption band around 500 nm, which arose in case of premixing of silver nitrate and tetrachloric acid solutions, became smaller and shifted to shorter wave lengths in case of increasing flow rate of PVA (Fig. 9). This effect was interpreted by more homogenous reaction conditions because higher polymer concentrations inhibit the seed growth significantly. The order of reactants addition is very important in case of macromolecular effectors, too. The influence of reaction order was tested by two different fluid arrangements using the 2+1+1 static micro mixer. First, the polymer solution was mixed with the reducing agent before gold (tetrachloroaurate) and then silver (silver nitrate) were added (Fig. 10, A). In the second case, the both metal supplying solutions were added to the polymer solution before the reducing agent was mixed with them in the last mixing step (Fig. 10, B). Concentrations and flow rates of all educt solutions were the same in both experiments. Nevertheless, the product solution of the first fluid arrangement showed a significant higher absorption in the range between 400 and 500 nm (difference spectrum in Fig. 11). It is to assume, that the interaction of growing seeds with the macromolecules is responsible for the difference. In the first case, the nucleation starts simultaneously with interaction of metal ions with the macromolecules, whereas in the second experiment, the metal ions could interact with the macromolecules before the start of the nucleation through mixing of metal salt solutions with reducing agent.
The interaction of metal particles with macromolecules was concluded from SEM images. Figures 12 and 13 show examples of aggregates of water-soluble polymers with Au/Ag-nanoparticles after the chemical amplification of gold seeds with higher concentrated silver nitrate solutions. Obviously, the macromolecules are responsible for the formation of the nanoparticle aggregates. Flow-through experiments with PVA addition in a 2+1+T-arrangement supply core-shell particles (Fig. 14).
Experiments with gold and silver salt concentrations in the same order of magnitude lead to nanoparticles with different composition, shape and size distribution. The colour of scattered light is a nice indicator for the features of obtained products. Figure 15 gives an example of a diverse composed mixture of AuAg nanoparticles obtained by reduction using ascorbic acid and following embedding of the nanoparticles in a PVA matrix. This example makes clear, that the observed spectra in the micro fluidic experiments must be discussed with respect to different optical properties and, therefore absorption behaviour of homogenous nanoparticle populations or as spectra of mixtures of them.

4. Conclusions

The formation of metal nanoparticles by reduction of metal ion containing solutions with ascorbic acid at room temperature is a good example for a comparatively fast chemical process, which is sensitive to the mixing conditions. Therefore, static micromixers were applied to investigate the influence of several parameters, since they provide reproducible and homogenous mixing. Differences in the product properties can easily be characterized by the optical spectra of product solutions.

Different product characteristics are not only obtained by changing the ratio of flow rates of the different educt solutions, but also in dependence on the absolute flow rate at same flow rate ratios or in dependence on mixing order, if all flow rates are kept constant. Beside mixing conditions and mixing order, the introduction of macromolecules effects the formed nanoparticle populations, additionally.

The differences in product properties in dependence on process parameters was discussed with respect to the interdiffusion of educts in the static micro mixers and to the mechanisms of nucleation and seed growth. It seems, that the outer part of the particles (shell) is mainly responsible for the optical absorption properties of the colloidal product solutions. The choice of flow rate and mixing order allow to control the nanoparticle formation more in the direction of gold or silver-dominated optical properties of the products at same total concentration of gold and silver in the product. TEM, AFM and single particle spectroscopy will be applied for the characterization of product solutions on the single particle level in further investigations.

Acknowledgement

We thank W. Fritzsche and A. Steinbrück (Jena) for helpful discussions. The support of J. Albert and M. Sossna (preparation of micro mixers, IPHT Jena) and technical support during the microfluid experiments by F. Möller (Ilmenau) is gratefully acknowledged. The authors thank for financial support by the German Federal Environmental Foundation (DBU).

References

[1] Henglein, A.: Chem. Rev. 89 (1989), 1861
[2] Mann, S.; Cölfen, H.: Angew. Chem. 115 (2003), 2452
[3] Krichevski, O.; Tirosh, E.; Markovich, G.: Langmuir 22 (2006), 867
[4] Niemeyer, C.M.:Appl. Phys. A 68 (1999), 119
[5] Niemeyer, C.M., Blohm, D.: Angew. Chem. 111 (1999), 3039 54.
[6] Esumi, K.; Hosoya, T.; Suzuki, A.; Torogoe, K.: J. Colloid and Interface Sci. 226 (2000), 346
[7] Gröhn, F.; Bauer, B.J.; Akpalu, Y.A.; Jackson, C.L.; Amis, E.J.: Macromolecules 33 (2000), 6042
[8] M. Mertig, L.C. Ciacchi, R. Seidel, R. Pompe, A. DeVita, Nano Letters 2 (2002), 841
[9] L.C. Ciacchi, M. Mertig, R. Seidel, W.Pompe, A. DeVita, Nanotechnology 14 (2003), 840
[10] Niemeyer, C.: Curr. Opinion Chem. Biol. 4 (2000), 60
[11] Möller, R., Csáki, A., Köhler, J.M., Fritzsche, W.: Nucleic Acid. Res. 28 (2000), 20 53.
[12] Taton, T.A., Mirkin, Ch. A., Letsinger, R.L.: Science 289 (2000), 175
[13] Köhler, J. M., Csáki, A., Reichert, J., Möller, R., Straube, W., and Fritzsche, W. Sensors and Actuators B 76, 166-172. 2001
[14] Fritzsche, W.: Rev. Mol. Biotechnol. 82 (2001), 37-46
[15] Park SJ, Taton TA, Mirkin CA, Science 195 (2002), 1503-1506
[16] Haruta, M.: Nature 437 (2005), 1098
[17] Hughes, M.D. et al.: Nature 437 (2005), 1132
[18] Shipway, A.N.; Katz, E.; Willner, I.: ChemPhysChem. 1 (2000), 18
[19] Ehrfeld, W.; Hessel, H.; Löwe, H.: Microreactors (Wiley VCH 2000)
[20] Schenk, R.; Donnet, M.; Hessel, V.; Hofmann, Ch.; Jongen, N.; Löwe,H.: IMRET 5, (2001), 489
[21] Hamley, I.W.: Soft Matter 1 (2005), 36
[22] Song, C.; Wang, D.; Lin, Y.; Hu, Z.; Gu, G.; Fu, X.: Nanotehcnology 15 (2004), 962
[23] P.M. Günther, J. Wagner, G.A. Groß, J.M. Köhler: 9th Internatl. Conf. On Miniaturized Systems μ-TAS (Boston, October 9-13, 2005), 918-920

[24] J. Wagner, T. Kirner, G. Mayer, J. Albert, J.M. Köhler, Chem. Eng. J. 101 (2004), 251
[25] J.M. Köhler, J. Wagner, J. Albert, J. Mater. Chem. 15 (2005), 1924
[26] Schwesinger, N.; Frank, T.; Wurmus, H. 1996. J. Micromech. Microeng. 6, 99
[27] Kirner T, Albert J, Günther M, Mayer G, Reinhäckel K, Köhler J M (2004), Chemical Engineering Journal 101: 65-74
[28] Köhler J M, Wagner J, Albert J, Mayer G, Hübner U (2005). Chemie Ingenieur Technik 77: 867-873
[29] Link, S.; Wang, Z.L.; El-Sayed, M.A.: J. Phys. Chem. B 103 (1999). 3529
[30] Link, S.; El-Sayed, M.A.: J. Phys. Chem. B 103 (1999), 8410
[31] Moskovits, M.; Srnová-Sloufoová, I.; Vlcková, B.: J. Phys. Chem. 116 (2002),10435
[32] J.Wagner, J. M. Köhler, Nano Letters 5 (2005), 685

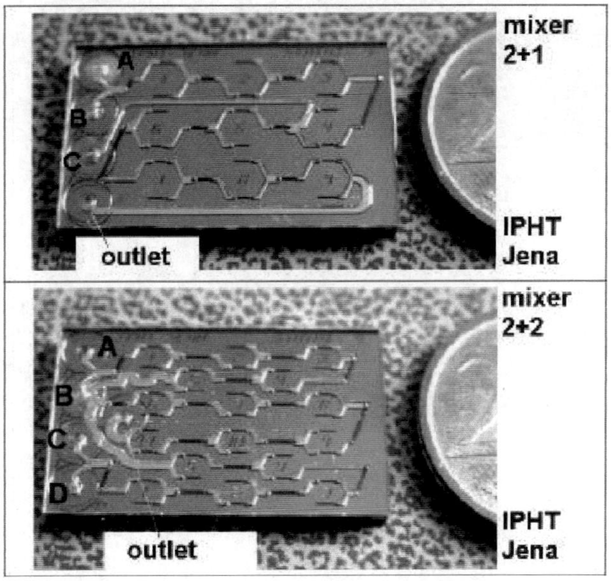

Fig. 1
Glass/Si/glass static micro mixers for two mixing steps (2+1, above) and three mixing steps (2+2, below)

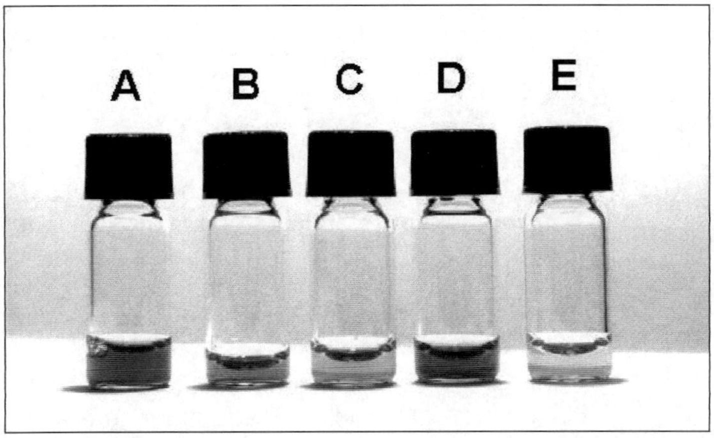

Fig. 2
Examples for product solutions containing colloidal Au/Ag nanoparticles; the differences in colour are caused by differences in Au/Ag ratio as well as by different mixing order of educts: in batch: A: 100 µL 0.2% PVA+12.5 mg SDS+450 µL 0.1 M ascorb. acid+450 µL 2 mM $HAuCl_4$; B: 40 µL 1 mM $AgNO_3$+160 µL 0.5 mM $HAuCl_4$+100 µL 0.1 M ascorb. acid; C: 250 µL 0.1 M $HAuCl_4$+250 µL 0.1 M ascorb. acid; D: 100 µL 0.5 mM $HAuCl_4$+100 µL 0.1 M ascorb. acid+100µl 1 mM $AgNO_3$ (after about 0.3 s); in micro flow ((2+1)+T), all 100 µL/min: E: PVA 5%+ 0.1M ascorb. acid + 0.5 mM $HAuCl_4$ + 1 mM $AgNO_3$

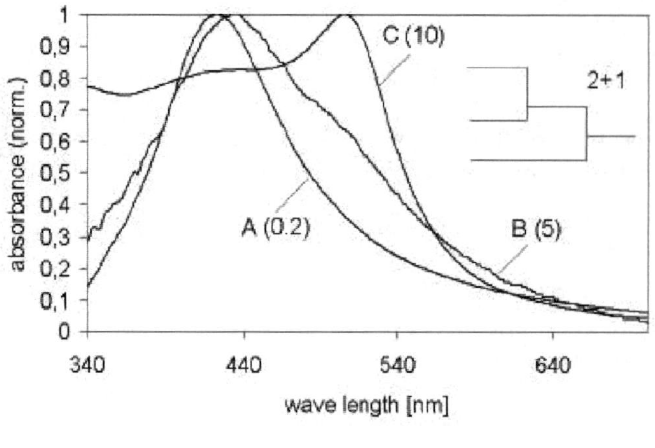

Fig. 3
Influence of flow rate ratios on product properties in micro flow-through experiments using a static micro mixer (2+1-arrangement); tetrachloroaurate solution and ascorbic acid solution at equal concentrations mixed first, addition of silver nitrate solution in the second mixing step; A: Au/Ag-ratio: 0.2, total flow rate: 70 µL/min; B: Au/Ag-ratio: 5, total flow rate: 110 µL/min; C: Au/Ag-ratio: 10, total flow rate: 105 µL/min;

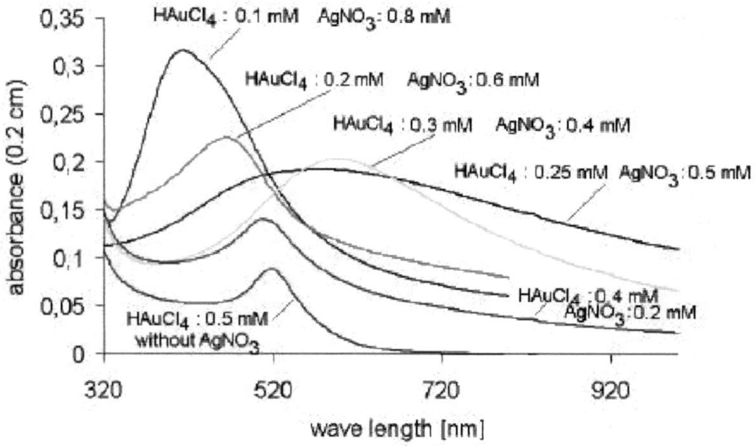

Fig. 4
Optical spectra of AuAg colloidal product soltuions from reference experiments in batch: premixing of tetrachloroaurate and silver nitrate in different concentration ratios (200 µL) and subsequent mixing with 0.1 M ascorbic acid solution (100 µL), product solutions dilluted 5:1, measured with an optical path length of 1 cm

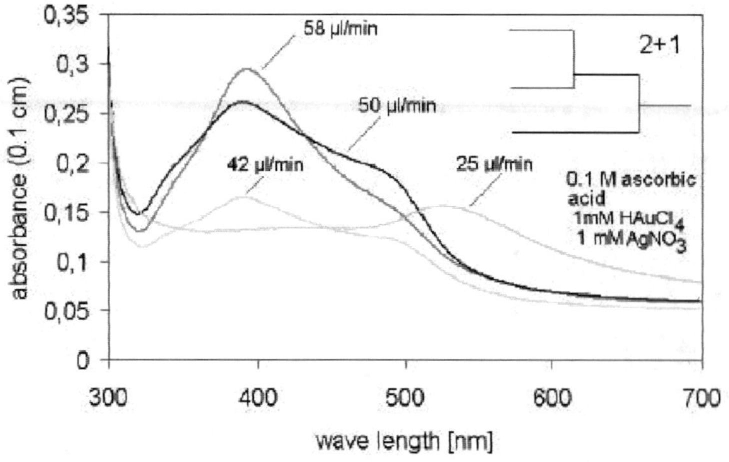

Fig. 5
Influence of total flow rate on nanoparticle formation in a 2+1 static micro mixing arrangement (Statmix) reflected by absorption spectra of product solutions. Concentration of educt solutions in all experiments: ascorbic acid: 0.1 mol/L, tetrachloroaurate: 1 mmol/L, silver nitrate: 1 mmol/L, premixing of tetrachloroaurate and ascorbic acid, equal flow rates in all three inlet channels (58, 50, 42, resp. 25 µL/min in easch channel) and, therefore, constant flow rate ratio (=1) in all experiments , product solutions diluted 10:1, measured with an optical path length of 1 cm

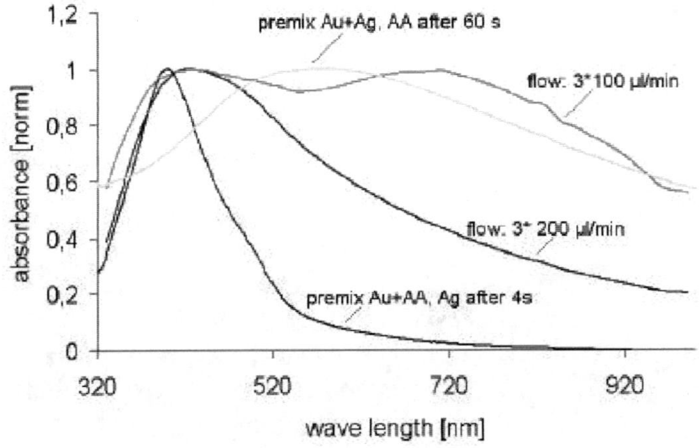

Fig. 6
Comparison of two micro flow-through experiments (same flow rate ratios, same order of mixing: premixing of tetrachlororaurate and ascorbic acid, followed by addition of ascorbic acid solution) with reference experiments in batch with different order of reactant mixing (premixing of tetracholoroaurate with ascorbic acid and premixing of tetrachloroaurate with silver nitrate), normalized absorbances

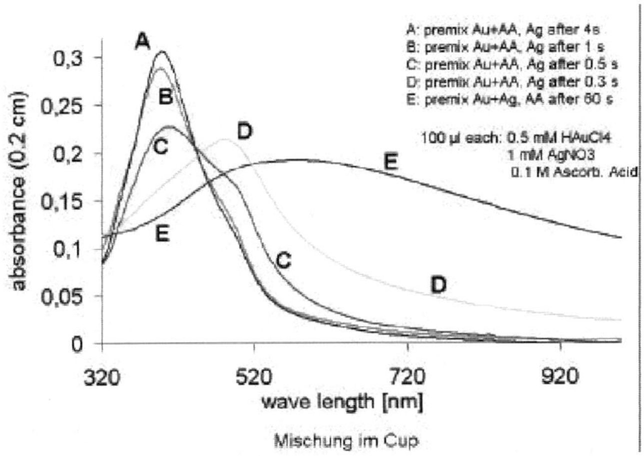

Fig. 7
Influence of time gap between the premixing of tertacholoroaurate and ascorbic acid and the addition of silver nitrate on the quality of nanoparticle product solutions characterized by optical spectra, A-D: time gaps of about 4s, 1s, 0.5 s, 0.3 s, E: changed educt mixing order, equal concentration ratios of educt solutions in all experiments, reference experiments in batch

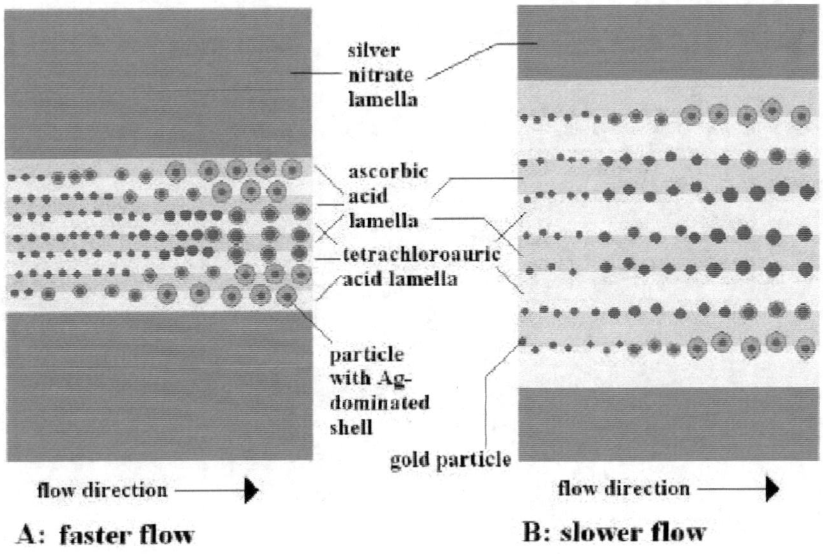

Fig. 8
Conditions of nanoparticle formation in a multi step micro mixer (hypothetical scheme):
parts of liquid cross section after some split-and-recombine units in the second mixing step for same residence time in the second mixing step: left: smaller liquid lamellas at higher flow rate, right: larger liquid lamellas at lower flow rate

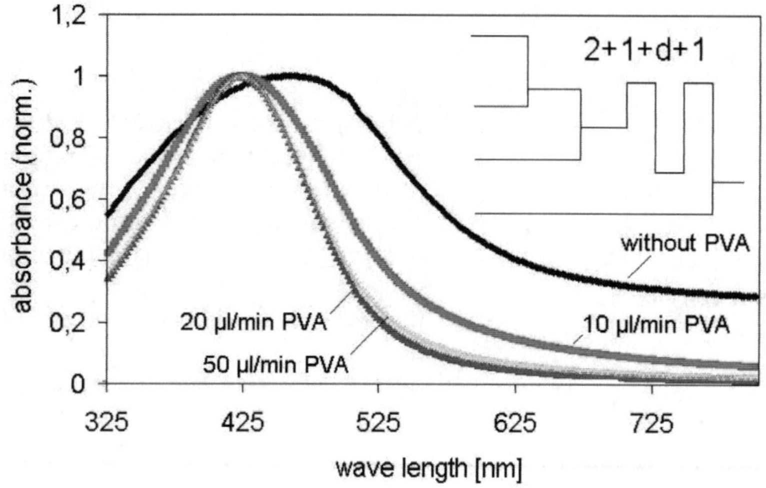

Fig. 9
Effect of polymer addition to the product spectra in case of AuAg nanoparticle formation in a micro flow-through experiment (2+1 mixer + residence loop and additional T-junction mixing), PVA (0,5 % in water) flow rates: 0, 10 µL/min, 20 µL/min, 50 µL/min, flow rate of other educts in all experiments: ascorbic acid (0.1 mol/L), tetrachloroaurate (0.5 mmol/L), silver nitrate (1 mmol/L): 50 µL/min

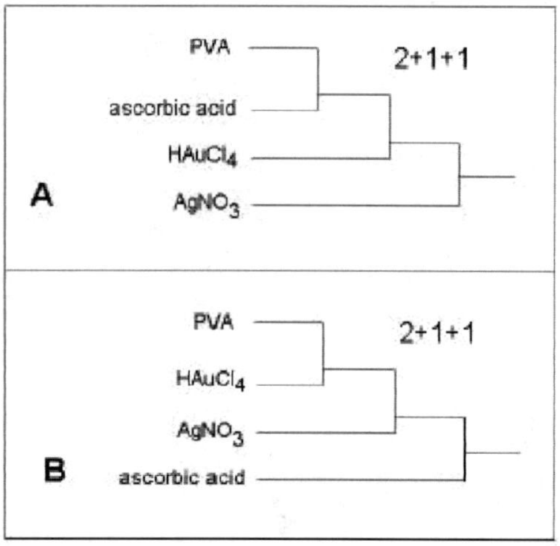

Fig. 10
Micro fluid arrangement for investigating the effect of mixing order on the product spectra using a 2+1+1 static micro mixer

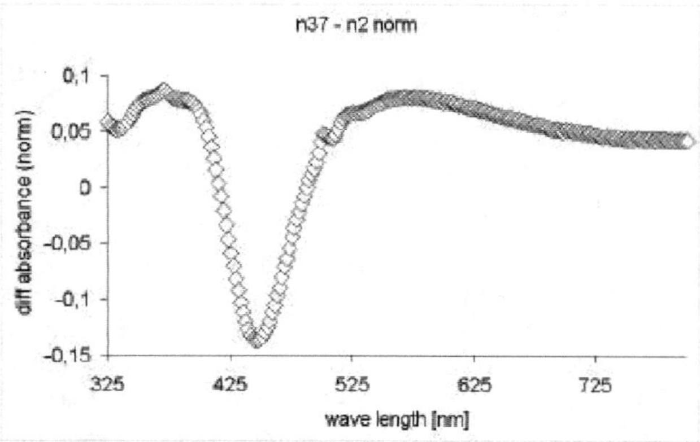

Fig. 11
Difference spectrum for the product solutions obtained in the experiments using the fluidic arrangements of fig. 10A and fig. 10 B, flow rates of all educts in both experiments: 50 µL/min, educt concentrations in both experiments: PVA: 0.5 %, ascorbic acid: 0.1 mol/L, tetrachloroaurate: 0.5 mmol/L, silver nitrate: 1 mmol/L

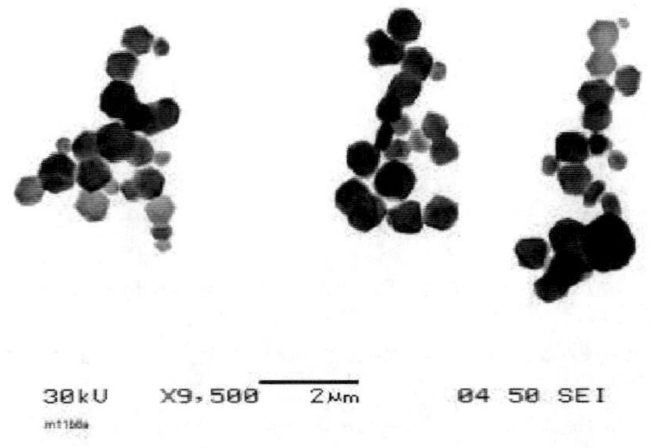

Fig. 12
SEM images of particle aggregates obtained by silver amplification of gold nanoparticles generated by the reduction of tetrachloroaurate with ascorbic acid in the presence of dilluted macromolecule solution (batch)

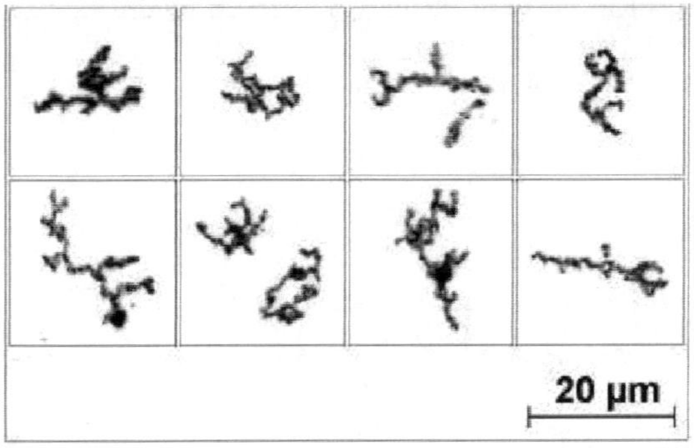

Fig. 13
Light microscopic images of particle aggregates obtained by silver amplification of gold nanoparticles generated by the reduction of tetrachloroaurate with ascorbic acid in the presence of dilluted macromolecule solution (batch)

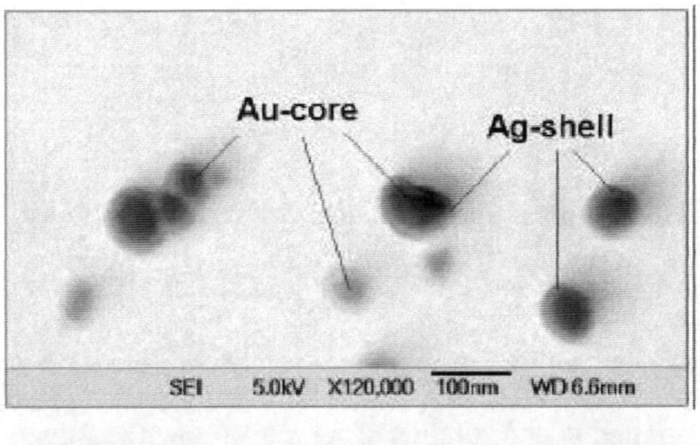

Fig. 14
Core shell particles obtained in a flow-through process using a 2+1 static micro mixer, a residence loop and an additional T-junction (compare Fig. 9): 0.5% PVA ; 0.1 M ascorb. acid; 0.5 mM $HAuCl_4$; 1 mM $AgNO_3$ (all components with a flow rate of 50 µL/min)

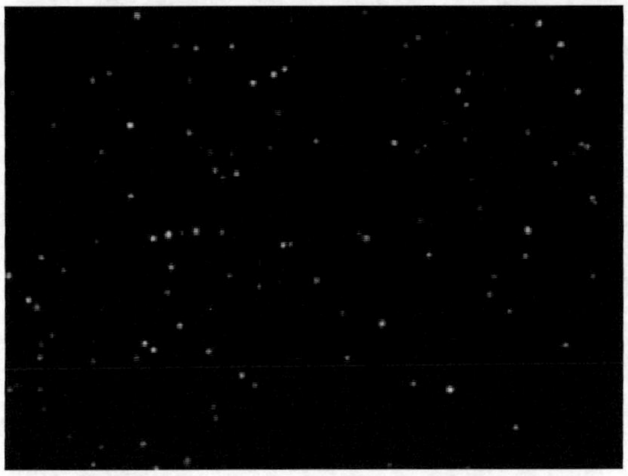

Fig. 15
Light microscopic image (scatterd white light) of a Au/Ag nanoparticle product solution after embedding in PVA showing a large variety of optical properties of the single particles (batch)

Preparation of $BaSO_4$ Nanoparticles in a Microchannel Reactor

Ying Ying, Guangwen Chen*, Jiansheng Chu, Quan Yuan

Dalian Institute of Chemical Physics, Chinese Academy of Sciences, 457 Zhongshan Road, Dalian 116023, China

Abstrat

Microchannel reactor has amazing performance for preparation of nanoparticles attributes to its high efficiency of micromixing. We can briefly synthesize nanoparticles through the microchannel without any additive agents. Neutralization synthesis of $BaSO_4$ from Na_2SO_4 and $BaCl_2$ solutions was carried out using microchannel reactors of both Y-type and T-type reactors. The size and morphology of particles were characterized by transmission electron microscope (TEM) images. The influences of concentration, flow rate, reaction temperature and molar ratio of reactants were studied and it turned out that the reactant concentration and flow rate had significant influence on the particle size and size distribution.

Key words: microreactor, micromixing, nanoparticles, barium sulfate, sodium sulfate, barium chloride

1. Introduction

Nanomaterials had attracted dramatically attention in the recent decades due to its excellent exhibition in optical, electronic and magnetic properties that bulk materials could not work out. Nanomaterial production is obviously a very important part in the whole nanomaterial industry. In all the methods to produce nanomaterials, the liquid-liquid chemosynthesis method is the most efficient and widely used one. In the liquid phase, the particles forming will go through the following three stages: nucleation, growth companying with Ostwald ripening, and aggregation [1]. For a single particle, these stages will perform step by step. In different stages, varied methods could be applied to control the particles to follow the way designed. For nanoparticle producing, we can increase the supersaturation to number up the nuclei in the first stage; provide an identical environment for growth stage; and remove the production from the reactor to prevent it from aggregation. But in traditional producing method, the process takes place in batch reactors. The particles in the batch comprise all of three stages. It's hard to control in such a complicated

* Corresponding author. Tel: +86-411-437-9031, Fax: +86-411-469-1570

E-mail address: gwchen@dicp.ac.cn (G. Chen).

situation. Therefore here is the good chance for novel method to overcome the disadvantages. Microchannel reactor had a good ability to control the process of nanoparticle formation.

One of the early reports about synthesizing compound semiconductor nanoparticles in microchannel reactor was by J. B. Edel et al [2]. The authors synthesize CdS in a microchannel reactor with the total volume of less than 600 nL by the reagents of Na_2S and $Cd(NO_3)_2$. Through the absorption spectrums obtained by Lambda 15UV-VIS spectrometer showed that the particle sizes range from 3.2 nm to 12 nm. The authors also made a good point that performing the reaction in a continuous flow discourages coalescence of the newly formed nanoparticles. It is a very significant advantage of microchannel reactors.

In the same year, H. Maeda et al. synthesized nanoparticle TiO_2 in a ceramic microchannel reactor with the channel of 360 μm wide, 200 μm deep and 90 mm long [3]. The channel had a Y-type inlet so that the reagents could go into the channel separately. They pumped the reagents by syringe pumps in such a slow velocity that the two immiscible reagents formed a steady interface on which the particles are designed to occur. The TEM image showed that titania particles of < 10 nm was produced without any surfactant.

The same group applied another kind of microchannel reactor to synthesize CdSe nanoparticles [4]. The channel was actually a silica glass capillary with width of 200-500 μm. The temperature was controlled by an oil bath in which the capillary was dipped. Finally the particles of 2-4.5 nm were successfully prepared. Using the same method, they prepared ZnS-coated CdSe composite particles [5]. In this experiment they show that the particle size and layer thickness could directly be controlled by flow rate. They also take the advantage of high reproducibility of microchannel reactor to study the particle growth kinetics [6].

A. Paul Alivisatos et al. at Berkeley studied more about the size-controlled growth of CdSe in a chip-based microfluidic reactor [7]. Microchannel reactors were isotropically etched into borosilicate glass wafers using standard micro-fabrication techniques. Reaction time, temperature, and precursor concentration were independently affecting the growth of CdSe nanocrystals. The authors had the confidence that such lab-on-a-chip technology allows for rapid, cost-effective, and environmentally friendly prototyping will accelerate the rational development and production of novel nanostructures.

Moungi G. Bawendi et al. have the same interesting in CdSe nanoparticles prepared in microchannel reactor [8]. The channel was a 250 μm ID glass tubing following a convective mixer in which the Cd and Se precursors mixing together before the heating section. They control the particle size by systematically varying the temperature, flow rate, and concentration and conclude that a combination of the RTD and intrinsic nucleation and growth processes place restrictions on the range of sizes.

2. Experimental details

The reagents used in this experiment were Na_2SO_4 (analytic grade, Shenyang federation reagent factory) and $BaCl_2{\cdot}2H_2O$ (analytic grade, Shenyang xingshun chemical reagent factory). They were dissolved in distilled water without any further purification.

A sketch map of the reactor is showed in Fig.1. The channels were fabricated on a polymethyl methacrylate (PMMA) substrate and enclosed by pressure bonding of a blank one. Two types of channel shape were studied in this paper, Y-type and T-type. The angle of the Y-type reactor is 60°, and the length of the three channels are 20 mm while the depth and width are 0.2 mm. For the T-type reactor, the length of the vertical channel is 20 mm with the width of 1mm and depth of 0.2 mm. And the other two channels are 10 mm long, 0.2 mm wide and 0.2 mm deep.

Sodium sulfate and barium chloride were pumped into the reactor separately through two inlets. They meet and nucleate in the contact, and the reaction continues when the cores grow larger through the channel. The nanoparticles were collected in a beaker through the outlet.

The morphology of the nanoparticles was characterized by transmission electron microscope (TEM) images.

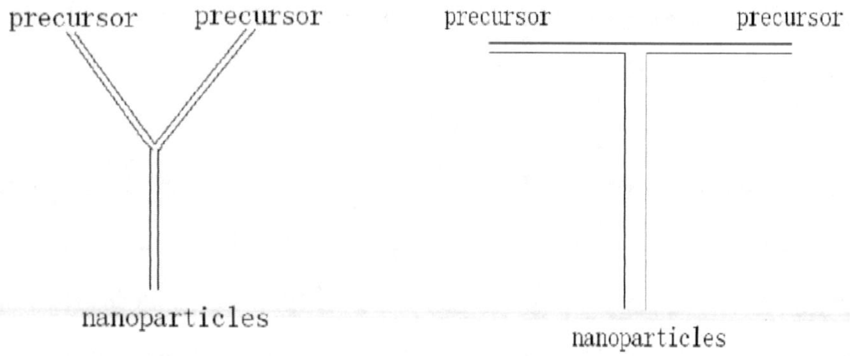

Y- type reactor | T- type reactor

width:0.2mm, depth:0.2mm | width of parallel channels:0.2mm, depth:0.2mm

length of vertical channel:2cm | width of vertical channel:1.0mm, depth:0.2mm, length:2cm

Fig. 1. Sketch maps of the reactors

3. Results and discussion

3.1 *Y-type reactor*

The TEM image of barium sulfate prepared by Y-type reactor is showed in Fig.2a. The particle diameters are mostly around 15nm, which is significantly smaller and more uniform than the particles prepared by traditional methods [9]. This would attribute to the great performance of micromixing in microchannels. The flow is totally in the turbulent state as the Re number is 3300 in the channel. This strong turbulent flow in the confined channel made the surroundings identically. Thus all the particles had the same chance to get new products to grow larger. Since the react between sodium sulfate and barium chloride is instantaneous [10], the uniform environment would result in uniform particles.

3.2 *T-type reactor*

Another T-type reactor was tested in the experiment. The channels near the inlets are 0.2 mm wide and 0.2 mm deep, and the one after the connection is 1mm wide and 0.2 mm deep. The major difference of the T-type reactor from the Y-type reactor is that the channel after the connection is much wider. The TEM images show that the particle sizes are almost the same. It suggests that the nucleation play a key role in the particle size controlling. Sodium sulfate and barium chloride turn into barium sulfate as soon as they meet in the connection. Instantaneously, the supersaturation of barium sulfate reaches to such a high level that they all burst into nuclei. The huge number of nuclei guarantees that there is not extra barium sulfate for all the nuclei to grow large. On the other hand, the homogeneous environment in the narrow channel provides all the nuclei the same chance to get material to grow large. These two facts work together to prevent the particles from growing large and make the size distribution narrow.

The influence of concentration also suggests the importance of the nucleation. Influence of concentrations of sulfate sodium of 0.1 mol/L, 0.5 mol/L and 1.0 mol/L were compared when the other parameters are the same. The particles are near 18 nm and uniform when the concentration is 0.5 mol/L as showed in Fig.2b. When reduce it to 0.1 mol/L, as showed in Fig. 2c the particles grow to above 20 nm, some of them even reach to 25 nm. This would attribute to the relatively small number of nuclei at the low concentration. Thus the material that each nucleus could get becomes abundant. When increase it to 1.0 mol/L, the smallest particles reach to 12 nm, however the uniformity gets slightly worse.

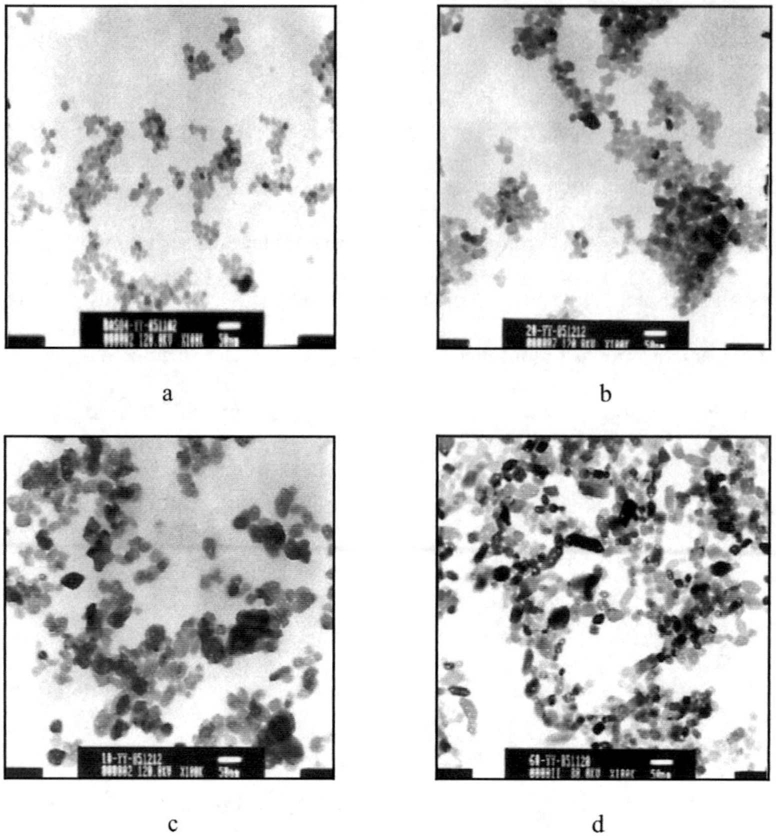

Fig.2. TEM images of $BaSO_4$ nanoparticles

number	a	b	c	d
Flow rate (mL/min)	40	40	40	10
$BaCl_2$ (mol/L)	0.85	0.425	0.085	0.425
$NaSO_4$(mol/L)	1.0	0.5	0.1	0.5

Another important parameter that would strongly affect the nanoparticles quality is the flow rate. As set it on 40 mL/min, the particles size and distribution show at the best state. While reducing it to 20 mL/min, the particles size becomes a little larger, and the size distribution gets wider. This trend becomes obviously when the flow rate slows down to 10 mL/min as showed in Fig. 2d. The large ones spread all over the view field. And the ellipse shape of the particles was elongated, which is not at all the same morphology with the large particles produced under the condition of 0.1 mol/L, though they all grow large. This phenomenon may attribute to Ostwald ripening [11]. Along with the reducing of the flow rate, the residence time becomes longer. In the middle of the channel the monomers of barium sulfate were all consumed. Then for the equilibrium reason, some of the barium sulfate particles were dissolving, and some of the particles grow even larger of the monomers from dissolving. Since the particles are all smaller than the 'critical radius' [11]. Ostwald ringing always sharpens the differences between small ones and large ones that enlarge the size distribution.

Other parameters such as temperature (20□，40□，60□) and molar ratio of $BaCl_2$ and $NaSO_4$ (0.85/1, 1/1, 1/0.85) were also tested in the experiment, but not significant different found. It shows that the influence of these parameters are not as important as either concentration or flow rate changing.

4. Conclusion

In this paper, a novel way for synthesize nanoparticles in liquid phase was introduced. The effects of concentration, flow rate, temperature and molar ratio were studied to determine the most important parameter in the experiment. Due to the small volume of the channel, the microchannel reactor could provide a homogeneous environment for the reaction system. It ensures that the nucleation and growth of all the particles are homogeneous which results in narrow size distribution. These advantages of microchannel reactor make the way for us to synthesize barium sulfate nanoparticles of 15 nm in a direct way without any additional reagents. This virtue can help for quite a lot of other nanomaterials especially for producing ultra-fine drug particles that have strictly restrictions of purity. Moreover we can easily control the residence time by changing the channel length to vary the growth time, which would help for size controlling and restrain Ostwald ripening.

The continuous synthesis in the microchannel reactor was a simple and efficient way to prepare composite nanoparticles. It can change the traditional batch process to continuous process, which would highly improve the producing efficiency and reduce the labor intention. Thus we could say that this novel way is an environmentally friendly and cost-effective process for nanomaterials production.

References

[1] John deMello, Andrew deMello, *Lab. Chip.,* 2004, 11N-15N

[2] J. B. Edel, R. Fortt, J. C. deMello, A. J. deMello, *Chem. Commun.*, 2002, 1136-1137

[3] Hongzhi Wang, Hiroyuki Nakamura, Masato Uehara, Masaya Miyazaki and Hideaki Meada, *Chem. Commun.*, 2002, 1462-1463

[4] Hiroyuki Nakamura, Yoshiko Yamagrchi, Masaya Miyazaki, Hideaki Maeda, Masato Uehara and Paul Mulvaney, *Chem. Commun.*, 2002, 2844-2845

[5] Hongzhi Wang, Xianying Li, Masato Uehara, Yoshico Yamaguchi, Hiroyuki Nakamura, Masaya Miyazaki, Hazime Shimizu and Hideaki Maeda, *Chem. Commun*, 2003, 48-49

[6] Hiroyuki Nakamura, Asuka Tashiro, Yoshiko, Yamaguchi, Masaya Miyazaki, Takanori Watari, Hazime Shimizu and Hideaki Maeda, *Lab. Chip.*, 2004, 237-240

[7] Emory M.Chan, Richard A. Mathies and A. Paul Alivisatos, *Nano Lett.* 2003, 199-201

[8] Brian K. H. Yen, Nathan E. Stott, Klavs F. Jensen and Moungi G. Bawendi, *Adv. Mater.*, 2003, 1858-1862

[9] Liu Youzhi, Li Junping, Yuan Rusheng, Liu Laishuan, *Journal of basic science and engineering,* 2001, 141-145

[10] Daniele L. Marchisio, Antonello A. Barresi, and Mirko Garbero, *AIChE Journal*, 2002, 2039-2050

[11] Nikos V. Mantzaris, *Chemical Engineering Science*, 2005, 4749-4770

Industrial Use of microstructured devices in Solids Technology

Dirk Kirschneck, Walter Linhart and Günter Tekautz
Microinnova KEG, Reininghausstraße 13a, 8020 Graz, Austria
tel: +43 316 908606-0, mail: office@microinnova.com, web: www.microinnova.com

The aim of solids technology is the production of disperse solid products. Disperse substances require to have specific characteristics in terms of production and application. The handling and production of solid substances is a mayor task in chemical industry because a vast majority of products are solid substances or at least pass through a disperse state in the course of their production.

The application of microstructured devices offers new opportunities for the production of solid chemicals. Microstructured devices usually contain structures below 1000µm. Their main advantage is their high surface to volume ratio which leads to a large specific surface area. In terms of solids technology microstructured devices can be used to obtain specific application-related product properties. Especially the designs of the regular structures offer possibilities for influencing the particle size distribution. However, the handling of solid substances in microstructured devices is always challenging.

In the current paper the Starlaminator (designed by Institut für Mikrotechnik Mainz) has been used to produce well-defined particles in the submicron area. The Starlaminator is based on the micromixing concept of flow multilamination. The solid products produced in the micro mixer were compared to those obtained from conventional stirred batch reactors. Special attention was also given to the influence of surface-active agents on the precipitation. Product properties and process conditions were measured to characterize the precipitation process. For example particle size analysis was performed using laser diffractometry. A concept for an industrial production has been done in a second development step. Viscosity effects have been studied as well as effects of extreme phase ratios like 1:10. The project ends with a setup for a production scale reactor with a throughput of 4 tons per hour.

Microfluidics and Reactors in Porous Materials

Rob G. H. Lammertink, Jorrit de Jong, Matthias Wessling

University of Twente, Science and Technology, NL-7500 AE, Enschede, The Netherlands

Microfluidics and reactors are very interesting systems for a wide variety of reaction systems, frequently involving fast and exothermic reactions. In microfluidics, fluid flow is laminar and therefore mixing progresses by diffusion only. A microfluidic channel can be regarded as a microreactor. Reagent enter the channel and due to the small dimensions have a large contact area to volume ratio. In order to enhance mixing between components, both static and dynamic micromixers have attracted a great amount of interest.

When dealing with multi phase systems, contacting principles become increasingly important. For instance, for gas-liquid reaction systems, large interfacial area is desirable. Such a gas-liquid interface, however, strongly depends on the hydrodynamic conditions. For certain gas-liquid flow ratios, so-called Taylor flow can be obtained. For certain other conditions even tubular flow can be obtained, where the liquid wets the channel wall while the gas phase is located at the center of the channel (surrounded by a continuous liquid film).

Here, we describe another concept for microreactors where we profit from porosity present in the matrix material. As such, it is possible to provide one phase through the walls of the microchannels contacting the other phase that resides within the microchannel. This concept is illustrated below in Figure 1.

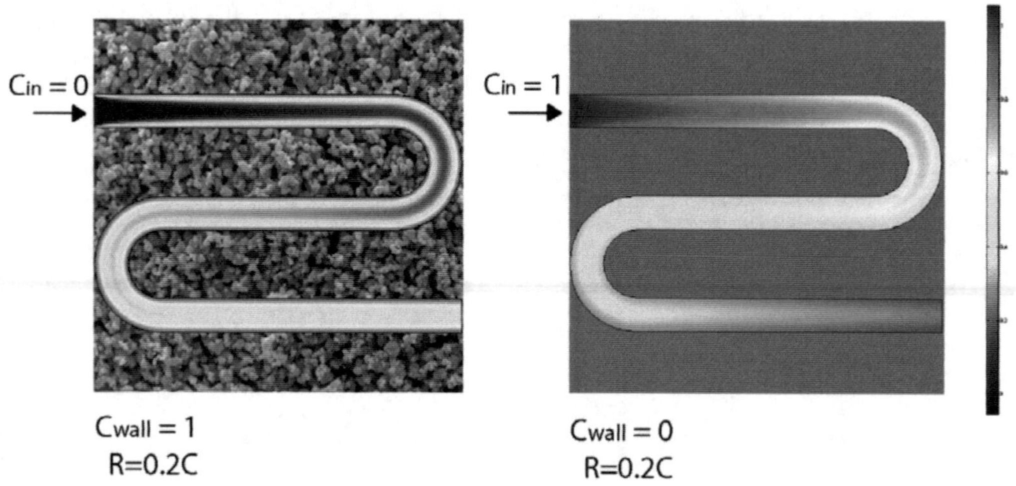

Figure 1 Schematic illustration of a porous microreactor, where the reactant is applied through the channel wall (left), resulting in a continuous reaction along the full channel length. On the right, a conventional microchannel is displayed where the reactant is present at the entrance, but is consumed by the reaction as it flows through.

A forthcoming advantage of such an approach is that the delivery through the microchannel wall can be externally controlled, and locally varied. In, for instance, slug flow systems, the gas phase can potentially be consumed and the driving force for reacting/dissolving can not be controlled once the bubbles have entered the channel. In the proposed gas-liquid contacting microreactor, the driving force for reaction/dissolution can be controlled locally.

Fabrication of porous microfluidic structures is a relatively simple process. The process is schematically depicted in Figure 2. A microstructured mold is needed to define the required channel structures. For this, standard cleanroom procedures, including lithography and reactive ion etching, can be used to obtain the molds in silicon, for example. Then, a polymer solution is cast onto the mold at a well defined thickness. This thickness is typically several hundreds of micrometers. The mold with the polymer solution is then immersed in a non-solvent bath. The solvent and non-solvent exchange and eventually the polymer precipitates out of the solution. A polymer film can be released from the mold and the mold can be reused. As such, the porosity in the polymer film is controlled by the precipitation conditions, including the type of solvent and non-solvent, the concentration, the temperature, the humidity, etc. The wide variety of recipes can be found in membrane fabrication literature. Membranes are made in a similar way and control of pore size (distribution) is of great importance.

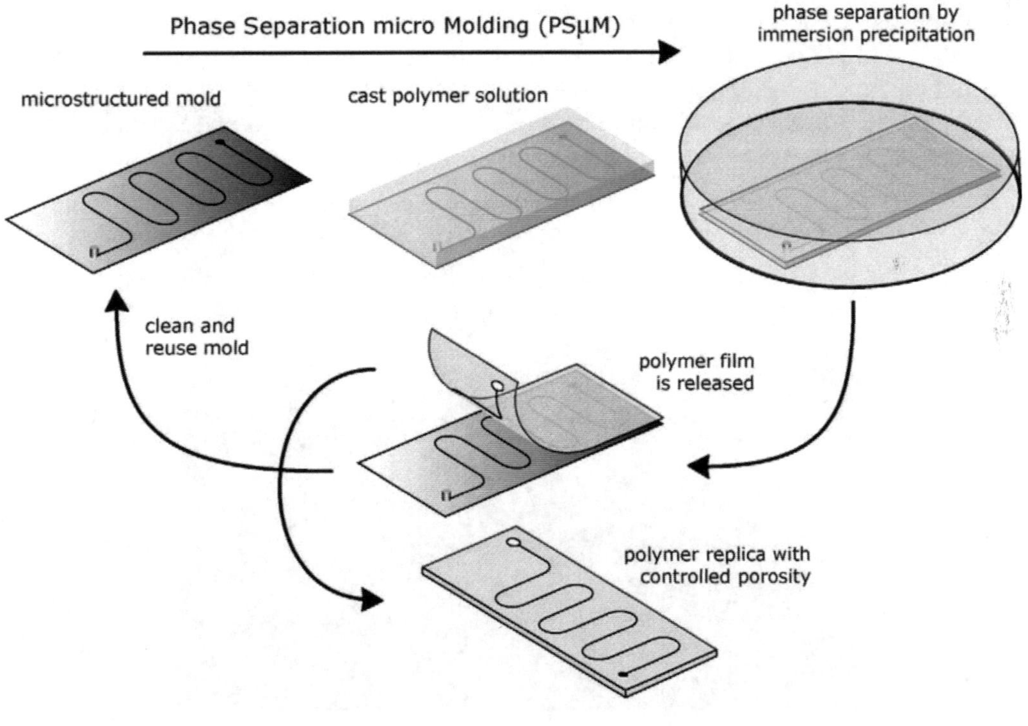

Figure 2 Illustration of the method for fabricating porous microfluidic structures.

Preparing porous polymer films by immersion precipitation is a well-known method for fabricating polymeric membranes. Therefore, a substantial amount of literature is present that describes the influence of process conditions (temperature, humidity, concentration, additives) on the resulting morphology.

The method offers great flexibility in terms of materials choice. In contrast to other replica molding methods, this method is not restricted to melt-processable polymers. Any polymer for which a solvent and non-solvent are present can be structured by this micromolding technology. The only additional requirement is that the solvent and non-solvent are (partly) miscible.

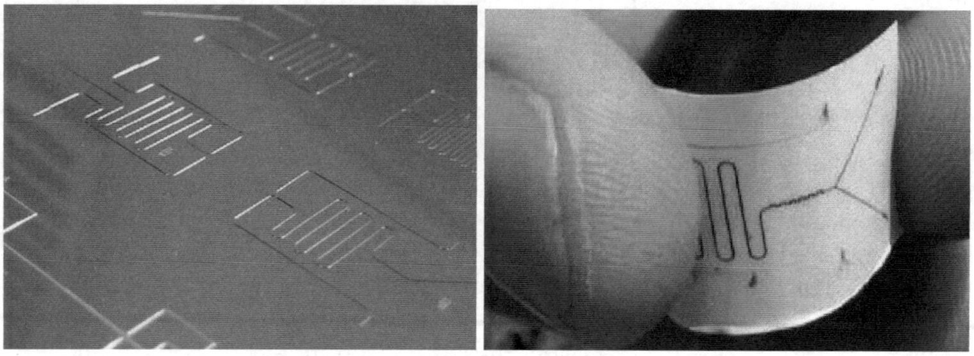

Figure 3 A microfluidic relief pattern in silicon (left) obtained by cleanroom technology, used to fabricate thin film microfluidics (right) by a replication method.

The porous fluidic structures must be sealed or stacked to form microfluidic channels. Stacking provides multiple distinct channels that are separated by a membrane with tunable porosity. Stacking and sealing methods are dependent on the used material. In Figure 4 an example of a sealed porous microfluidic channel is displayed. This device consists of a porous PMMA layer, sealed with a dense PMMA layer (so optical examination is possible).

Figure 4 SEM image of a cross-section of a sealed porous microfluidic channel.

With the introduction of porosity in microfluidic channels, membrane functions can be easily integrated. We have demonstrated this concept with a gas-liquid contacting device in a microfluidic system. In such a system, the liquid inside the microfluidic channel is contacted with a gas phase through the porous polymer matrix. The extend of contacting can be tuned by the porous matrix material as well as the matrix morphology. Even the use of materials with relatively low gas permeability can then result in significant gas transport, by adjusting the material morphology towards a more open and porous structure.

Figure 5 Optical micrograph of a microfluidic channel in a porous matrix. The color change in the channel is due to acidification as a result of carbon dioxide uptake.

Besides the wide variety of polymeric materials, it is possible to structure inorganic and metallic materials as well. In such an approach, ceramic precursor or metallic particles are mixed into the polymer solution prior to structuring. After the immersion precipitation, a thin polymer-particle composite film is obtained. The polymer matrix is removed by means of pyrolysis, while the inorganic or metallic particles are sintered. An example of a metallic porous microfluidic layout is shown in Figure 6.

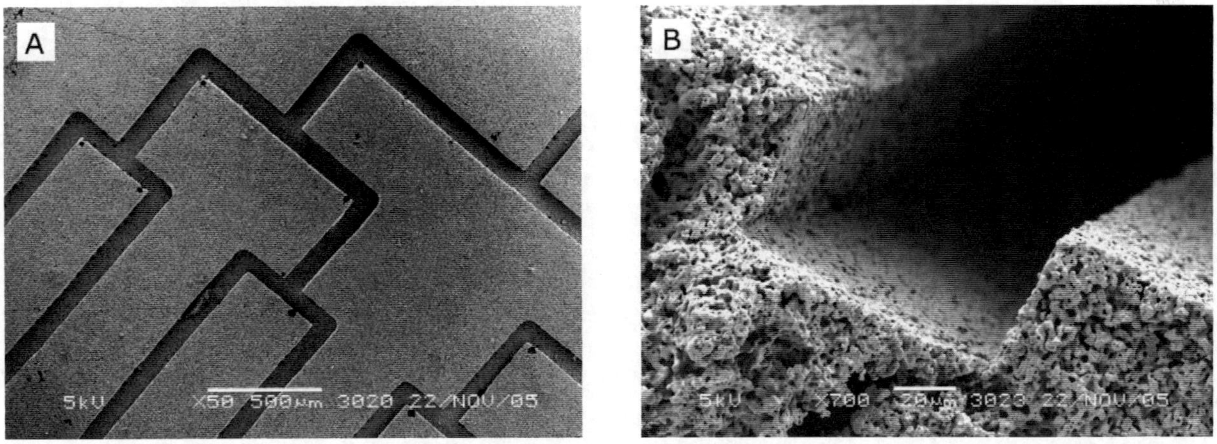

Figure 6 SEM images of Fe microchannels, obtained after phase separation micromolding and sintering. Channel dimensions are approximately 100 micrometer wide and 40 micrometer high.

The integration of membranes within microfluidic and microreactor structures is very promising for different applications. One could make use of membrane functionality such as gas-liquid contacting, filtration, emulsification, etc. We believe that the presented microstructuring method, phase separation micromolding, is a very versatile method. It can be applied to fabricate thin microstructures with tunable porosity. Such films are therefore ideal for stacked devices as well.

References

1. de Jong, J.; Ankoné, B.; Lammertink, R. G. H.; Wessling, M. "New replication technique for the fabrication of thin polymeric microfluidic devices with tunable porosity" Lab on a Chip 2005, 5, 1240-1247.
2. Vogelaar, L.; Lammertink, R. G. H.; Barsema, J. N.; Nijdam, W.; Bolhuis-Versteeg, L. A. M.; van Rijn, C. J. M.; Wessling, M. "Phase separation micromolding: a new generic approach for microstructuring various materials", Small 2005, 1, 645-655.
3. Vogelaar, L.; Barsema, J. N.; van Rijn, C. J. M.; Nijdam, W.; Wessling, M. "Phase separation micromolding - PSµM", Advanced Materials 2003, 15, 1385-1389.

Assembling of Nanoparticles Using Ice Crystals

[*1] *Shin R. Mukai,* [*2] *Kazuhiro Mitani,* [*3] *Hirotomo Nishihara and* [*2] *Hajime Tamon*
[*1] *Graduate School of Engineering, Hokkaido University, Sapporo, Japan*
[*2] *Graduate School of Engineering, Kyoto University, Kyoto, Japan*
[*3] *Institute of Multidisciplinary Research for Advanced Materials, Tohoku University, Sendai Japan*

Introduction

Various inorganic metal oxides can be synthesized through the sol-gel method. Such metal oxides can be synthesized in the form of fibers by freezing their parent hydrogels "unidirectionally."[1] We found that fibers obtained through this method are generally porous, which indicates the possibility of using them as adsorbents and/or catalysts.[2] Moreover, we showed that the diameters of the fibers, as well as the sizes of the pores within them, can be controlled separately, and such controlling can be conducted simultaneously by simply adjusting preparation conditions.

The unidirectional freezing method was originally developed to obtain long and strong fibers, so thoroughly aged hydrogels were usually used as the precursor. However, we found that by applying this method to hydrosols or freshly gelled hydrogels, gels with unique morphologies, such as lamellar sheets, flat fibers and microhoneycombs, can also be obtained.[3,4] We also found that the morphology which appears during unidirectional freezing can be controlled by adjusting the firmness of the precursor sol or gel. As this method practically uses ice crystals as the template, we named this method the "ice templating method."

Microhoneycombs obtained through the ice templating method have micrometer sized straight channels, and the thicknesses of the walls which form the channels are only up to a few micrometers. This means that the pressure drop when fluids are passed through them will be extremely low even though the diffusion path lengths within the porous walls are extremely short, a feature that cannot be realized in packed columns. Therefore, this method is expected to provide porous materials with ideal morphology for microsystems such as microreactors and microseparators. However, the applicability of this method is limited to materials which can be synthesized through the sol-gel method.

In this work, we attempted to widen the range of applicability of the ice templating method. We applied the ice templating method to mixtures of SiO_2 hydrogels and various nanoparticles possessing various functions. SiO_2 is expected to act as the binder and also to provide a large surface area to the resulting material. Therefore this modified ice templating method is expected to provide materials with unique functions and unique morphologies.

Experimental

Commercial sodium silicate solutions (Wako Pure Chemical Industries) were diluted with ion-exchanged and distilled water, and the SiO_2 concentration in them were adjusted. Then the pH of the solutions was adjusted using an ion-exchange resin. To these solutions, particles with various functions were added. The particles used in this work were TiO_2 nanoparticles (p25, Degussa) a typical photocatalyst, high silica Y-type zeolite particles (HSZ, Tosoh) a typical catalyst used in various petrochemical processes, and graphite particles (Wako Pure Chemical Industries) a chemically stable and electric conducting material. They will be respectively denoted as p25, HSZ and GRA, hereafter. First, the particles were dispersed in the solutions using an ultrasonic homogenizer, and the resulting slurries were poured into polypropylene tubes (i.d.: 10 mm, L: 100 mm) and were aged for different periods. Next the tubes were dipped at a constant rate of 8 cm/h into a cold bath maintained at 77 K. After the tubes were completely frozen, they were taken out from the bath and thawed at 323 K. Then the water in the tubes was exchanged with *t*-butanol, and the samples were freeze-dried. Some samples were heat treated in an air atmosphere (a nitrogen atmosphere for samples including GRA) at 1273 K for 4 h.

The morphology of the samples was directly observed using a scanning electron microscope (SEM). The compression strengths of honeycomb type samples were measured using a load cell. The porous properties of the samples were evaluated through nitrogen adsorption-desorption experiments. First, nitrogen adsorption-desorption isotherms of the samples were measured at 77 K. Next, BET surface areas of the samples were calculated using the BET method. The pore size distributions and mesopore volumes of the samples were calculated using the Dollimore and Heal method. The samples were also analyzed using an X-ray Diffractometer (XRD) to check whether the included particles maintained their structures.

Results and Discussion

Sample Morphology

In the original ice templating method we found that the morphology which appears during freezing depends on the firmness of the precursor. A microhoneycomb structure appears when a fairly soft hydrogel is used, and a firm hydrogel gives a fibrous structure. The firmness of the precursor can be easily controlled by adjusting the time allowed for it to age. Therefore, in this work we adjusted the aging time of the particle-SiO_2 mixtures, and attempted the synthesis of microhoneycombs and fibers including various particles with different functions.

Figure 1 shows SEM images of typical samples obtained in this work. It was confirmed that microhoneycombs and fibers of SiO_2 including p25, HSZ and GRA particles can

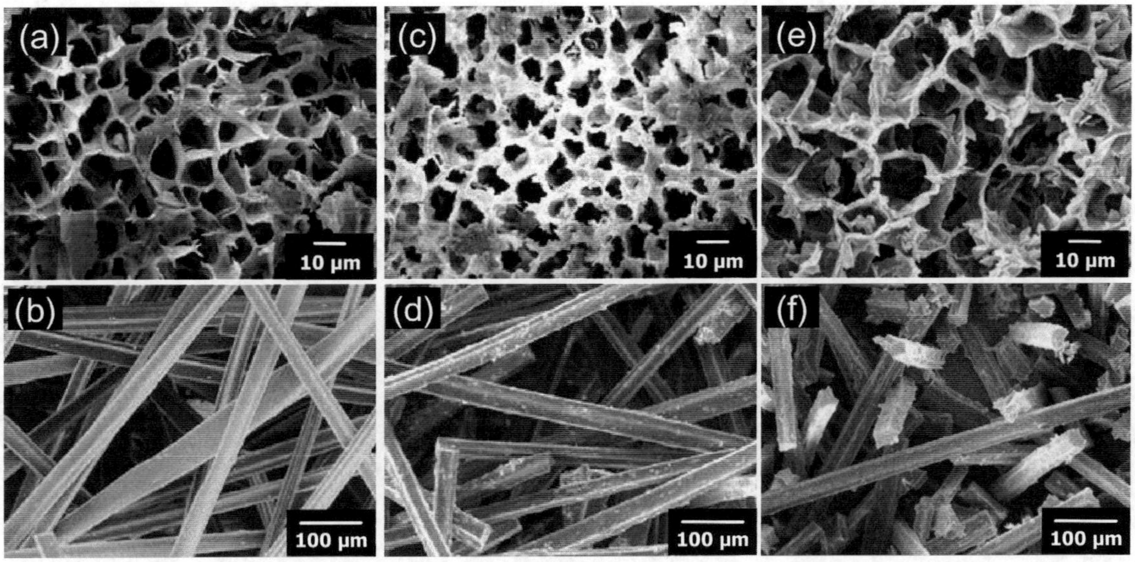

Fig. 1 SEM Images of Typical Samples Obtained in this Work
(a) SiO_2 Microhoneycomb Including p25 (b) SiO_2 Fibers Including p25 (c) SiO_2 Microhoneycomb Including HSZ (d) SiO_2 Fibers Including HSZ (e) SiO_2 Microhoneycomb Including GRA (f) SiO_2 Fibers Including GRA

be synthesized using the modified ice templating method. As expected, the morphology which appears after freezing could be controlled by adjusting the time allowed for the precursor to age prior to freezing.

The thickness of the walls of the obtained microhoneycombs seemed to depend on the sizes of the included particles which increase in the order p25 (about 20 nm), HSZ (about 300 nm) and GRA (over 1 μm). The channels of the microhoneycombs were not as straight and as regularly ordered as those in the microhoneycombs obtained through the original ice templating method. However, it was confirmed that the channels run from one side to the other side of the honeycomb through ink penetration tests. Fibers including particles tended to be shorter than those synthesized without particle inclusion.

As porous fibers can be obtained through other methods, hereafter we will focus on the microhoneycomb samples.

Compression Strength

As mentioned before, the walls of the honeycombs obtained through the original ice templating method are extremely thin. Accordingly, the diffusion path lengths within such porous walls become extremely short, which is extremely advantageous for porous materials to be used as adsorbents and catalysts. However, the thinness of the walls also limits the strength of such honeycombs.

In order to verify the effect of particle inclusion on the strength of the microhoneycombs, the compression strengths of the samples were tested using a load cell.

Table 1 Compression Strength of Typical Samples

Sample	Strength (As-prepared) [kg/cm^2]	Strength (Heat-treated) [kg/cm^2]
SiO_2 Microhoneycomb	0.21	0.38
SiO_2 Microhoneycomb Including p25 (p25/ SiO_2 = 0.6 g/g)	0.29	0.64
SiO_2 Microhoneycomb Including GRA (GRA/ SiO_2 = 0.6 g/g)	0.42	Over 6.6

Table 2 Porous Properties of Typical Samples

Sample	S_{BET} [m^2/g]	V_{meso} [ml/g]
SiO_2 Microhoneycomb	1024	0.28
SiO_2 Microhoneycomb Including p25 (p25/ SiO_2 = 0.6 g/g)	795 (658)	0.29 (0.18)
SiO_2 Microhoneycomb Including HSZ (p25/ SiO_2 = 0.6 g/g)	927 (965)	0.33 (0.18)
SiO_2 Microhoneycomb Including GRA (GRA/ SiO_2 = 0.6 g/g)	757 (655)	0.20 (0.18)

*Values in parenthesis are calculated by simply adding the values of the SiO_2 microhoneycomb and the particles multiplied by their inclusion ratios

The results are summarized in Table 1. Interestingly, the compression strengths of the samples were found to increase with the inclusion of particles. This increase is thought to arise from the thickening of the walls due to particle inclusion. Heat treatment of the samples led to further increase in compression strength, and this increase was significant when GRA particles were included.

In the original ice templating method, the thickness of the walls depends on the concentration of the starting solution. However, a starting solution with a high concentration tends to transform to a gel and harden quickly. Therefore the starting concentration is limited to moderate values in the original ice templating method. This modified method allows the synthesis of microhoneycombs with thick walls, which leads to the reinforcement of the structure.

Porous Properties

The porous properties of the samples were evaluated through nitrogen adsorption experiments. Table 2 summarizes the BET surface areas (S_{BET}) and mesopore volumes (V_{meso}) of typical samples. p25 and GRA are nonporous materials, and HSZ is a fairly porous material but its BET surface area and mesopore volume are lower than those of typical SiO_2 microhoneycombs. Therefore it is natural that the BET surface areas of the samples

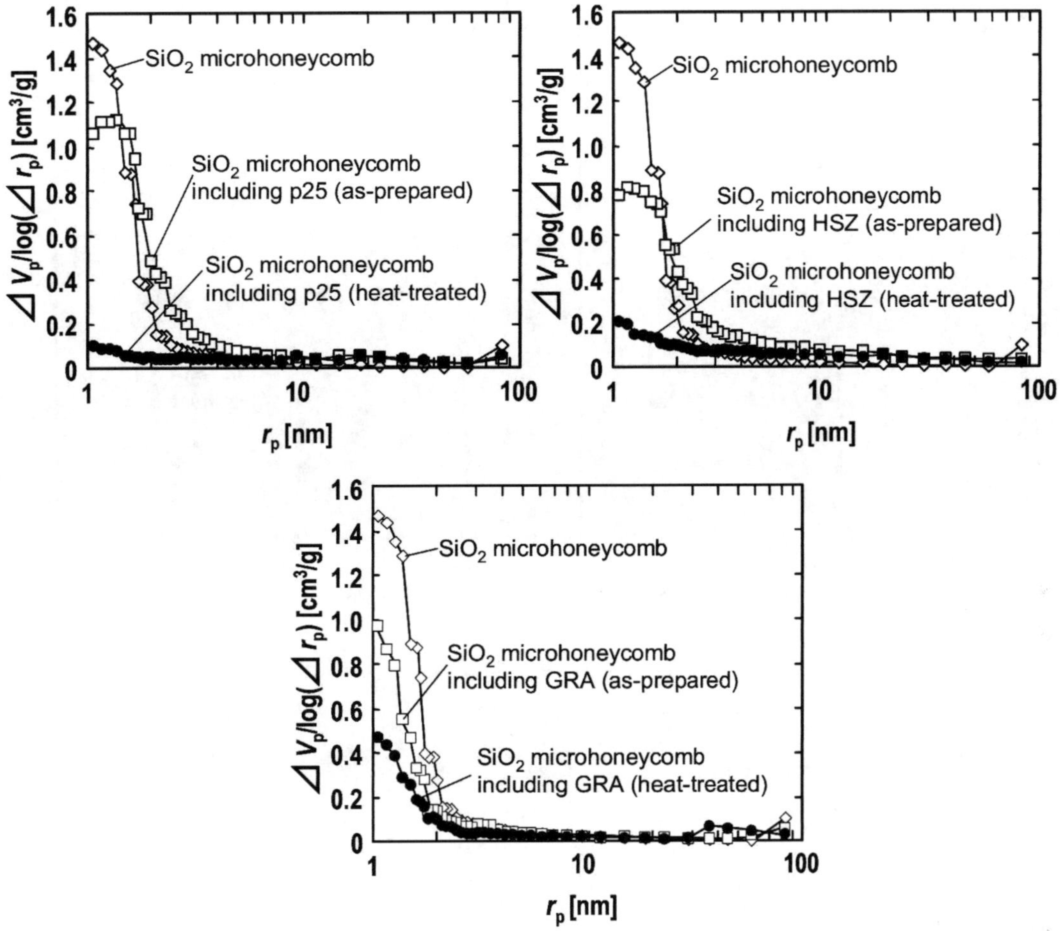

Fig. 2 Pore Size Distributions of Typical Samples Obtained in this Work (Particle/SiO_2 ratio 0.6 g/g)

including particles will decrease. However, it was confirmed that the BET surface areas and mesopore volumes of the samples were larger than the values calculated by simply adding the values of the SiO_2 microhoneycomb and the particles based on their amounts in the microhoneycomb.

Figure 2 shows the pore size distributions of typical microhoneycomb samples. It can be noticed that as expected, the mesopore volumes decrease due to particle inclusion and heat treatment. The sizes of the mesopores of samples with and without GRA inclusion seem to be quite similar, but it is slightly shifted to larger values when p25 and HSZ are included.

Considering the results obtained through compression and adsorption experiments, it can be assumed that in SiO_2 microhoneycombs including particles, the particles are dispersed fairly homogeneously in the SiO_2 matrix.

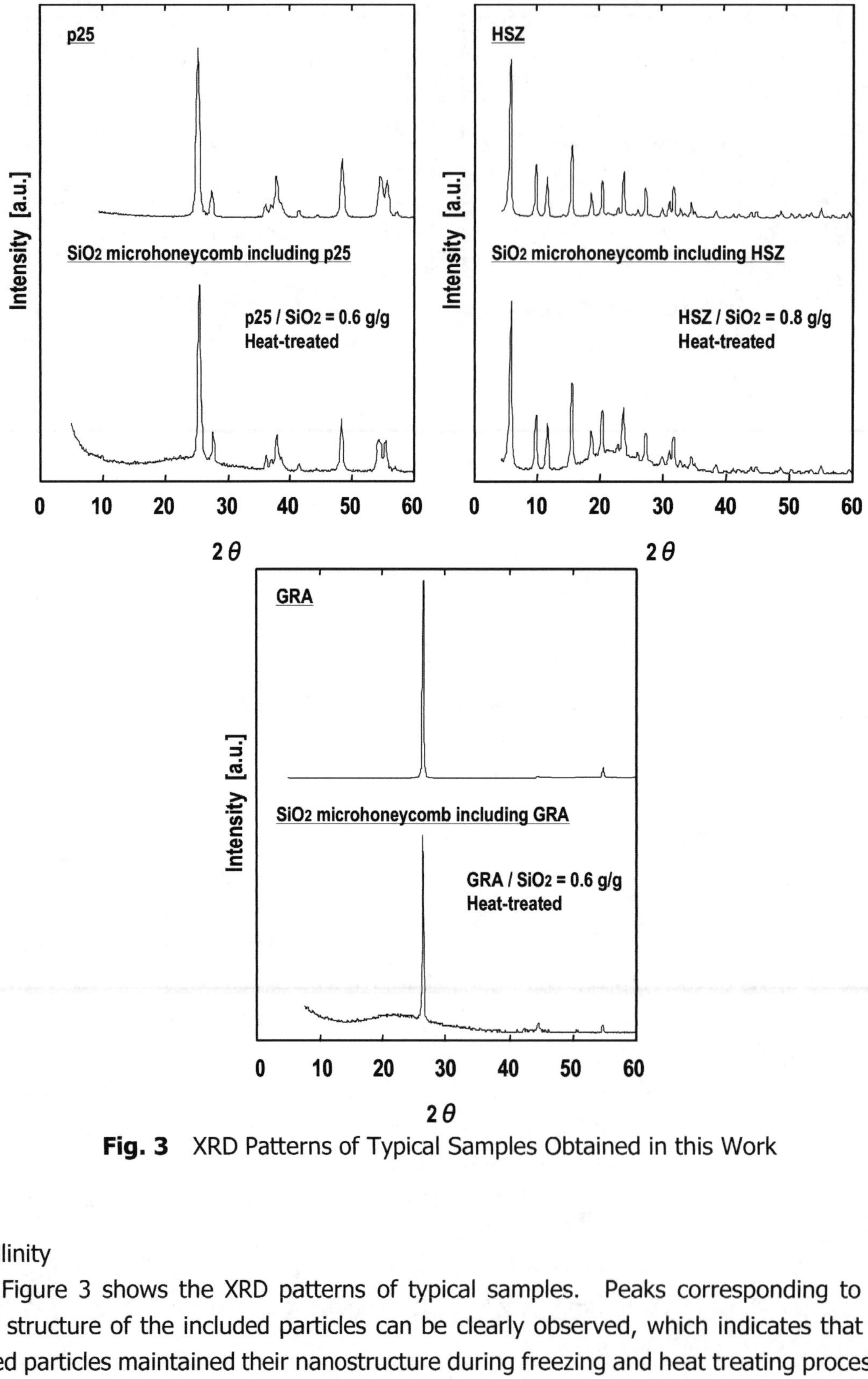

Fig. 3 XRD Patterns of Typical Samples Obtained in this Work

Crystallinity

Figure 3 shows the XRD patterns of typical samples. Peaks corresponding to the crystal structure of the included particles can be clearly observed, which indicates that the included particles maintained their nanostructure during freezing and heat treating processes. Therefore, the particles are assumed to maintain their unique functions, such as catalysis and adsorption abilities, after processing. Therefore, microhoneycombs and fibers with unique combined functions can be easily synthesized using this method.

Conclusion

In this work, we attempted to widen the range of applicability of the ice templating method by applying this method to mixtures of SiO_2 hydrogels and various nanoparticles possessing various functions. It was experimentally shown that microhoneycombs and fibers of SiO_2 including various nanoparticles can be synthesized using this modified ice templating method. This indicates that this method can also be applied to materials which cannot be obtained through sol-gel transition, as long as they are combined with sol-gel derivable materials such as silica. It was also found that the morphology which appears during freezing can be controlled by adjusting the firmness of the precursor. From the results of structural analysis, it is assumed that the included particles are uniformly distributed in the SiO_2 matrix. It was also confirmed that the particles maintained their structure, and that particle inclusion hardly affects the structure of the SiO_2 matrix. Therefore, microhoneycombs and fibers with unique combined functions can be easily synthesized using this method.

We believe this is a promising method to provide microhoneycombs and fibers which can be used as adsorbents and/or catalysts in various microdevices, such as microreactors and microseparators.

References

1. Mahler, W. and Bechtold, M. F. (1980), "Freeze-formed silica fibers," *Nature*, 285, pp. 27-28
2. Mukai, S. R., Nishihara, H. and Tamon, H. (2003), " Porous Properties of Silica Gels with Controlled Morphology Synthesized by Unidirectional Freeze-Gelation," *Microporous and Mesoporous Materials, 63*, pp. 43-51
3. Mukai, S. R., Nishihara, H. and Tamon, H. (2004), "Formation of Monolithic Silica Gel Microhoneycombs (SMHs) Using Pseudosteady State Growth of Microstructural Ice Crystals," *Chemical Communications*, pp. 874-875
4. Nishihara, H., Mukai, S. R., Yamashita, D. and Tamon, H. (2005), " Ordered Macroporous Silica by Ice-Templating," *Chemistry of Materials*, 17, pp. 683-689

139a Applying Microreactor Technology in Chemical Product Research and Process Development

Jaap C. Schouten, Chemical Engineering and Chemistry, Eindhoven University of Technology, Eindhoven, 5600 MB, Netherlands

The particular features of microreactors, such as the large driving forces for heat and mass transport and the high surface-area-to-volume ratio, stimulate the use of these devices as new tools in chemical product research and process development. For example, microreactors are well suited for rapid screening of materials and catalysts, for measuring intrinsic reaction kinetics, and for discovering new process chemistries. An additional advantage is that micro structured chemical systems may be easily integrated with dedicated instrumentation, sensors and actuators, in small bench top units, requiring less fume hood space and fewer utilities. Nowadays microreactor systems are also more and more combined with modern measurement technologies for real-time analysis of the system's performance and of the chemistry conducted in these reactors. Scale-up to production levels by replication of microreactor units used in the laboratory could subsequently eliminate costly redesign and pilot plant experiments, thereby shortening the development time from laboratory to commercial production.

At present three areas of measurement needs can be identified: (1) the characterization of the microreactor itself in terms of its performance characteristics (e.g., flow distribution, temperature distribution), (2) the characterization of the chemistry being conducted in the microreactor (conversion, selectivity, product yield), and (3) the characterization of banks of microreactors which would be found in commercial production.

In the lecture, various examples will be given of the characterization of the performance of microreactors as well as determining the kinetics and mechanism of various exothermic reactions (e.g., ammonia oxidation, preferential CO oxidation, ethane ammoxidation). Important issues will be covered such as flow distribution, isothermal operation, product gas sampling, etc. The characterization of different types of microreactors will be discussed, such as microchannel reactors and parallel-plate microreactors. It is demonstrated that real-time analysis in well characterized microreactor systems leads to better understanding of the chemistry which ultimately will lead to more efficient chemicals production.

139b Characterization of Microchannel Microreactors as Tools for Laboratory Measurements of Reaction Kinetics

Daniel A. Hickman and Daniel D. Sobeck. Engineering & Process Sciences, The Dow Chemical Company, 1776 Building, Midland, MI 48674

The high specific surface areas and small dimensions in microchannel microreactors can enable essentially ideal isothermal, plug flow performance in small scale continuous systems. Thus, these features enable characterization of reaction systems under carefully controlled conditions, with the potential to extract accurate reaction kinetic parameters using a simple mathematical model. To confirm the plug flow assumption for a particular commercially available microreactor, we conducted experiments to characterize the residence time distribution and characteristic mixing time. Experiments in this microreactor gave a characteristic mixing time as low as 6 seconds for a model hydrolysis reaction at 2.4 mL/min total flow, while reaction-diffusion calculations for this device predict minimum mixing times near 1 second at flow rates of 10 mL/min or higher. In contrast, the literature reports mixing times as low as 0.01 seconds for other microchannel systems and 0.002 seconds for a Mettler RC1 calorimeter. We hypothesize that one or more of the following three factors caused the relatively slow mixing in this reactor. (1) The initial mixing of the interdigital feeds takes place while the lamellae are being thinned from 4500 μm down to 33 μm in a 0.095 mL volume. (2) Each of the mixing and reaction channels contains only 6 lamellae, so the characteristic mixing length approaches that of the channel height for reactions operated near the stoichiometric ratio. (3) For reactions that are extremely sensitive to stoichiometry, uneven distribution of the two feeds to the multiple channels can lead to anomalously high apparent mixing times. Building on the foundation of the microreactor characterization case study, we developed a system to characterize homogeneous liquid phase reactions using serial screening in a microreactor. The system used a single microchannel reactor with reactants injected as finite pulses. Fundamental reaction engineering principles (including equations and constraints for heat transfer, mixing, axial dispersion, and pressure drop) were used to design or select the system components to enable analysis of the reactor effluent at undiluted, steady-state concentrations while injecting only microliters of the reactant solutions per experiment. The custom single channel reactor was shown to give the expected low axial dispersion of injected reactants, consistent with the analysis of axial dispersion of Newtonian fluids in circular tubes. However, experiments with different materials of construction demonstrated a variable lag time between the injection of a reactant pulse and its arrival at the analytical sample point as a function of the material of construction. We hypothesize that this effect is the result of adsorption or absorption of the reactants on or in the tubing walls. The impact of this observation is that only some materials of construction may be used for high throughput serial screening by injection of reactant pulses, dependent upon the chemistry employed.

139c Analytical Sensors, Instrumentation, and Laboratory Applications of Nessi

Dave Veltkamp, University of Washington, Center for Process Analytical Chemistry (CPAC), : Chem Lib Bldg, Rm 137, Box 351700, Seattle, WA 98195-1700

While originally conceived and developed as an alternative to traditional sampling systems for process analyzers, NeSSI also provides a platform for development of compact sensors and analyzers for both process and laboratory applications. This talk will provide a survey of some of these development efforts. Examples using NeSSI as an analytical sensor development tool will be presented and will highlight some of the advantages of NeSSI in the research laboratory and teaching settings.

139d Online Analysis in Microreaction Technology: Suitable Tools for Process Screening and Optimization

Wolfgang Ferstl, Wenka Schweikert, Maud Schwarzer, and Stefan Loebbecke. Fraunhofer Institut für Chemische Technologie ICT, Joseph-von-Fraunhofer-Strasse 7, Pfinztal, D-76327, Germany
Microreaction Technology provides well-known benefits such as improved heat and mass transport for the continuous processing of chemical reactions. The precise control and regulation of process parameters in microreaction processes allow an in-depth analysis of mixing sensitive and strong exothermic reactions on basis of Design-of-Experiment (DoE) procedures to identify promising routes for process optimization. Here we report on the implementation of appropriate online spectroscopic techniques such as UV/Vis, Near IR, Mid IR and Raman spectroscopy to microreaction processes for screening and optimization purposes. Suitable optical cells, analytical interfaces and quenching modules have been developed that can be adapted to a microreaction process. If necessary, chemometric techniques can be applied for qualitative and quantitative real-time monitoring of microchemical processes. In combination with experimental design it can help identifying crucial factors of process intensification and speeding up optimization procedure.

Determination of Temperature Profile within Continuous Micromixer-Tube Reactor Used for the Exothermic Addition of Dimethyl Amine to Acrylonitrile and an Exothermic Ionic Liquid Synthesis

Patrick Löb[1,], Holger Löwe[1,2], Volker Hessel[1,3], Suzanne May Hubbard[4], Gabriele Menges[1], Monika Balon-Burger[1]*

[1] *Institut für Mikrotechnik Mainz GmbH, Carl Zeiss Strasse 18-20, 55129 Mainz, Germany*
[2] *Johannes-Gutenberg University Mainz, Department of Chemistry, Pharmaceutics and Earth Sciences, Institute of Organic Chemistry, Duisbergweg 10-14, 55128 Mainz, Germany*
[3] *Eindhoven University of Technology, Department of Chemistry and Chemical Engineering, PO Box 513, 5600 MB Eindhoven, The Netherlands*
[4] *Europa Fachhochschule Fresenius, Department of Chemistry and Biology, Limburger Strasse 2, 65510 Idstein, Germany,*

* *Author for correspondence; Tel.: +49 6131 / 990 377, Fax: +49 6131 990 205, E-mail: loeb@imm-mainz.de*

Abstract

Continuous micromixer-tube reactors are versatile and widely used tools to explore the benefits of microreactor processing for organic synthesis. Due to enhanced mass and heat transfer capabilities and mechanical stability with regard to temperature and pressure their use can lead for given chemical protocols to process intensification and as important additional aspect enable process chemistry with novel features not feasible so far because of limits of conventional equipment. Naturally, the heat transfer capabilities of the tube section within continuous micromixer-tube reactors are - albeit enhanced - limited especially since they are often rather meso- than microstructed. In this context the temperature profiles within the tube section of a continuous micromixer-tube reactor was determined for two exothermic reactions, the addition of dimethly amine to acrylonitrile and an ionic liquid synthesis, in order to determine the occurrence and extent of hot spot formation for different process conditions and to gain thereby also a deeper understanding of the processes. Seven temperature sensing elements were therefore integrated into the tube section. Furthermore, it is outlined how these experiments can contribute to the targeted adaptation of the reactors.

1. Introduction

The use of microstructured reactors with their benefits especially concerning enhanced mass and heat transfer for organic synthesis can lead for given chemical protocols to process intensification and as important additional aspect enable process chemistry with novel features not feasible so far because of limits of conventional equipment.[1-7] Novel features hereby can be for example[8]:

- direct routes from hazardous elements
- routes at increased concentration or even solvent-free
- routes at elevated temperature and/or pressure (see also high-p,T processing concept [9])
- routes mixing the reactants all at once
- routes using unstable intermediates
- routes in the explosive or thermal runaway regime

- routes omitting the need of catalyst and auxiliary agents

Continuous micro-reactor rigs including conventionally interconnected micro-flow devices are thereby versatile tools because of their simplicity, reliability, and flexibility – as opposed to tailored integrated micro-reactor concepts. This includes combinations of micromixers and heat exchangers and in particular capillary or tube sections (being rather meso- than microstructured). A generic overview over possible flow sheets for such rigs targeting at different reaction requirements is given elsewhere [10].

Recently, the authors investigated two different reactions types in a continuous micromixer-tube reactor addressing some of the above mentioned novel features which are now also inspected with regard to temperature profile.

The first reaction type is the addition of secondary amines to α,β unsaturated carbonyl compounds and nitriles, such as acrylic acids and their esters or acrylonitrile.[11] This reaction type is a common synthesis path in organic chemistry Using a batch procedure these reactions give good yields, in some cases even exceeding 85%, but at long reaction times up to 24 hours. The latter, however, is not caused by the intrinsic kinetics of the reaction, but relates in many cases to their great exothermicity. In order to enable safe removal of the heat created by the reaction, it is necessary to add the α,β unsaturated compound quite slowly to the amine. To overcome these limits the process has been shifted to a continuous one using a micromixer-tube reactor embedded into a thermostat bath.[11] Systematic investigations were performed for different systems using as amine either dimethyl amine, diethyl amine, piperidine and either acrylonitrile or arylic acid ethyl ester as α,β unsaturated carbonyl compound. The set-up allowed a direct contacting of the reactants under controllable conditions. Determination of yield as function of temperature and total flow rate (residence/reaction time) for the different combination possibilities allowed to compile the reactivity row of the amines and the α,β unsaturated carbonyl compounds (piperidine > dimethyl amine > diethyl amine; acrylonitrile > acrylic acid ester). Furthermore, as expected the microreactor processing lead to higher space-time yields.[11] Out of the range of investigated reactions, here the exothermic addition of aqueous dimethyl amine to acrylonitrile was selected for further studies focusing on the temperature profile within the micromixer-tube reactor.

The second reaction type is the synthesis of an ionic liquid carried out in a micromixer-tube reactor recently in the framework of the EU-project IMPULSE with Solvent Innovation GmbH, Cologne, Germany as partner.[10, 12] It is a solvent-free exothermic process with aggressive reactants which exhibits heat-transfer sensitivity. Too high temperatures will lead to formation of unwanted side products decreasing product quality which is already visually observable by the yellow colouring of the otherwise clear product. The concrete chemistry cannot be disclosed at that time due to intellectual properties rights of the industrial user. Nevertheless, some generic conclusions can be extracted from the performed experiments. Some more information about the process is given below in the context of the description of the supplementary experiments targeting at the temperature profile.

2. Experimental set-up for temperature profile determination

The experimental set-up used for the temperature profile determination was the same for both reactions – only the micromixer was changed. The set-up was composed of a micromixer for mixing the two reactants followed by a PTFE tube with 1/8´´ outer diameter and of 7.5 ml inner volume as residence time section, both embedded in a thermostat bath.

The reactants were preheated by tube sections in front of the mixer also immersed into the thermostat bath. The reactor tube was divided in several sections connected via T-junctions allowing the embedment of thermocouples in the reactant flow. The thermocouples were positioned directly after the mixer, after 5%, 10%, 15%, 25%, 50% and 100% of the tube length. Therewith the determination of the temperature profile along the tube section was possible. For the ionic liquid synthesis an interdigital micromixer (type SIMM-V2 Ag40, Institut für Mikrotechnik Mainz GmbH, Germany) and for the amine addition a caterpillar mixer (type CPMM-V1.2-R600/12, Institut für Mikrotechnik Mainz GmbH, Germany) was used. At the end of the tube a valve allowed to build up pressure. After the valve the reaction mixture was cooled down by a micro heat exchanger (type CRMH, Institut für Mikrotechnik Mainz GmbH, Germany). The reactants were fed by two syringe pumps (PN 1610 syringe dosing systems, Postnova Analytics, Landsberg, Germany).
The internal volume of the caterpillar mixer (interdigital mixer) is about $5.9x10^{-3}$ ml (about $1.5x10^{-3}$ ml). The inner volume of the connection between mixer outlet and inlet tube section with first thermocouple was estimated to be about $80x10^{-3}$ ml. The complete experimental set-up is shown in Figure 1, a more detailed view of the reactor embedded in the thermostat bath is shown in Figure 2.

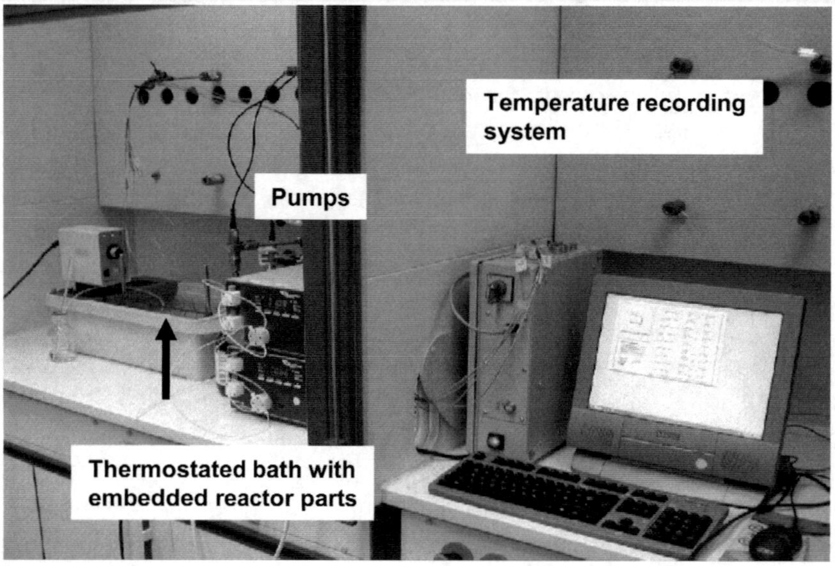

Fig. 1: Experimentally set-up used for determination of the temperature profile in continuous micromixer-tube reactor.

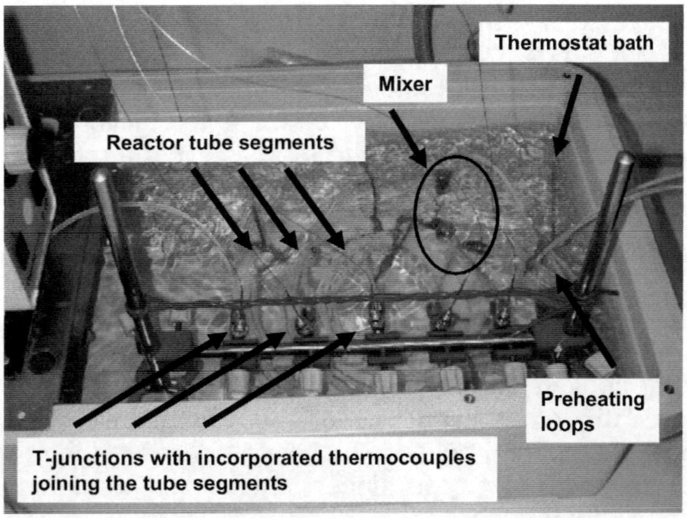

Fig. 2: Reactor part from the experimental set-up used for the determination of the temperature profile in continuous micromixer-tube reactor embedded in thermostat bath. Visible are single tube sections connected via T-junctions with integrated thermocouples.

3. Temperature profile determination for the exothermic addition of dimethyl amine to acrylonitrile

The reaction was carried out using undiluted acrylonitrile and dimethyl amine dissolved in water (40 wt% dimethyl amine) in a molar ratio of 1 : 1.1. At first, temperature profile was determined for a total flow rate of 4.68 ml/min referring to a residence time of 1.6 min for different tempering conditions: thermostat bath without filling (i.e. air cooling, room temperature), thermostat bath operating at room temperature or thermostat bath operating at 60 °C. The obtained temperature profiles are shown in Figure 3.

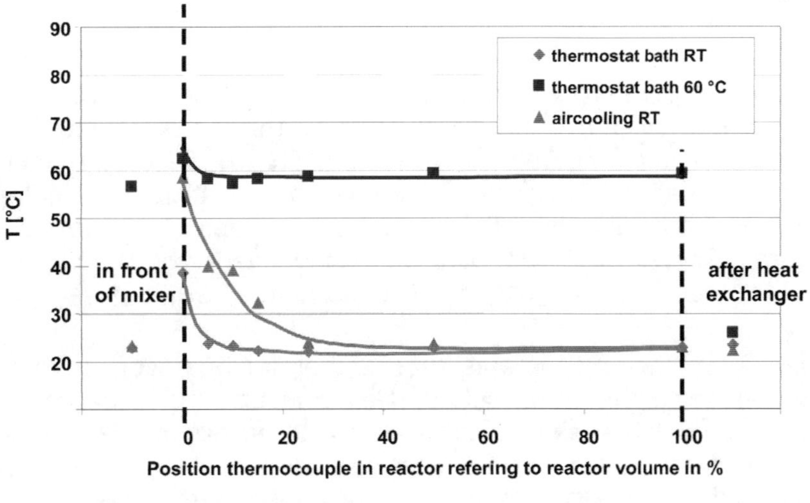

Fig. 3: Experimentally determined temperature profile in continuous micromixer-tube reactor used for the exothermic addition of dimethyl amine to acrylonitrile for a total flow rate of 4.68 ml/min for different tempering options.

For all three tempering option the highest temperature rise above bath temperature is measured at the beginning of the tube sections. The highest temperature rise was detected when starting at room temperature and only air cooling was applied; not unexpected because of the missing active cooling by a thermostat filling. Detected temperature rise at a thermostat temperature of about 60 °C is quite low. A higher temperature rise was expected because of higher reaction rate at higher temperature. It is possible that there is a higher temperature rise in front of the first thermocouple, i.e. in the connection between mixer and tube section. Bath temperature is reached again in the case of the room temperature and 60 °C processing already after 5% of the reactor length. Without thermostat filling the heat transfer capability of the set-up is much lower and therefore room temperature is reached later (after about 25% of the reactor length). The obtained yields of the addition product are given in Figure 4 for all the experiments performed. The reaction is selective. No significant amounts of side products were detected. Therefore, yield and turnover of acrylonitrile is almost identical. For a total flow rate of 4.68 ml/min and therewith a residence time of 1.6 min almost complete conversion was achieved when working at a thermostat temperature of 60 °C. For the other two tempering options conversion is lower. The conversion for the air cooling tempering option is higher compared to the room temperature thermostat processing since in the first case higher reaction temperatures were obtained due to more pronounced hot spot formation.

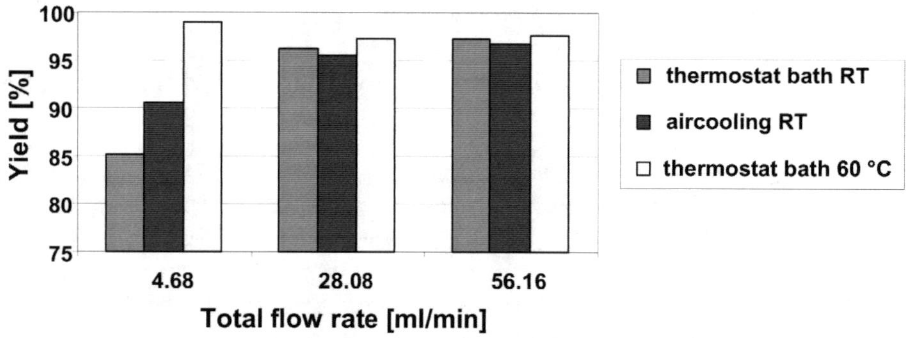

Fig. 4: Yields of the addition of dimethyl amine to acrylonitrile as function of total flow rate for different tempering options.

At first sight it is surprising that the yield/conversion at a thermostat bath temperature of room temperature increases with increasing total flow rate from 4.68 ml/min to 28.08 ml/min, i.e. decreasing residence and reaction time from 1.6 min to 0.3 min. But looking at the corresponding temperature profiles (see Fig. 5), the explanation is that for the higher flow rate the reaction heat cannot be removed efficiently leading to a more pronounced hot spot formation and therewith higher reaction temperature and higher conversion. Figure 5 shows also that with increasing flow rates the hot spot is shifted downwards in the tube and the absolute temperature rise above thermostat bath temperature rises.

It is clear from the experiments that isothermal operation conditions were not reached in this case with the used set-up. Nevertheless, the experiments have been performed under a controllable manner, certainly not to be expected for a direct, one step contacting of the reactants in a batch. Furthermore, the selectivity of the reaction was not significantly influenced by hot spot formation, i.e. albeit isothermal operation conditions are not reached at all, the set-up is sufficient to get the product in right quality. This is no longer true for the synthesis of the ionic liquid.

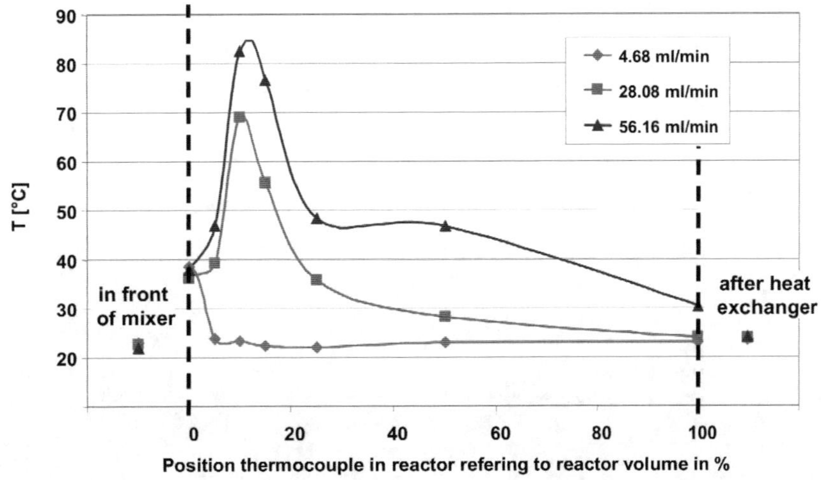

Fig. 5: Experimentally determined temperature profile in continuous micromixer-tube reactor used for the exothermic addition of dimethyl amine to acrylonitrile for a thermostat temperature of 22 °C for different total flow rates.

4. Temperature profile determination for the exothermic ionic liquid synthesis

The reaction was carried out using an imidazole derivative and an alkylating agent both without any solvents in almost stochiometric ratio. Similar observation as in the case for the amine addition can be made.[11] Exemplarily, Figure 6 shows the determined temperature profiles for a thermostat temperature of 50 °C for different total flow rates. For all flow rates yield is here almost 100% (side products already at a very low level lead to a colouring of the product). As in the case of the amine addition with increasing total flow rate hot spot becomes more pronounced and moves downwards the tube. While for the low total flow rate (1.74 ml/min) the observed temperature rise is only in the range of acceptable 6°C, hot spot formation with temperature rises up to 50°C over thermostat temperature occurred for higher thermostat temperatures. The increasing amount of formed coloured side products can be attributed to the increased temperature rise and high temperature duration as found for the higher flow rates. Obviously, the reactor or the selected reactor dimensions are not capable of removing reaction heat in a sufficient manner for the higher flow rates.

In order to improve heat removal for the higher flow rates, one could use finer structures, e.g. 1/16´´ tubes instead of 1/8´´ tubes. Naturally, the smaller tubes impose an increase of pressure drop which may become a limiting operational parameter. Therefore, the determination of temperature profiles becomes so important by locating the reactor section where smaller tubes have to be used and therewith to minimise the use of smaller tubes to where necessary. As can be seen from the temperature profiles, the heat removal capacity of the 1/8´´ tubes is sufficient in large parts of the reactor, e.g. for a total flow rate of 3.48 ml/min beginning at a reactor volume of 15%.

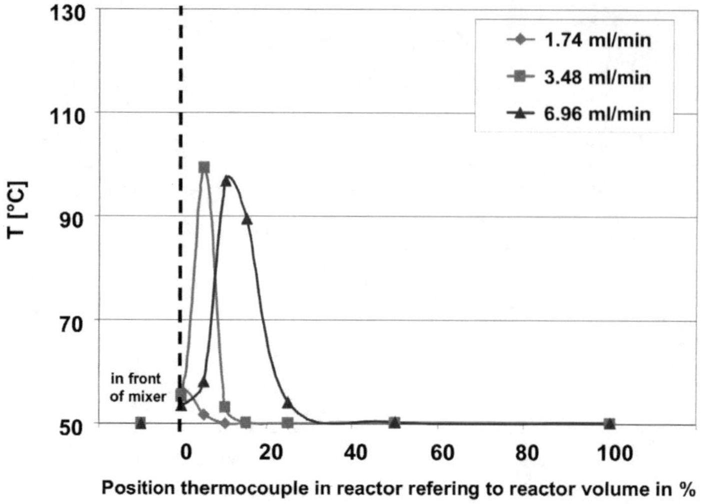

Fig. 6: Experimentally determined temperature profile in continuous micromixer-tube reactor used for an ionic liquid synthesis working at a thermostat temperature of 50 °C.

Therefore, the experimental set-up has been modified.[11] The first two reactor sections where exchanged by 1/16´´ tubes with same internal volume as the replaced 1/8´´ tubes. Exemplarily, the obtained temperature profiles with such a set-up for a total flow rate of 3.48 ml/min are given in Figure 7 with the corresponding profile of the set-up with all 1/8´´ tube sections as comparison. The maximum temperature rise above thermostat temperature was reduced from 50°C to 10°C with this reactor modification, yielding a clearer product. For the highest flow rate (6.96 ml/min) the modification did not prevent hot spot formation, since a good portion of the reaction is occurring after the first two tube sections and therewith not affected by smaller dimensions in the first two tubes (see Figure 6).

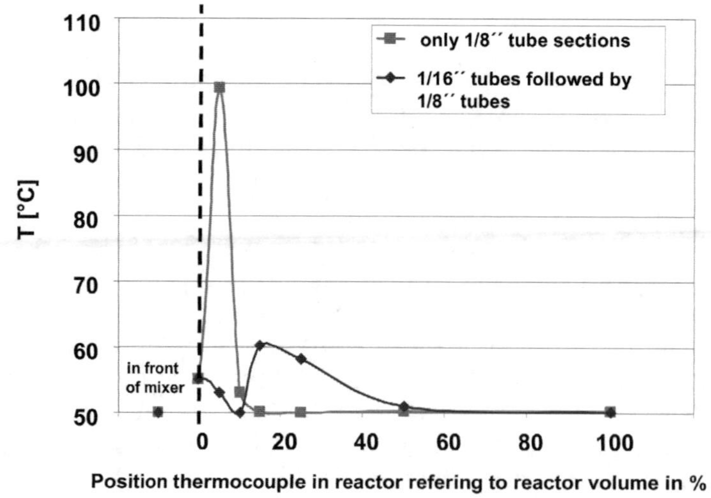

Fig. 7: Experimentally determined temperature profile in continuous micromixer-tube reactor used for an ionic liquid synthesis working at a thermostat temperature of 50 °C for two reactor configurations.

6. Summary

The determination of temperature profiles within the tube section of a continuous micromixer-tube reactor was demonstrated for two exothermic organic reactions. Occurrence and extent of hot spot formation can be monitored, allowing better interpretation of otherwise not obvious results. The determination of the temperature profile also allows an experimental access to reaction data necessary for specifying adapted reactor design when no kinetic and thermodynamic data are available.

Acknowledgements

Parts of the presented works were funded by the EU-project IMPULSE.

References

[1] Hessel, V., Löwe, H., Müller, A., Kolb, G., *Chemical Micro Process Engineering - Processing and Plants*, Wiley-VCH, Weinheim (2005).
[2] Hessel, V., Hardt, S., Löwe, H., *Chemical Micro Process Engineering – Fundamentals, Modelling and Reactions,* Wiley-VCH, Weinheim (2004).
[3] Gavriilidis, A., Angeli, P., Cao, E., Enhong, K.K., Wan, Y.S.S., *Trans. Inst. Chem. Eng.* A, **2005**, 80, 3
[4] Jensen, K.F., *AIChE J.*, **1999**, 45, 2001
[5] Jensen, K.F., *Chem. Eng. Sci.*, **1999**, 56, 293
[6] Haswell, S. T.; Watts, P. *Green Chem.*, **2003**, 5, 240.
[7] Fletcher, P. D. I.; Haswell, S. J.; Pombo-Villar, E.; Warrington, B. H.; Watts, P.; Wong, S. Y. F.; Zhang, X. *Tetrahedron*, **2002**, 58, 4735.
[8] Hessel, V., Löb, P., Löwe, H., *Current Organic Chemistry*, **2005**, 9, 765.
[9] Hessel, V., Hofmann, C., Löb, P., Löhndorf, J., Löwe, H., Ziogas, A., *Organic Process Research & Development*, **2005**, 9, 479.
[10] Löb, P., Hessel, V., Krtschil, U., Löwe, H., *Chimica Oggi / Chemistry Today*, **2006**, submitted for publication.
[11] Löwe, H., Hessel, V., Löb, P., Hubbard, S., *Organic Process Research & Development*, **2006**, submitted for publication.
[12] Hessel, V., Löb, P., Löwe, H., Uerdingen, M., *Fine Chemical and Initial Ionic-liquid Syntheses in Microstructured Reactors*; to be presented during ACS Symposium Atlanta, 2006.

139f Use of Catalytic Micro-Reactor to Improve Selectivity of Gas Sensors :

christophe Pijolat[1], Marilyne Roumanie[2], Guy Tournier[2], and Jean-Paul Viricelle[2]. (1) EMSE, 158 cours Fauriel, Saint-Etienne, 42023, France, (2) SPIN, EMSE, 158 cours Fauriel, Saint-Etienne, 42023, France

The development of gas micro-reactors able to work at high temperature is an important current objective. There are several requests for catalytic reactions of mixtures of gases. We have recently work in this sense in order to produce a micro-reactor able to work at high temperature up to 500 or 600°C. We have firstly focussed this development on a micro-production of hydrogen using the catalytic reaction of deshydrogenation of MCH into toluene. For this goal, it was necessary to use a catalyst (platinum supported on alumina). The objective of the present paper is to use this micro-reactor for gas analysis in the sense of Micro-TAS development. The micro-reactor is associated to a conventional gas micro-sensor (semi-conductor type using electrical properties of tin oxide). As the main defect of such gas sensor is its lack of selectivity, it is interesting to place a micro-reactor before the sensor in order to have a pre-conditioning action on the gas mixture. The catalytic reactor can be used to change the composition of the gas mixture and so to improve the selectivity of the device. A Silicon micro-reactor has been developed with micro-technologies. Channels are etched in silicon substrates with classical deep reactive ion etching process. These channels of 100mm depth are structured with pillars to increase the surface and thus improve the overall reaction rate compare to a channel without pillar. The structured microreactor is molecularly bonded with a silicon cover to form an airtight devices. In order to work at high temperatures, it was necessary to include in this system some heaters to have an autonomous microreactor. Thick film resistances of platinum have been deposited by the screen printing and used as microheaters. The tests of microheaters are realised using the infrared thermographic camera. Special gas connexions able to work at 400 or 500°C have been specially developed with metallic tubes (300 microns diameter) and sealing process with ceramic material. The second part of the study concerns the deposition of the catalyst directly in the closed silicon channels by fluidic process. Firstly, the alumina is deposited by wash-coat associated to a vacuum step and a sintering process. The platinum is then deposited by fluidic impregnation. Physical and characterisations of the catalyst in the channels are presented (DRX, MEB, TEM, XPS, AFM…). The catalytic activity of the micro-reactor is also evaluated on a special catalytic bench developed for such very low flows. The bench is also equipped with gas analyses and especially a mass spectrometer in order to measure to gas composition before and after the micro-reactor. Finally, the micro-reactor is used in association with a gas sensor. The SnO_2 sensor is placed in a small cell after the micro-reactor. As such sensor presents response both to CO and CH_4, the micro-reactor is used to transform CO into CO_2 (no response with SnO_2). As platinum has no action on CH_4 at medium temperature, the device become selective to detect CH_4 in a mixture of CO + CH_4. The influence of the temperature is also evaluated for such selective responses. On the base of such feasibility study, new types of pre-processing of gas mixtures are currently investigated for more complex gas detection, for example in NOx detection. Another application is also under investigation for the development of a pre-concentrator for gas analysis. For this objective, the microreactor must be filled with absorbent materials. Current work are performed for the moment the achieved such objectives.

140a Superior Emulsion Formation in Microchannel Architecture

Laura Silva, Business Development, Velocys, Inc., 7950 Corporate Boulevard, Plain City, OH 43064, Anna Lee Tonkovich, Technology Development, Velocys, Inc., 7950 Corporate Boulevard, Plain City, OH 43064, and Jan J. Lerou, Experimental Operations, Velocys, Inc., 7950 Corporate Boulevard, Plain City, OH 43064.

The characteristics and performance of emulsified products are the result of a combination of the chemical formulation and the processing technology employed. Traditionally, emulsions are formed under high shear conditions using static mixers, ultrasound devices, homogenizers, or rotor/stator mixers. The shear field, temperature history, and composition profiles during the process are key factors in the final form of the emulsion. Emulsion compositions and processing conditions are typically over-designed to ensure product requirements are met. This formulation and processing approach leads to the use, and often overuse, of expensive surfactants, overshearing sensitive materials, high energy consumption, and can ultimately lead to less than optimum product performance.

The microchannel emulsification process adds discontinuous phase droplets to the continuous phase, one droplet at a time, in multiple parallel channels. The resulting droplet size distribution is controlled by a variety of processing conditions, including substrate pore size, stream pressures and temperature. The microchannel approach permits precise control of mixing energy, mixing time, and heating/cooling energy, all of which stay constant during process scale-up. This tight process control offers many advantages. First, the droplets are formed one at a time by pushing the discontinuous phase through a specially designed substrate. This allows for high shear at the wall, but low bulk shear to avoid damaging the sometimes fragile emulsion components. Second, tight droplet size distribution possible in microchannel architecture enhances mixture stability. Third, in-line heating and cooling permits gentle reagent heating as well as rapid product quenching, which helps produce very stable emulsions.

Examples will demonstrate the innovative approach to emulsification allowed by the microchannel architecture. Furthermore, the paper will discuss its potential use to explore processing conditions to enable previously inaccessible product formulations with shear-sensitive materials and controlled drop size distribution. Microchannel emulsification could be beneficial for a wide range of industrial and consumer applications, including pharmaceuticals, food products, adhesives, and personal care products.

Droplet Formation by Collision of Two Aqueous Solutions in an Organic Phase and Application to Ag Particle Formation

Ken-Ichiro Sotowa, The University of Tokushima, Tokushima, Japan
Keisuke Irie, Kyushu University, Fukuoka, Japan
Takanori Fukumori, The University of Tokushima, Tokushima, Japan
Katsuki Kusakabe, Fukuoka Women's University, Fukuoka, Japan
Shigeru Sugiyama, The University of Tokushima, Tokushima, Japan

Introduction

In typical microreactors, two or more reactants were brought into contact in a microchannel. Since the characteristic length of the microchannel is in the order of 100 micrometers or less, mixing can be completed quickly. When microreactors are used for reaction crystallization, fast mixing leads to higher supersaturation. This may allow us to produce finer particles. A major problem associated with the crystallization in microchannel is fouling. Some nuclei formed at the interface between two reactant fluids adhere to the surface of the microchannel. Also some crystals deposit on the channel wall directly. The crystals on the wall then begin to grow in size and this eventually leads to fouling. In an attempt to suppress fouling, a co-axial microchannel has been proposed [1]. By making the co-axial flow pattern, the interface appears to have a cylindrical shape and thus do not touch the channel wall. This prevented solid deposition on the channel wall.

In the present study, a new microchannel device was proposed. The channel consists of main channel and two nozzles. Organic phase was fed to the main channel, and the reactants were forced into the stream of organic phase through the nozzles. The reactants fed to the main channel collide each other and form droplets, in which nucleation and growth take place. Since the nucleation takes place in the dispersed phase, which hardly touches the channel wall, the problem of fouling can be relieved.

The performance of the new microchannel device was experimentally evaluated. As a first example, the device was used to produce calcium carbonate particles, which are prone to deposit on the channel wall. The microreactor was then applied to silver particle synthesis. The fluid behavior within the microreactor was also examined by means of computational fluid dynamics simulation.

Experimental

The microreactor developed in this study consists of a microchannel and two stainless steel tubes. A reactant is fed to the microchannel through one of the tubes. As the organic phase, soybean oil was used and fed to the main channel at 20 mL min-1. In the experiment to form calcium carbonate, a calcium chloride solution (0.05 mol L-1) and sodium carbonate solution (0.05 mol L-1) were injected to the stream of organic phase at a rate of 10 micro liters per minute. The two solutions merged at the tip of the nozzles and give calcium carbonate particles. The size of the particles was analyzed using SEM.

In silver particle synthesis, the microreactor was placed in a water bath that was kept at 297 K. The tube connected to the outlet of the microreactor was immersed in another water bath that was kept at 317 K. The silver precursor solution was a mixture of 5mL of silver sulfate (5mM) and 4 mL of aniline (0.02M), diluted with de-ionized water to 44mL. The pH of the solution was adjusted to 9.5 by dropping ammonia solution. The reducing agent was a hydrazine solution (0.5 wt%), diluted to 50 times. The silver particles were synthesized by mixing these solutions in the microreactor.

Results and Discussion

Figure 1 shows the droplet formed at the tip of the stainless steel tubes. The organic phase flows from left to right of the figure. It was revealed that even though the distance between the two nozzles was below 1 mm, the two reactants do not collide and form independent droplets alternately, unless the flow rates of organic and aqueous phases are within an appropriate range.

An SEM photograph of calcium carbonate particles obtained using the microreactor is shown in Fig. 2. The size of the particles ranged from 1.3 to 3.0 micrometers, and the average was approximately 1.8 micrometers. For comparison, the calcium carbonate particles were synthesized in a beaker from the same reactant solutions. The average particle size was found to be 5.1 micrometers. It can be said that as compared with a stirred tank, a microreactor is more suited for producing finer calcium carbonate particles.

The longest experimental run in the current study was 9 hours. During the run, no particle deposition was observed on the channel wall.

Figure 3 shows a SEM photograph of silver particles synthesized in the study. The size of the particles was approximately 100 nm. The size of the silver particles synthesized in a beaker ranged from 30 to 150 nm. Thus, for the present case, the microreactor is not always effective in producing finer particles. However, it was found that the size distribution of the silver particles was narrower.

Figure 1. Droplet formed by collision of two reactants injected from the two nozzles

Figure 2. Calcium carbonate particles obtained in this study.

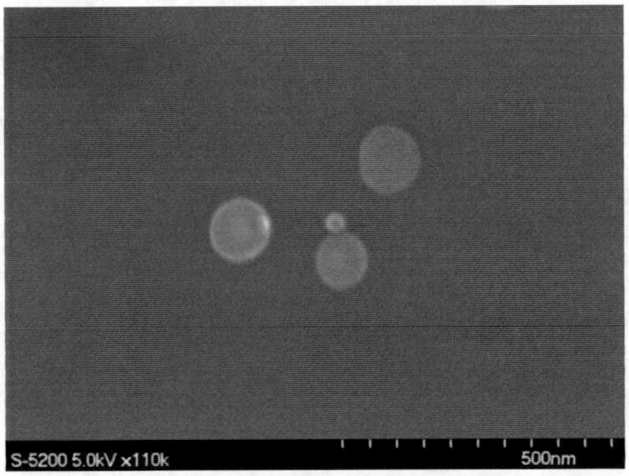

Figure 3. Silver particles obtained in this study.

Numerical simulation

A numerical simulation was carried out to examine the fluid behavior in the microreactor. The mixing rate within the droplet was evaluated for different channel configuration and flow rates.

The fluid behavior in the calcium carbonate formation experiment was examined. The simulation results predicted internal recirculation in a droplet, which can enhance the mixing rate. However, due to the laminar nature of the flow, the each reactant forms an independent recirculation path. One of the reactants forms closed streamlines in one hemisphere of the droplets and the other reactant on the other hemisphere.

Further simulation study was carried out to examine the effect of flow rates on the flow pattern in the microchannel. It was predicted that when the flow rates of the reactants are different from each other, the interface between the two reactants is formed off the center of the droplet and the internal recirculation can enhance mixing of the two reactants.

Concluding remarks

A new microreactor suited for particle synthesis was developed. It was confirmed that no fouling was observed even after 9 hours of experimental run. This is because nucleation and crystal growth take place in the aqueous phase, which hardly touches the channel wall. The experimental study showed that it could be used to obtain particles with small average size or narrower distribution.

The simulation study indicated that the flow pattern and mixing behavior is strongly affected by the flow rates of the reactants. Further experimental study will be conducted to verify the results.

References

1. H. Nagasawa, H. Mae (2003), Proc. of the 7th International Conference on Microreaction Technology (IMRET-7), pp.127-129.

140c Direct Nucleation Studies of Colloidal Systems inside Microfluidic Droplets

Tieying Gong[1], Jingyi Shen[1], Manuel Marquez[2], and Zhengdong Cheng[3]. (1) INEST Program, Philip Morris USA; Artie McFerrin Department of Chemical Engineering, Texas A&M University, 200 Jack E Brown BuildingTexas A&M University, Texas A&M University, College Station, TX 77840, (2) New Technology Research Department, Phillip Morris USA, 4201 Commerce Rd., Richmond, VA 23234, (3) Texas A&M University, Chemical Engineering Department, College Station, TX 77843
Even though the nucleation of crystalline materials is extensively studied, many aspects such as the dynamics of nucleation remain under hot debate, partly due to the limitations of experimental evidence. In this paper, emulsion droplets containing gel particles of a thermo-responsive polymer, poly-N-isopropylacrylamide(PNIPAM), has been used as an experimental model system for the study of nucleation. The polymer suspension is injected into a microfluidic channel together with two immiscible oil streams inside the outer channels to fabricate PNIPAM suspension droplets in oil. The concentration of PNIPAM gel particles is varied by temperature to determine the thermodynamic phase and the droplet size is varied using different flow rates of oil and colloidal suspensions to tune the nucleation volume. Here we present quantitative measurements of the nucleation process of colloidal systems using UV-Vis spectroscopy, light scattering and confocal microscopy. Effects of particle size and surface charge on the nucleation are also investigated.

140d Study of the Phase Transition Behavior of Self-Oscillating Pnipam Particles

Jingyi Shen[1], Srinivasa Pullela[2], Manuel Marquez[3], and Zhengdong Cheng[2]. (1) Artie McFerrin Department of Chemical Engineering, PMUSA INEST Program at Texas A&M University, 409 Jack E Brown Building, Texas A&M University, College Station, TX 77843-3122, (2) Artie McFerrin Department of Chemical Engineering, Texas A&M University, College Station, TX 77843-3122, (3) New Technology Research Department, Phillip Morris USA, 4201 Commerce Rd., Richmond, VA 23234
A self-oscillating gel was synthesized via copolymerization of N-isopropylacrylamide (NIPAM) and a Ru catalyst crosslinker. In the presence of substrates of the Belousov-Zhabotinsky (BZ) reaction, the size of the gel particles oscillates due to the periodic change in polymer hydrophilicity induced by the chemical oscillation. We report here the preparation of catalytic PNIPAM particles of relatively uniform size, and the study of their phase behavior by use of UV-Vis spectroscopy and light scattering techniques.

Miniaturization of Metallic Micro-Separator/Classifier and its Application to Emulsions

Shinichi Ookawara, Takumi Ishikawa, Kohei Ogawa

Department of Chemical Engineering, Graduate School of Science & Engineering, Tokyo Institute of Technology, 2-12-1, O-okayama, Meguro-ku, Tokyo 152-8552, Japan

Abstract

Applicability of a micro-separator/classifier, whose structure was originally proposed at IMRET7 (Ookawara, et al., 2003, 2004), to an oil-in-water emulsion is experimentally examined in this paper. The metallic micro-separator/classifier, whose width and depth of the main microchannel were originally 200 and 150 microns in the previous study (Ookawara et al., 2005), is further miniaturized for performance intensification. All the dimensions of the functional parts are proportionally reduced by a 1/2. The new device, which is fabricated by means of photolithography and diffusion bonds (Dai Nippon Printing Co., Ltd.), has six built-in curved microchannels each with a bifurcation at one end. For the feed emulsion, a small amount of vegetable oil is dispersed in deionized water with surfactant using an ultrasonic homogenizer. The emulsion is fed into the microchannels and is discharged from the inner and outer branches that form the bifurcations. The numerical size distribution (NSD) of oil droplets in the feed and discharge is measured by a particle size analyzer. The polydisperse NSD of the feed emulsion typically falls into the range of 0.5 to 50 microns. In this study, it is assumed that total droplet number density is proportional to the absorbance of the emulsion regardless of the NSD. The reduced grade efficiency and the grade purity coefficient are evaluated based on the droplet number density, NSD and flow rates in the inner and outer branches. It is found that both cut-size and pressure loss are much smaller than those of existing hydrocyclones. It can be concluded, therefore, that the miniaturized micro-separator/classifier is applicable to liquid-liquid separation of emulsions with the intensified separation efficiency.

Introduction

Liquid-liquid separation, especially oil-water separation, is a major operation in the chemical and petrochemical industries. The hydrocyclone is often adopted for the liquid-liquid separation to take advantage of low maintenance costs due to the absence of moving parts, easy set-up and operation, space-saving and so on. It is often the case, however, that the cut-size of hydrocyclones in liquid-liquid separation is rather large, mainly due to the small density difference between the phases. It is stated that the smallest oil droplet size with 100% efficiency d_{100} is 100 microns, although there is commercial evidence of efficiencies of over 99% down to 20 microns (Bennett and Williams, 2004).

Motivated by the low efficiencies and limitations of hydrocyclones, the applicability of a micro-separator/classifier, which was originally proposed for solid-liquid separation at IMRET7 (Ookawara, et al., 2003, 2004), to the liquid-liquid separation is experimentally examined in the present study. The recent numerical work suggested that the original microdevice brought about solid-liquid separation in the entry section of a curved microchannel, viz., arc within 30 degrees, due to the lift force induced by steep shear rates (Ookawara et al., 2006). Since the miniaturization of the device could augment the shear rate and centrifugal force with a constant Reynolds number in the curved channel, the higher separation efficiency can be expected in spite of the small density difference between the phases. Although the miniaturization causes larger pressure drops, the short arc length with just 30 degrees makes its operation practical. Therefore, the metallic micro-separator/classifier with arc of 30 degrees (Ookawara et al., 2005), is miniaturized for the performance intensification.

The performance of the miniaturized device is evaluated by a grade purity coefficient (GPC) (Grady et al., 2003) as well as by a reduced grade efficiency (RGE). The breakage and coalescence of oil droplets in the liquid-liquid separation could cause a RGE value higher than unity. On the other hand, the GPC is straightforward to interpret, regardless of breakage and coalescence. This study evaluates the RGE for the performance comparison with the original device, associated with its re-definition.

Fabrication of miniaturized micro-separator/classifier

The width, depth and curvature radius of the curved microchannel, which is the functional part of the metallic micro-separator/classifier, were originally 200 microns, 150 microns and 20 mm, respectively (Ookawara et al., 2005). All the dimensions of functional part are proportionally reduced by a 1/2. In this study, only an arc with 30 degrees is examined for the operation with low pressure loss.

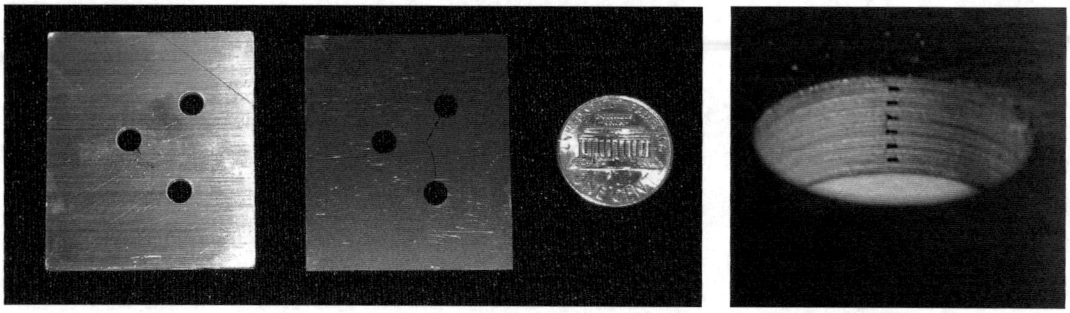

(a) (b) (c) (d)

Figure 1 Pictures of (a) the numbering-up device created by diffusion-bonds, (b) the slit formed by photolithography, (c) one cent coin, and (d) microchannel inlets in the inlet through-hole.

Fig.1 shows pictures of the device fabricated in the present study. The curved slit with width is 112 microns, the curvature radius is 10 mm and the arc angle of 30 degrees is formed by photoetching both sides of an SUS304 plate with a thickness of 80 microns as shown in Fig.1 (b). The SUS304 plates with and without the slits are stacked in turn and diffusion-bonded as shown in Fig. 1 (a), where the top plate without the slit is seen. As a result, the new device has six built-in curved microchannels with a bifurcation at one end (fabricated by Dai Nippon Printing, Co., Ltd.). Fig. 1 (d) shows the position of the six inlets of the microchannels from a photograph taken of the inlet through-hole.

Experiments and analysis

In order to prepare the feed emulsion, a small amount of vegetable oil is dispersed in deionized water with surfactant using an ultrasonic homogenizer. The density of the oil is 930 kg/m^3 at 30 °C. The intensity of the homogenizer is adjusted so that the numerical size distribution (NSD) of oil droplets falls into the range of 0.5 to 100 microns. The emulsion is fed into the inlet and discharged from the outlets of either the outer or the inner bifurcation branches. The device and sample containers are immersed in a constant temperature bath kept at 30 °C. The emulsions collected from the inner and outer exits are weighed to determine the ratio of flow rates in the inner and the outer branches (Q_{in} and Q_{out}) as well as the flow rate in the curved main channel (Q_{feed}). The absorbance of those emulsions is also measured to evaluate the oil concentrations (C_{feed}, C_{in} and C_{out}) in terms of the number density. In this study, it is assumed that the total number density of oil droplet is proportional to the absorbance regardless of the polydispersity of the NSD. The NSD, viz., frequency fraction of oil droplet with diameter of d_p ($f_{feed}(d_p)$, $f_{in}(d_p)$ and $f_{out}(d_p)$) are measured by a particle size analyzer. Subsequently, the grade purity coefficient at the droplet diameter of d_p ($G_{in}(d_p)$) is given as;

$$G_{in}(d_p)=\frac{C_{feed}f_{feed}(d_p)-C_{in}f_{in}(d_p)}{C_{feed}f_{feed}(d_p)} \qquad (1)$$

The $G_{in}(d_p)$ becomes unity when the oil droplet with diameter of d_p cannot be found in the emulsion collected from the inner branch. The smallest droplet size with GPC of unity, viz., 100% efficiency, is defined as d_{100}. When the number concentration of oil droplets with diameter of d_p in the inner branch is same as that of the feed, the $G_{in}(d_p)$ is zero, but it is not necessarily a minimum. This is because the breakage and coalescence of oil droplets due to high shear rates in the microchannel could give minus values of GPC by increasing the concentration in the inner branch compared to that in the feed. On the other hand, the RGE ($R_c(d_p)$) can be calculated as;

$$R_c(d_p)=\frac{Q_{out}C_{out}f_{out}(d_p)/\{Q_{in}C_{in}f_{in}(d_p)+Q_{out}C_{out}f_{out}(d_p)\}-Q_{out}/(Q_{in}+Q_{out})}{1-Q_{out}/(Q_{in}+Q_{out})} \qquad (2)$$

The NSD of the emulsion could vary considerably during its passing through the arc due to the breakage and coalescence aforementioned. Therefore, the amounts related to feed in the equation are substituted by the sum of amounts collected from inner and outer branches as shown in Eq. (2). The R_c given here implies the recovery of oil droplets with a particular size reduced by water recovery, from just before the bifurcation point to the outer branch. The R_c becomes unity when all droplets with a particular size report to the outer exit while it takes a minus value when the droplets are concentrated into the emulsion in the inner branch rather than in the outer branch. This value is calculated for the performance comparison with the original metallic micro-separator /classifier with the double size of the present miniaturized device (Ookawara et al., 2005).

Results and discussion

Fig. 2 shows a typical result of the grade purity coefficient $G_{in}(d_p)$ and the reduced grade efficiency $R_c(d_p)$ at Re = 510 (De = 35). Reynolds number Re and Dean number De are defined as;

$$Re = \frac{\rho U D_h}{\mu} \tag{3}$$

$$De = Re\sqrt{\frac{D_h}{2R}} \tag{4}$$

where μ and ρ is viscosity and density of water, U is the mean velocity in the main channel, D_h and R is equivalent diameter and radius of the arc channel, respectively. The weight concentration of oil and surfactant in feed are 1 wt% and 0.5 wt%, respectively. The feed pressure required for the present condition is about 200 kPa.

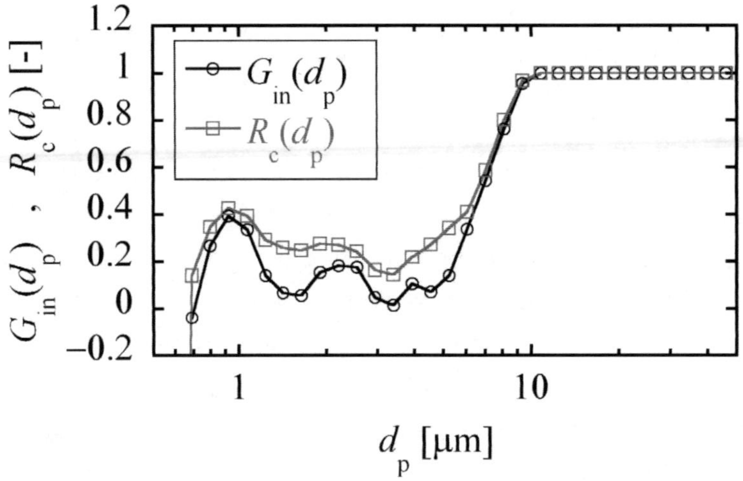

Figure 2 The grade purity coefficient $G_{in}(d_p)$ and the reduce grade efficiency $R_c(d_p)$ at Re = 510 (De = 35).

As shown in Fig. 2, both efficiencies take nearly the same values in the diameter range beyond 6 microns and the efficiency range beyond 0.4. Therefore, it is possible to compare the performance indices, such as the smallest oil droplet size with 100 % efficiency, the size with 80 % and 50% efficiencies between devices and conditions in terms of either GPC or RGE. At first, it should be noted that the size with 100 % efficiency is just about 10 microns, which is one tenth or even half of that achievable by hydrocyclones (Bennett and Williams, 2004).

The original micro-separator/classifier with the double size gave 13 and 19 microns as the cut size with 80 % efficiency in terms of RGE at *De* = 45 for resin particles with density 1190 kg/m^3 by the arc angle of 180 and 30 degrees, respectively (Ookawara et al., 2005). Based on the work of Cheng (1976), the pressure loss of water in the arc section with 180 and 30 degrees was estimated as 198 and 33 kPa, respectively. The feed pressure is probably larger due to the additional pressure loss in the straight inlet section and branches. On the other hand, the Kelsall type hydrocyclone requires 276 kPa as the feed pressure to separate 21 microns of similar resin particle with 80 % efficiency (Kelsall, 1953). In spite of small arc angle of 30 degrees, a smaller density difference and a smaller *De* (*Re*) number, the present miniaturized device gives much smaller cut size with 80 % efficiency such as 8 microns. Further, the pressure loss in the arc with 30 degrees of the present miniaturized device is comparable to that in the arc with 180 degrees of the original device with the double dimensions. Therefore, the miniaturized device could be a choice where the priority is given to the cut size. In case the desired cut size is not so small, on the other hand, the original device with arc 30 degrees is a better choice due to the considerably smaller pressure loss.

Further, it should be noticed that the lighter oil phase is concentrated into the outer branch in the present study as well as the denser solid particles as shown in the previous study (Ookawara et al., 2005). This implies that the separation mechanism of the micro-separator/classifier is not due to the centrifugal force enhanced by the miniaturization, but due to the lift force caused by the steep shear rates in the microchannel (Ookawara et al., 2006). Therefore, there is a high possibility that the lift force could concentrate the dispersed phase into the higher velocity region, which is shifted to the outer wall due to the centrifugal force and mostly flow into the outer branch, regardless of either lighter or denser density compared with the continuous phase. In addition to the separation mechanism for the lighter phase, it is of further interest to examine whether a completely neutrally-buoyant particle or droplet could be separated.

It is usual that the GPC and the RGE curves for solid particles monotonously converge to zero with the decreasing diameter. As a result, the efficiency curve for solid particle often becomes a simple S-shaped curve. However, the trend of the GPC and the RGE curves obtained in this study are considerably different in the small diameter range. It fluctuates and even takes a minus value for the smallest diameter of emulsified droplets. The minus value of GPC happens when the oil concentration in the inner branch is larger than that in the feed as aforementioned. It is quite reasonable that the breakage of oil droplets occurs and generates smaller droplets because the shear rate in the curved microchannel can be estimated as 3.8×10^5 1/s. Therefore, it is of interest for a population balance approach to be applicable in order to explain this trend. Before reaching a conclusion, however, the trend should be carefully investigated in future, associated with the interpretation of the RGE trend.

The pressure loss of 200 kPa in the miniaturized device to operate a given emulsion seems to be higher than that expected for water. This could be because the deformation of the oil phase and its breakage require additional energy. The weight concentration of oil is rather small and the relative viscosity of the emulsion is expected to be practically unity for normal flow conditions. The extremely high shear rate could cause the augmentation of pressure loss. This issue also will be addressed based on the additional experiments in future.

Conclusions

In the present study, a new micro-separator/classifier miniaturized compared with the original device (Ookawara et al., 2005) was applied to oil-water separation. The miniaturization apparently brought about the performance intensification. As a result, the smallest oil droplet size with 100 % efficiency became one tenth of the commonly suggested size for hydrocyclone (Bennett et al., 2004). It was confirmed that numbering-up device of 6 layers worked well due to the structure suitable for the methodology. The concentration of the lighter phase into the outer branch as well as of denser solid particle implied that the separation mechanism was due to the lift force induced by the steep shear rate rather than the centrifugal force. Further works to be addressed in future are suggested, such as fluctuations of the GPC and the RGE curves in the smaller diameter range and the additional pressure loss, both of which could be explained by the deformation and breakage of oil droplets.

Acknowledgements

This research was partially supported by a Grant-in-Aid for Scientific Research (A) (No. 17206079) from the Japan Society for the Promotion of Science (JSPS). The authors are most grateful to Dr. David Street for valuable advice on the present study.

References

Bennett, M. A., and Williams, R. A., 2004. Monitoring the operation of an oil/water separator using impedance tomography. Minerals Engineering, 17, 605-614.

Cheng, K. C., Lin, R. C., Ou, J. W., 1976. Fully developed laminar flow in curved rectangular channels. Transactions of the ASME. J. Fluids Eng. March, 41-48.

Grady, S. A., Wesson, G. D., Abdullab M. A., Kalu, E. E., 2003. Prediction of 10-mm hydrocyclone separation efficiency using computational fluid dynamics. Filtration & Separation, 40, 41-46.

Kelsall, D. F., 1953. A further study of the hydraulic cyclone. Chemical Engineering Science. 2, 254-272.

Ookawara, S., Higashi, R., Street, D., Ogawa, K., 2003, 2004. Feasibility study on concentration of slurry and classification of contained particles by microchannel. Book of Abstracts of 7th International Conference on Microreaction Technology, Lausanne, Switzerland, 231-233, (2003), Chemical Engineering J. 101, 171-178 (2004).

Ookawara, S., Oozeki, N., Ogawa, K., 2005. Experimental Benchmark of a Metallic Micro-Separator/Classifier Compared with Representative Hydrocyclone. IMRET8 in 2005 AIChE Spring Meeting, Atlanta, USA, 129g.

Ookawara, S., Street, D., Ogawa, K., 2006. Numerical Study on Development of Particle Concentration Profiles in a Curved Microchannel. Chemical Engineering Science, in press.

Mixing Enhancement by Internal Circulation Flow Using Liquid-Liquid Slug Flow in Microreactors

*Wiroon Tanthapanichakoon, Kazuo Matsuyama, Nobuaki Aoki and Kazuhiro Mae**
Dept. of Chemical Engineering, Kyoto University, Kyoto 615-8510, Japan.
**E-mail: kaz@cheme.kyoto-u.ac.jp*

Abstract

Mixing characteristics inside a microfluidic liquid slug using the Computational Fluid Dynamics (CFD) simulations are reported. Slug-based microfluidics offers rapid mixing by internal circulation and transport with narrow residence time distribution, making it suitable for precise reaction and mixing operations. The simulation results show that the circulation flow occurs in the slug and stagnant regions exist in the front and tail parts and in the center of recirculation vortices. This circulation flow greatly reduces mixing time as compared with the mixing time when mixing is driven by only molecular diffusion. The results also provide insights into the influences of operating parameters on slug-based mixing rates. Based on the simulation results, the modified Peclet number, $Pe^* = U_s d_s^2/lD$, is proposed for designing mixing in liquid slugs. A novel method using Pe^* to estimate mixing rates and design liquid slugs to obtain desired mixing rates is discussed and confirmed experimentally. In conclusion, design of microfluidic liquid slug for measuring kinetics of fast reactions and mixing operations is made possible by realizing instant mixing and narrowed residence time distribution.

Keywords: Multiphase flow; Laminar flow; Simulation; Mixing; Internal circulation flow; Modified Peclet number, Dimensionless mixing rate

Introduction

Flow regime in microchannels is laminar flow, enabling precise control of liquid-liquid slug flow. Since slugs in the channel move like a periodic plug flow, narrow residence time distribution can be achieved precisely [1]. Internal circulation flow from the friction between the channel wall and fluid in the slug is expected to enhance the performance of mixing, which is mainly driven by molecular diffusion in laminar flow. If the mixing of the reactant in the slug is instant, we can apply this liquid-liquid slug flow to reaction kinetics measurements. From this viewpoint, we estimate quantitatively the internal mixing in the slug by Computational Fluid Dynamics (CFD) simulations. Based on the simulation results, the modified Peclet number, $Pe^* = U_s d_s^2/lD$, is proposed for designing mixing in liquid slugs. Furthermore, the experimental operation for designing slug to reach a desired mixing performance is also shown, and the validity of the use of Pe^* for estimating mixing rate is verified experimentally.

Simulation Methods

A commercial CFD code, Fluent 6.2.16, was used in the simulations. Each slug was modeled as either two-dimensional (2D) or three-dimensional (3D) single-phase flow domain. Fig. 1 illustrates an internal circulation flow and the settings of slug boundary conditions to generate the internal circulation flow in a 2D slug. To generate an internal circulation flow inside the slug, the slug traveling at velocity U_s in the positive x-direction can be simulated by moving the sidewalls in the negative direction with velocity $U_{wall} = -U_s$. We considered mixing of two reactants (A and B) that have an equal density and volume in the slug. The diffusivity D was set at 10^{-9} m^2/s or 10^{-8} m^2/s. The density and viscosity were set at 998.2 kg/m^3 and 0.001 Pa·s, respectively. Two initial arrangements of reactant segments, axial arrangement (Fig. 2a) and radial arrangement (Fig. 2b), were investigated. Moreover, influences of slug length l and diameter d_s were investigated.

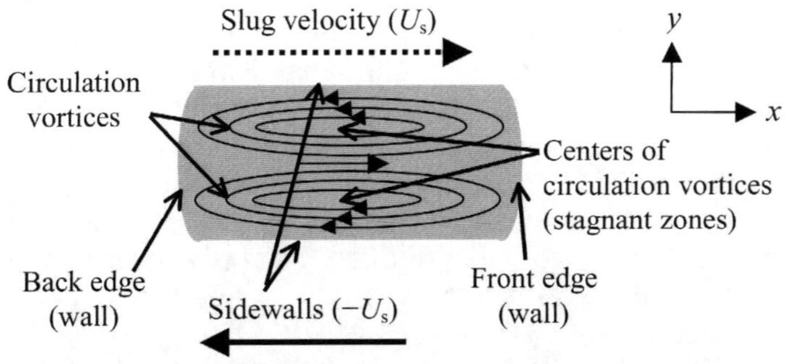

Fig. 1. A slug as a single-phase flow domain where internal circulating flow was generated by moving the slug sidewall in the negative *x*-direction.

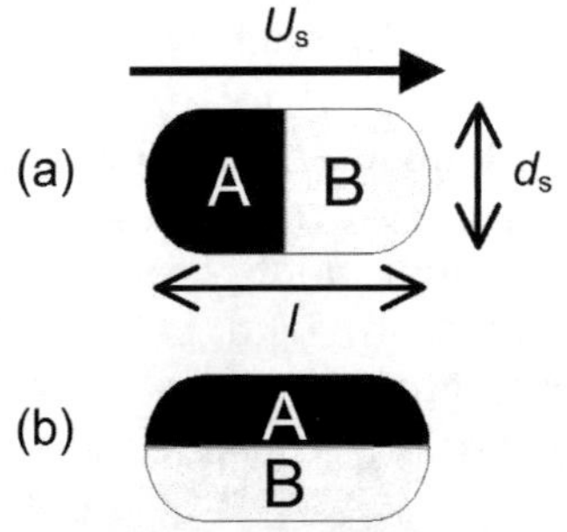

Fig. 2. Initial arrangement of reactants in slug: (a) Axial arrangement, (b) Radial arrangement.

Evaluating method of mixing rates

We evaluated mixing by tracing mass fraction deviation against flow time. Mass fraction deviation is an often-used indicator for evaluating mixing rate and extent of mixing [2,3]. Decreasing rate of the mass fraction deviation is used to indicate mixing rate, and mass fraction deviation δ at any time is compared with the initial mass fraction deviation δ_0 to obtain extent of mixing at that time.

Simulation Results and Discussion

Influence of initial arrangement of reactant segments inside a slug

We first discuss effects of the two initial arrangements of reactant segments shown in Fig. 2. The slug (l = 100 μm, d_s = 50 μm) traveling at 0 (=stationary), 1, 50, and 500 mm/s was used in the simulations. Fig. 3 shows relative mass fraction deviation δ/δ_0 against flow time t_m for initially axial and radial arrangements of reactants at various slug velocities. Mixing rates are much faster in moving slugs than stationary slugs, suggesting that the internal circulation flow is important in enhancing mixing rates. Accordingly, the operations using the internal circulation flow are shown effective for instant mixing. Though mixing rate increases greatly with U_s, the channel length (= slug moving distance) required for obtaining the same extent of mixing also increases greatly with U_s.

We next discuss differences between the two initial arrangements. For stationary slugs, the radial arrangement gives faster mixing rate than the axial arrangement because the initial diffusion length between the reactants for the former arrangement is

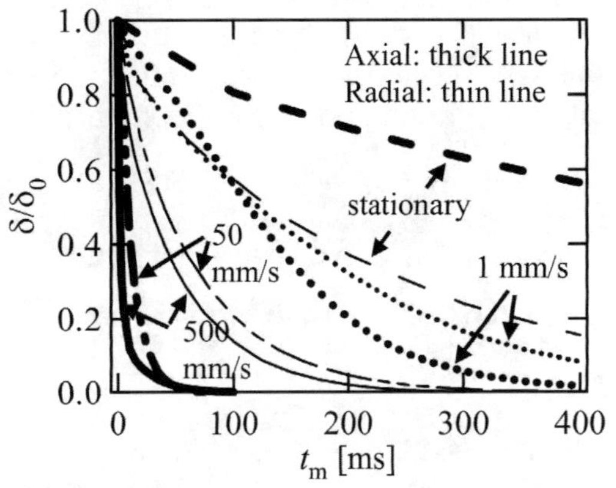

Fig. 3. Relative mass fraction deviations δ/δ_0 for radial and axial arrangements of reactants at various slug velocities.

shorter than that for the later. For moving slugs, the axial arrangement gives much faster mixing rate than the radial arrangement except for U_s = 1 mm/s, where mixing rate in the initial state for the radial arrangement exceeds that for the axial arrangement. Mixing rate depends on the direction in which shear stress is applied to the initial reactant segments inside the slug. Fig. 4a shows mass fraction distributions at various flow times for the axial arrangement. In this arrangement, the shear-direction is perpendicular to the interface of the initial reactant segments of initial reactant segments, enhancing the deformation of the initial reactant segments. The deformation helps create new concentration distributions in the radial direction by interlayering, resulting in shorter striation thickness and faster mixing rate. Fig. 4b shows mass fraction distributions at various flow times for the radial arrangement. In this arrangement, the shear-direction is parallel to the interface of the initial reactant segments, making it difficult to deform the initial reactant segments. With almost no deformation of the initial reactant segments, almost no concentration distributions are newly formed in the axial direction, leading to slow mixing rates.

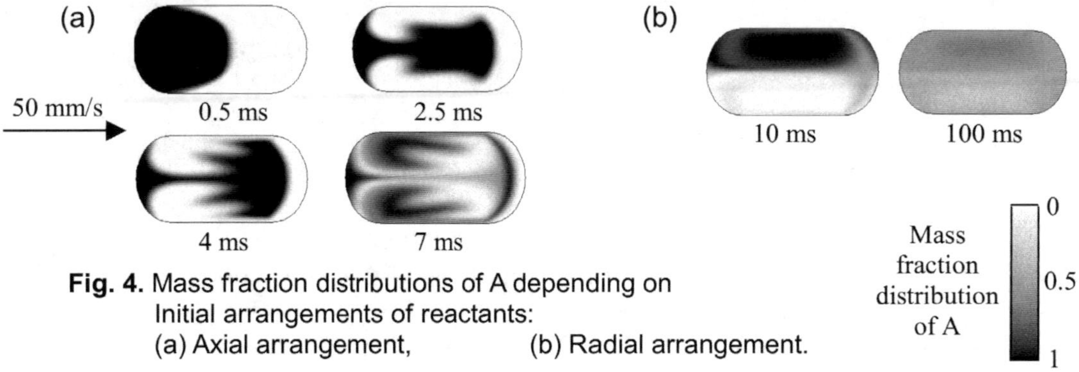

Fig. 4. Mass fraction distributions of A depending on Initial arrangements of reactants:
(a) Axial arrangement, (b) Radial arrangement.

Verification of the simulated circulation patterns in slugs by experiments

The circulation patterns in Fig. 4a and 4b at U_s = 50 mm/s are then compared with experimental results. An axial arrangement of reactants inside a slug was formed by generating two kinds of slugs (blue-dyed water and undyed water) flowing alternately in a microchannel (500 μm × 500 μm) and then forcing these two kinds of slugs to merge. Fig. 5a shows alternately flowing slugs before merging and circulation patterns in the slug soon after merging to form axial arrangement (U_s = 27 mm/s). A radial arrangement of reactants in a slug was formed by merging slugs at a junction of a cross mixer [4]. Fig. 5b shows the formation of a slug with radial arrangement and circulation patterns in the slug after leaving the junction (U_s = 40 mm/s). The circulation patterns in Fig. 4 at U_s = 50 mm/s agree qualitatively with those in Fig. 5 for each initial arrangement of reactants, showing that the simulation methods and assumptions are valid.

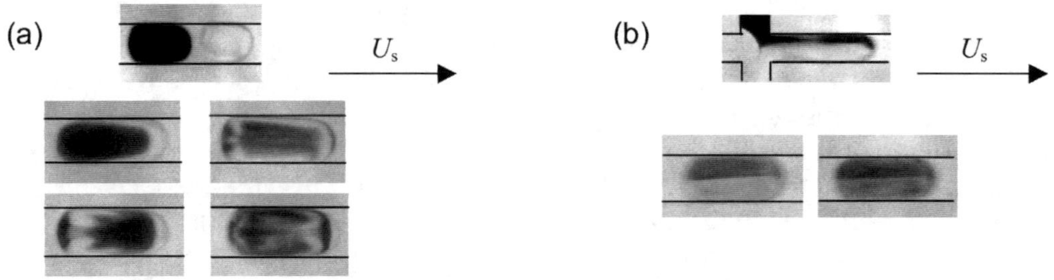

Fig. 5. Photos of distributions inside a slug from experiments:
(a) Axial arrangement (U_s= 27 mm/s), (b) Radial arrangement (U_s= 40 mm/s).

Pe based on two mixing regimes in slugs*

According to the circulation patterns showed in the previous section, two mixing regimes in slugs with internal circulation flow, which are the radial diffusion caused by the radial concentration distributions and the axial convection caused by the internal circulation flow, can be considered for estimating actual mixing rates. The diffusive timescale after new concentration distributions are formed in the radial direction is proportional to d_s^2/D, and the convective transport timescale is proportional to the time required to displace the slug by one slug length, l/U_s. Therefore, the modified Peclet number, $Pe^* = U_s\, d_s^2/\, lD$, which is the ratio of diffusive timescale to convective transport timescale in a slug with internal circulation flow, can be used to determine whether mixing is diffusion-dominated or convection-dominated. It is showed in the following sections that the mixing rates can be quantitatively estimated using Pe^* and that slug mixing can be designed based on D, l, d_s, and U_s.

*Estimation and design of mixing rates in slugs using Pe**

A method to estimate and design mixing rates using Pe^* ($= U_s\, d_s^2/\, lD$) is proposed. The case of axial arrangement is considered. Mixing rates are determined from δ/δ_0 curves plotted against flow time t_m. Fig. 6 shows typical δ/δ_0 curves for mixing with axial reactant arrangement (d_s = 50 μm, U_s = 50 mm/s). Mixing rate decreases as mixing proceeds because mixing rate is generally fast (steep slope) in the initial stage, but becomes much slower (gentle slope) in the final stage. The δ/δ_0 curves at small t_m (mixing rates before the curves lie down) are fitted by equation $y = -mx + 1$, and m is defined as mixing rate with a dimension of s^{-1}.

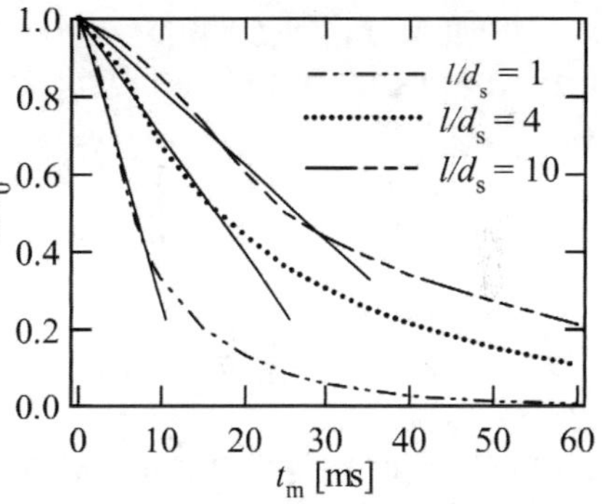

Fig. 6. Determination of mixing rates for axial arrangement by curve fitting (solid lines):
d_s = 50 μm, U_s = 50 mm/s.

The simulation results of mass fraction deviation δ at various l, d_s, and U_s were used to verify whether Pe^* can be used to estimate mixing rates for axial arrangement. Fig. 7a shows the relation between Pe^* and mixing rates, m. Mixing rates increase linearly with Pe^* for a double logarithmic plot when the slug diameter is fixed, but the lines differ greatly from each other when the slug diameter is different. However, since the mixing rate has the dimension of s^{-1}, it should to be changed into a dimensionless parameter, using a referential mixing rate.

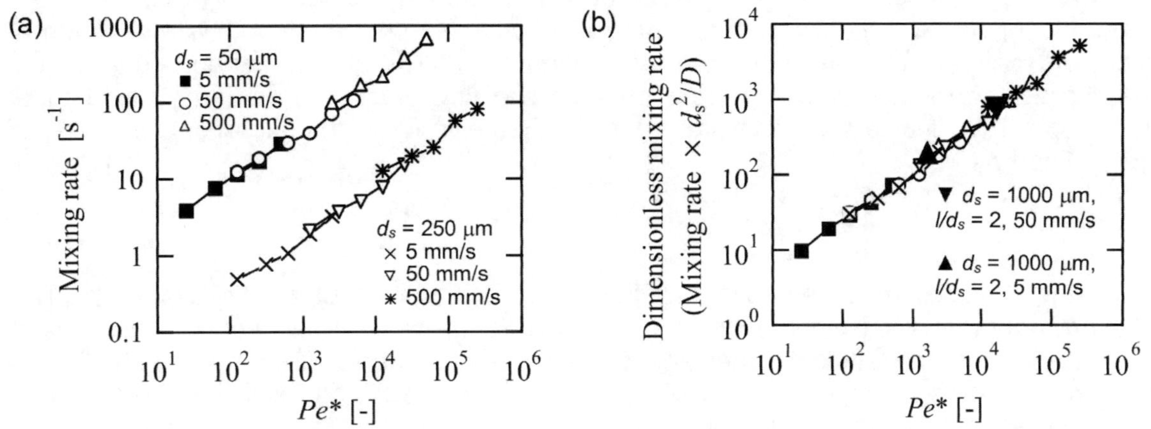

Fig. 7. Mixing rate estimation:
(a) Mixing rate vs. Pe^*, (b) Dimensionless mixing rate vs. Pe^*.

Thus, we can use the mixing rate divided by diffusion rate (= D/d_s^2) as reference. Fig. 7b shows the relation between Pe^* and dimensionless mixing rates ($m \times d_s^2/D$). The dimensionless mixing rates even for different slug diameters nearly superpose. Therefore, mixing rates in slugs of different slug diameters, lengths, and velocities can be quantitatively estimated using Pe^* and the relation showed in Fig. 7b.

Experimental validation of applying *Pe to estimate mixing rate**

For the purpose of validation of applying Pe^* to estimating mixing rate, we evaluated mixing performance of slug flow having the reactant arrangements shown in Fig. 2 using a competitive parallel reaction. We also compared mixing performances of flows with and without internal circulation. Moreover, mixing rates for the axial and radial arrangements are compared using Pe^*.

Evaluation of mixing performance by a competitive parallel reaction

Mixing performance can be evaluated by the following competitive parallel reactions (Dushman reaction). The Dushman reaction [5] consists of instantaneous neutralization and redox reactions:

$CH_3COO^- + H^+ \rightleftarrows CH_3COOH$ (instantaneous neutralization) (reaction 1),

$5I^- + IO_3^- + 6H^+ \rightleftarrows 3I_2 + 3H_2O$ (rapid redox reaction) (reaction 2).

The Dushman reaction occurs when mixing an HCl aqueous solution with an aqueous solution mixture of CH_3COONa, KI, and KIO_3. When the two solutions mix rapidly enough to produce a reaction-controlled condition, H^+ will be consumed only by reaction 1. However, when mixing is slow, reaction 2 also proceeds, resulting in I_2. Iodine then reacts with I^- to form I_3^-. Since I_3^- has a maximum UV absorption peak at 352 nm, the absorbance at λ = 352 nm increases with decreasing mixing rate. Therefore, we can evaluate mixing performance by this Dushman reaction by measuring the absorbance at 352 nm.

Experimental procedures

The concentration of the HCl solution was 0.138 mol/L. The concentration of the solution mixture of CH_3COONa, KI, and KIO_3 was [KI] = 0.032 mol/L, [CH_3COONa] = 1.33 mol/L, [KIO_3] = 0.0064 mol/L, and [CH_3COONa] = 1.33 mol/L.

A micro cross-mixer (I.D. 500 μm × 500 μm, made of PDMS) was used in the experiments. Enhancement of mixing performance by internal circulation flow was confirmed by producing a parallel flow without internal circulation and a slug flow with internal circulation. Silicone oil was used as the continuous phase and the prepared aqueous solutions as the dispersed phase (slug phase). The flow was observed using an optical microscope. Total flow rate V_t and water fraction wf were varied in the experiments. V_t and wf were defined by

$$V_t = V_{w1} + V_{w2} + V_o,$$
$$wf = (V_{w1} + V_{w2})/V_t.$$

Fig. 8 shows the schemes to form the flows with and without internal circulation. The radial arrangement was formed by flowing the oil while merging the two aqueous streams at the cross junction and letting the merged aqueous streams snap off into a droplet as shown in Fig. 8a. Lengths of slugs increase with wf when V_t is fixed. The axial arrangement was formed by generating two slugs flowing alternately in a microchannel and then forcing these two slugs to merge as shown in Fig. 8b. The formation of alternately flowing slugs becomes possible by increasing Capillary number *Ca*. The parallel flow without internal circulation was produced by merging the two aqueous streams at the cross junction without flowing the oil (V_o = 0) as shown in Fig. 8c. Values of the absorbance at 352 nm (ABS) of the three flows were then measured.

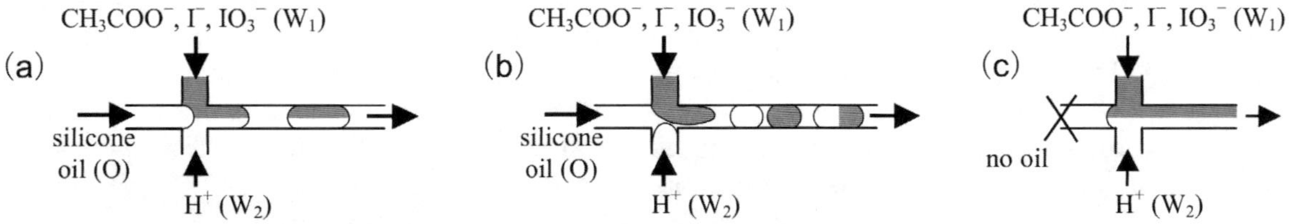

Fig. 8. Experimental schemes: (a) to form the radial arrangement, (b) to form the axial arrangement, (c) to form the parallel flow without internal circulation.

Confirmation of mixing enhancement effect by internal circulation flow

Fig. 9a shows ABS versus V_t for the three flows in Fig. 8. Mixing performance increases with decreasing ABS. Mixing is better when internal circulation occurs than when it does not. For the radial arrangement at fixed V_t, the short slug (low *wf*) gives better mixing than the long slugs (high *wf*). At V_t = 0.2 mL/min, The axial arrangement mixing gives the best mixing performance. However, mixing performances of these two arrangements cannot be compared simply by ABS because of the difference in slug lengths between the axial and radial arrangements. Thus, the mixing performances are instead compared using *Pe**. Fig. 9b shows 1/ABS versus *Pe** for the axial and radial arrangements. Since ABS is proportional to a concentration of I_3^-, 1/ABS increases with mixing rate. At the same value of *Pe**, the axial arrangement gives faster mixing rate than the radial arrangement. Furthermore, for the radial arrangement, 1/ABS and *Pe** show a linear relation on a double logarithmic plot. This agrees qualitatively with the simulation results in Fig. 7b when d_s and D are constant. Therefore, the experimental results give support to the suggestions by simulation, which are the effects of internal circulation flow, the effects of initial arrangement of reactants, and the estimation mixing rates using *Pe**.

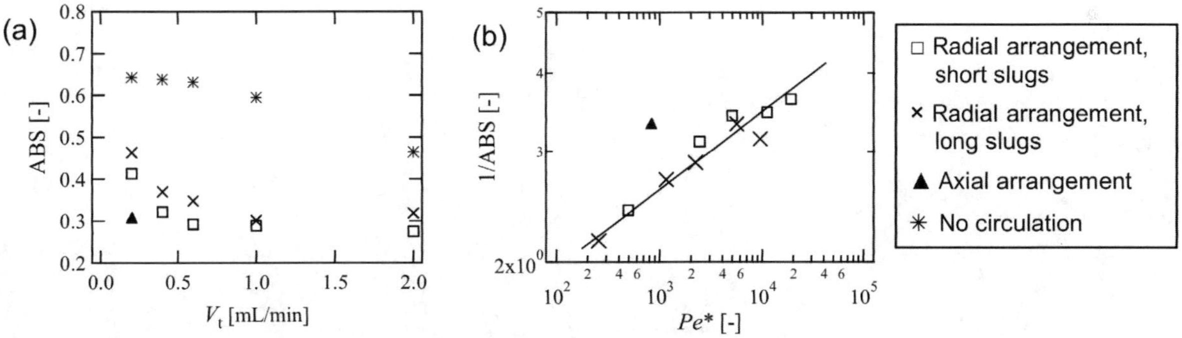

Fig. 9. Experimental results: (a) ABS vs. V_t, (b) 1/ABS vs. *Pe**.

Conclusion

Computational Fluid Dynamics (CFD) simulations were used to investigate mixing characteristics in microfluidic liquid slugs. The results show that the internal circulation flow in a slug is important in enhancing mixing rates and that slug traveling velocity and initial arrangement of reactants inside a slug are important parameters affecting mixing. Mixing performance improves with increasing slug traveling velocity. Axial arrangement gives much faster mixing rate than radial arrangement. To estimate and design mixing rates in a slug, we then proposed a new dimensionless number Pe^*, which can be used to determine whether mixing is diffusion-dominated or convection-dominated. According to the simulation results for axial arrangement at various l, d_s, and U_s, the quantitative relations between mixing rates and Pe^* were established. Moreover, in experimental investigations, the effects of internal circulation flow and initial arrangement of reactants have been confirmed, and the validity of the use of Pe^* for estimating mixing rate has been qualitatively verified. A relation between mixing rate and Pe^*, thus, allows us to design reactant arrangement and sizes of microfluidic liquid slugs for measuring kinetics of fast reactions and mixing operations by realizing instant mixing and narrowed residence time distribution.

References

1. Song, H., Tice J.D., Ismagilov, R.F. (2003) "A Microfluidic System for Controlling Reaction Networks in Time," *Angew. Chem. Int. Ed.*, 42, 768–772.
2. MacInnes, J.M., Chen, Z., Allen, R.W.K. (2005) "Investigation of alternate-flow mixing in microchannels," *Chem. Eng. Sci.*, 60, 3453−3467.
3. Xia, H.M., Wan, S.Y.M., Chew, Y.T. (2005) "Chaotic micromixers using two-layer crossing channels to exhibit fast mixing at low Reynolds numbers," *Lab Chip*, 5, 748−755.
4. Zheng, B., Tice, J.D., Ismagilov, R.F. (2004) "Formation of droplets of alternating composition in microfluidic channels and applications to indexing of concentrations in droplet-based assays," *Anal. Chem.*, 76, 4977−4982.
5. Ehrfeld, W., Golbig, K., Hessel, V., Lowe, H., Richter, T. (1999) "Characterization of Mixing in Micromixiers by a Test Reaction: Single Mixing Units and Mixer Arrays," *Ind. Eng. Chem. Res.*, 38, 1075−1082.

140g Microfluidic Networks for Complex Emulsions : from Single Active Elements to Highly Parallelised Systems

Georgios Tetradis-Meris, Corporate Research, Unilever, Colworth House, Sharnbrook, bedford, United Kingdom

Microfluidics systems have been demonstrated to allow precision control over engineered of complex particulate at micron length scale. Examples include complex (multilayered) encapsulates with precise drop-in-drop structure for protection and controlled release of actives. For many of Unilever's foods and HPC products, this precision control is important for desired product properties including both sensorial and functional. A challenge remains to produce in enough sample quantity to characterise. One way to produce these quantities is to parallelise microfluidic channels. An important step to understand how to parallelise microfluidics devices to form highly controlled micro structures is to fully comprehend the behaviour of a single microdevice and the manifold characteristics when running several microdevices in parallel. A bench scale rig prototype was designed and built in-house for the purpose to investigate the numbering up of microfluidics devices. This manifold rig is able to fully characterise and/or control the process conditions of this systems (e.g. pressure, droplet size, flowrate), which includes micro pressure sensors, optical sensors and accurate feeding system. We measured the resistances for each microdevice for both phases, and correlated the data with the existing models, for both Newtonian and Non-Newtonian liquids. Further investigation was carried out to understand the behaviour of running microdevices in parallel. The validation of existing models was achieved. Scale Up rules have been derived by quantifying the process parameters, fluid characteristics, and microdevice geometry, dimension & tolerances. Bench-top systems have been designed using the knowledge of the manifold rig to demonstrate the feasibility of a simple micro-fluidic emulsifier utilising massive parallelism. The demonstrator will be capable of producing samples of dispersion assemblies such as o/w and/or w/o emulsions with controlled dispersity. Results of the demonstrator will be shown in the conference.

Micro Reaction Technology As A Means Of Chemical Process Intensification

S. Becht, H. Hahn, Degussa AG, Hanau

Biography

HENRIK HAHN, born in 1968. Diploma in Process Engineering, University of Hannover, Germany, 1993. Ph.D. in Fluid Dynamics, University of Essen, Germany, 1999. 7 years industrial experience in process technology and cost controlling. Head of Multi-Purpose Pilot Plant Fluid Processing, 2001-2004. Project Manager for a strategic program in the field of process intensification, 2003-2004. Director Project House Process Intensification, since 01.01.2005. Member of the Executive Board of DECHEMA's Subject Division "Process Intensification", founded 07.09.2005
+49-6181-59-6137, henrik.hahn@degussa.com

SIMON BECHT, born in 1974. Since January 2005, Simon Becht has been responsible for projects in the area of micro process engineering and has been project manager for the POLYTREND Color Infusion Plus™ technology in Degussa's Process Intensification Project House. Member of executive committee of DECHEMA's Industrial Panel Micro Process Engineering – MicroChemTec and of the European IMPULSE initiative. While studying mechanical engineering at the University of Kaiserslautern, Becht worked as a student trainee at British Aerospace Systems and at DaimlerChrysler AG. After completing his studies, he began working for Degussa AG in the Process Technology & Engineering Service Unit in 2001. He is also completing his MBA at the University of Pforzheim while employed at Degussa
+49 6181 59-6374, simon.becht@degussa.com

Introduction

In addition to chemical synthesis, the apparatus technology and plant engineering used for reaction and product processing ultimately define the efficiency of chemical production processes. The defining factor is how successful chemists and engineers are in bringing starting materials and products to the right place at the right time and in controlling the heat and mass transport.

Competitive forces in the specialty chemicals sector today and tomorrow

Rethinking chemical processes in the above mentioned way becomes an issue of constantly growing importance as in many cases the traditional way of process optimization as well as the related management methods seem to have reached boundaries. Raising the last percent points of plant or process efficiency is in the speciality chemicals sector quite often prohibited.

Restrictions are either based on technical matters or are simply defined by an unproportional increase in the required expenses in plant assets as well as required R&D efforts.
However, apart from the limitations we discover within the organisation we also face an ever faster changing business environment. Regions with considerably lower labour costs as well as overall manufacturing costs are entering our markets with in many cases comparable products for the same customers.
Furthermore, we encounter a trend toward shortening product live cycles and of commoditisation even in the speciality chemicals sector. The competitive forces with special respect to the risk of being substituted are therefore very high.

From a strategic point of view this situation leaves a worldwide well established manufacturer of chemical intermediates with not too many options. Normally only two of them remain in the mid- to long-term perspective. Depending on the current position of the organisation, being defined by the specific competitive advantage as well as the existing product portfolio and the market attractiveness, two ways are available.
Based on an analysis of the internal and external environment the organization either redraws from the market due to economic considerations or considerable investments are made for sustainable technical and economic development.

If the investment option is the preferred one – at least in this case redrawing is not an option – basically two strategies are available in order to cope with the competitive situation on the international marketplace. Either the organisation decides for a 'differentiation strategy' which is normally combined with a premium price for a premium product or it decides for the 'cost-leadership strategy' with the latter option being chosen more frequently. A very broad scope of business activities will be maintained in both cases but the strategies will differ in the range of product differentiation.

However, both strategies require investments in order to achieve sustainable development. This is especially true for the 'cost-leadership strategy' since this goes at least for the established chemical manufacturers from Western Europe and the NAFTA region hand in hand with technology-leadership.

Process intensification – technologies and methods

Process intensification in this context is regarded as a basic approach to revolutionize both process and plant design, where novel chemical technologies and synthesis routes play the leading role.

For example, it is possible to realise chemical processes and plants with less investment and operational costs via a corresponding "technology push" such as through modular engineering and micro process technology. It is our understanding at Degussa that PI is not just limited to new apparatus technologies but includes also aspects and methods of process development and plant design.

Many investigations in the past decade evaluated prospects and limitations of micro reactors for the production of chemicals. Various advantages of micro reaction technology have been identified and demonstrated on the laboratory scale. Those potentials include higher space time yield (i.e. smaller reactor), higher selectivities (i.e. higher feedstock efficiency), full conversions (i.e. no recycle stream), or safety benefits (i.e. safe operation with explosive mixtures). PI from a purely technological perspective addresses all these aspects. Intensification of chemical processes is, using a very generic definition, the intensification of mass and heat transfer which in turn is used for minimization of limitations due to the reaction kinetics.

Combining the possible increases in process efficiency and safety with the flexibility of e.g. a modular plant design will – at least in theory – lead to considerable reductions in the time duration between the conception and the start-up of a plant. Time to market in combination with flexibility in terms of location and production capacity will be key competitive advantages for our future business environment. However, there are still considerable obstacles to a brought acceptance.

Obstacles to implementation

It has to be taken into account that there are at least two obstacles to a fast and glorious introduction of the PI idea.

- Firstly, PI in terms of technology and managing methodology is still considered as being new and unpredictable in its results. Experience is still very limited.
- Secondly, PI based on new, e.g. micro process technologies must be compared and evaluated on a strict "accounting basis" with alternative conventional technologies and in combination with different economic environments.

PI is not necessarily limited to the use of new technologies but can in many cases apply available conventional apparatuses and process methods. Only if a robust advantage can be derived from employing the new management approach and new technologies this combination will result in a positive perception and implementation of the PI idea.

It is the perception of people working in production as well as in R&D departments of major chemical manufacturers and engineering companies that micro process technology is still lacking market acceptance, references and guidelines. Even though it obviously has gained considerable momentum in the R&D areas of the chemical industry during the last decade, only a limited number of industrial applications are known so far.
This might be due to two apparent entrance barriers:

- Firstly, in many cases novel micro reactor technology solutions for existing production processes have to compete with already depreciated plants. This significantly affects the overall economic attractiveness.

- Secondly, the multiscale concept, referring to the interplay of micro reactors with conventional plant technology, is an issue which is still not fully resolved.

Degussa's approach to process intensification

In order to overcome the current status of "being in test", Degussa pursues new process technologies like micro reaction technology as a major but not the only means for intensification of chemical processes. Process intensification hereby is considered as a comprehensive management approach resulting in a new or at least modified, highly sustainable production environment for the existing and future product portfolio.

The presentation will deal with Degussa's experience in and perception of PI. Several examples will give an idea of the chosen approach which is and will be developed by the Degussa project house "Process Intensification" during a three years period.
The focus of the presentation is first of all on applications of micro process technology in the R&D area where high aspect ratios are used in order to minimize mass or heat transfer limitations.

It will be illustrated that a prerequisite for intensifying chemical processes is that not only a singular process step is considered but the complete process design. This in many cases includes not only aspects of improved mass or heat transfer but focuses on more superior issues like the adjustment of the overall process logistics and process design. A transfer from batch to semi-batch or even to continuous operation might be required in order to design the process according to the reaction and not the other way round.
This aspect is carried on by the multiscale process design approach. Being able to design a process based on multiscale process units will ensure the company's ability to meet future requirements for highly flexible, highly efficient plant concepts. However, such aspects and concepts will only be supported if a corresponding feasibility study at least reaches a development stage where the future plant design, either a 'greenfield investment' or a retrofit of an existing plant, is assessable e.g. by using NPV or DCF analysis methods. Only such objective figures will leave us with the opportunity to compare different scenarios for production. In any case the feasibility study will encompass the whole value chain starting from chemical synthesis and progressing until the plant design is known (basic or detail engineering).

Examples used in the presentation are regarded as a contribution to reduce the barrier between laboratory and industrial production:

Miniemulsion polymerization using a multiscale design approach

This project is part of the EU funded **IMPULSE** (Integrated Multiscale Process Units with Locally Structured Elements) project and aims at the implementation of a new, continuously operated production process for high quality polymer nanoparticles. Nanoparticles (in this case < 100 µm) are synthesized using technologies for emulsification, reaction and downstream processing that are considered as being new to the polymer sector. Micro process technology is expected to permit isothermal conditions during the polymerization step as the intensive heat release of the reaction would otherwise lead to uncontrolled polymerization and plugging.

Microreaction technology for industrial gasphase reactions

This project focuses on the development and commercial implementation of microstructured reactors for highly exothermal / endothermal gas phase reactions. In addition to the reactor itself highly active catalysts are developed in order to fully exploit the thermal capabilities (well controllable isothermal reaction conditions) of the reactor. The identified range of applications for such reactor concepts are world scale plants for bulk building blocks.

Ozonolysis of alkenes in microstructured reactors for highly selective synthesis

This project is part of the public funded project **µ.Pro.Chem** (Micro process technology for process intensification in chemical production) of the German BMBF. A full feasibility study is conducted starting with the analysis of the business environment, continuing with the development of the chemical synthesis steps and ending with the complete process implemented in a pilot plant set up.

Apart from those synthesis or product oriented examples of applications of micro process technology the modular engineering approach will be addressed.

From a technological point of view modular engineering is regarded as a generic approach which should be applicable for many processes, no matter whether micro process technology is used or not. However, it faces considerable economic constraints which must be taken into account.

The economic decision for an investment in world scale production capacity for huge chemical products is always based on a detailed analysis of the expected market development. It is a fact that today the time span between the economic decision and a 'plant on stream' is in most cases measured in years. This is due to an often very high complexity of the process and the time required for basic and detail engineering, procurement, construction and start of operation. In order to be in the market 'on time' the economic decision is often made before the market is duly assessed.

The uncertainty that is aligned with this approach can be considerably reduced with a modular engineering concept for chemical plants.

Such modular concepts are already known from the pharmaceuticals industry and are successfully applied in order to considerably reduce the time between clinical tests, legal approvals and start of production. Patent protection is another very important issue in the pharmaceuticals business since in many cases the patents are already close to expiry when production starts.

An aspect which is of considerable importance in the specialty chemicals business is that the modular approach facilitates the concept of 'growing plant capacity'. Due to the macro- and microeconomic analysis of the company's capabilities as well as of the business environment it might be preferable to start with smaller production capacities. Only when the market develops over time the capacity is adjusted simply by adding identical modules ('numbering up'). From the capital costs perspective this might even allow for a higher absolute investment compared with a conventional plant design.

180b Chemical Synthesis in Micro Reactors

P. Watts, University of Hull, Hull, United Kingdom

The miniaturisation of chemical reactors offer many fundamental and practical advantages of relevance to the pharmaceutical industry, who are constantly searching for controllable, information rich, high throughput, environmentally friendly methods of producing products with a high degree of chemical selectivity. Micro reactors used for chemical synthesis usually consist of a network of micron-sized channels etched into a solid substrate. For solution-based chemistry, the channel networks are connected to a series of reservoirs containing chemical reagents to form the complete device with overall dimensions of a few cm. Reagents can be brought together in a specific sequence, mixed and allowed to react for a specified time in a controlled region of the channel network using either electrokinetic or hydrodynamic pumping.

A concerted effort has now begun to establish the benefits that micro reactors can bring to the field of synthetic chemistry. For example, the ability to manipulate reagent concentrations in both space and time within the channel network of a micro reactor provides an additional level of reaction control which is not attainable in bulk stirred reactors where concentrations are generally uniform. Consistent with this notion, many single-step reactions have been demonstrated to show altered reactivity, product yield and selectivity when performed in micro reactors as compared with conventional bench top glassware. A selection of such reactions will be reviewed within this presentation.

Micro reactor chemistry is currently showing great promise as a novel method on which to build new chemical technology and processes. Reactions performed within a micro reactor invariably generate relatively pure products in high yield in short periods of time and in sufficient quantities to perform full instrumental characterisation. One of the immediate and obvious applications is therefore in combinatorial chemistry and drug discovery, where the generation of compounds with either different reagents or under variable conditions is an essential factor. In order to prepare complex products, of the type required by pharmaceutical companies, it is necessary to conduct multi-step reactions. In order to achieve such results it is important to be able to manufacture and use reagents in situ, examples of such reactions will be discussed.

The use of solid supported reagents adds even greater diversity to the range of reactions that may be achieved within such systems. It will be demonstrated that the dimensions of reactors may be increased in size while maintaining the classic advantages associated with miniaturisation. In such systems significant quantities of analytically pure compound may be prepared without additional purification.

180c Process Optimization of a Catalyzed Bleach Oxidation for the Production of Functionalized Aldehydes Using Microreaction Technology

Stefan Loebbecke[1], Wolfgang Ferstl[1], Maud Schwarzer[1], Elke Fritz-Langhals[2], and Juergen Stohrer[2]. (1) Fraunhofer Institut für Chemische Technologie ICT, Joseph-von-Fraunhofer-Strasse 7, Pfinztal, D-76327, Germany, (2) Consortium für Elektrochemische Industrie GmbH, a subsidiary of Wacker-Chemie GmbH, Munich, D-81379, Germany
Microreaction technology was used to investigate the continuous oxidation of 2-butoxyethanol to form 2-butoxyacetaldehyde. Major objective was a total conversion at high yields by suppressing the unwanted by-product butoxyacetic acid. The process is a liquid-liquid reaction with an oxidant based on bleach and bromide in the presence of catalytic amounts of 4-hydroxy TEMPO. The well-defined isothermal conditions in a microreaction system help to obtain information about necessary requirements for a continuous macroscopic system. For optimization, different mixing principles were examined. A variation of residence time unit in combination with a subsequent continuous quench gave a good estimation of the macroscopic reaction time within such a system. The parameter study also included investigations about the influence of temperature and stoichiometry.

180d High-P,T Micro-Reactor Processing for the Aqueous Kolbe-Schmitt Synthesis of Hydroquinone and Phloroglucinol

Volker Hessel[1], Christian Hofmann[2], Patrick Loeb[2], Holger Loewe[2], and Maria Parals[2]. (1) Institut für Mikrotechnik Mainz GmbH, Carl-Zeiss-Straße 18-20, Mainz, D-55129, Germany, (2) Chemical Process Technology, Institut für Mikrotechnik Mainz GmbH, Carl-Zeiss-Straße 18-20, Mainz, D-55129, Germany

Recently, the high-p,T (high pressure, high temperature) concept was introduced as one original, tailored micro-reactor operation, among some others which were specified as well [1]. The first validation was done for the Kolbe-Schmitt synthesis using resorcinol with water as solvent to give 2,4-dihydroxy benzoic acid [2,3]. Later, a detailed study on the Michael addition with alkylic amines and alpha, beta unsaturated compounds followed [4]. Further, some brief investigations on the radical side-chain bromination of toluene and m-nitro toluene were given [5].

By the high-p,T concept, liquid-phase reactions can be processed under so far mainly unknown conditions much above the boiling point of the process solvent, typically >200°C or more [1]. In this way, even formerly sluggish reactions can be speeded up so that the correspondingly increased heat and mass transfer issues demand for the use of microstructured reactors. The operational range of the Kolbe-Schmitt synthesis using resorcinol was so extended by about 120°C towards 220°C, as compared to the standard batch protocol under reflux conditions (100°C) [2,3]. Among other kinds of process intensification, a 500-times increase in reaction rate compared to the batch processing at comparable yield was found.

In this paper, the high-p,T Kolbe-Schmitt synthesis of phenolic substrates with position isomerism relative to and more hydroxy groups than resorcinol was investigated.

The aqueous Kolbe-Schmitt synthesis of the isomer 1,4-dihydroxy benzene (hydroquinone) gave only very low yields <1% of the 2,5-dihydroxy-benzoic acid. This is explained by a favoured oxidative route to benzoquinone. This is line with the missing aqueous-based batch process counterpart which is only known under non-aqueous high-pressure conditions. Thus, the high-p,T process function here is only to speed up routes, but it cannot enable routes which are not practicable by conventional chemistry.

Detailed investigations were made for the Kolbe-Schmitt synthesis of 1,3,5-trihydroxy benzene (phloroglucinol) to the 2,4,6-trihydroxy benzoic acid which turned out to be successful as continuous micro-reactor variant.

The synthesis needed to be done at lower phenolic content (27 g/l; 61.7 mmol) as compared to the original resorcinol protocol (89 g/l; 203 mmol), but using the same carbonate-to-phenolic excess ratio of about 4. This was done to avoid the solids' deposition during the process and the analytical problems encountered, due to the experiences stemming from the resorcinol synthesis. Indeed, reproducibility and repeatibility of the experiments were much better now, and more consisted results could be obtained.

Yields were determined for a wide temperature range at 80, 100, 110, 120, 140, 160, 180, 200, 210, 220, and 240°C. These experiments showed two main trends. The reaction time to reach maximum yield decreases considerably with temperature. It takes 750 s at 110°C, 140 s at 160°C, and 50 s at 200°C. Up to 200°C, the yields remain fairly constant for reaction times still longer than needed for maximum yield within the range investigated (typically up to 20 min). Decomposition and product generation seem to be in balance here. From there on, a decrease of the yield is observed, which is most pronounced at 240°C.

The second trend observed was that the degree of the maximum yield is strongly dependent on temperature. While at 80°C only 15% yield is achieved, the is given at 120°C with 55% yield. At 220°C, the best trade-off between T and maximum yield falls down to again only 16%. Thus, the phloroglucinol synthesis can be undergone at much lower temperatures (120°C; 465 s; 55% yield) as compared to the resorcinol one (180°C; 65 s; 43% yield) which is owing to the higher substrate reactivity, i.e. the phloroglucinol core is more electron-rich. However, longer residence times by more than one order of magnitude are needed, which reduces the increase in space-time yield compared to laboratory-flask operation from 440 (resorcinol) to about 20 for phloroglucinol. The decrease in reaction time is as high as by a factor of 144 (at 180°C) as compared to batch synthesis (resorcinol: ~500). The highest phlorglucinol productivity was 12 g/h at (200°C; 1620 ml/h; 20 s) as compared to the 68 g/h manufacture of resorcinol at (200°C; 2000 ml/h; 16 s). This is mainly due to the lower concentration used and the steeper decrease in yield at the high temperatures.

Several side and follow-up products of the phloroglucinol synthesis such as the diverse i,j-dihydroxy benzoic acids and phenolic substrates were identified by GC analysis with standards. This allows for mechanistic speculations on the diverse reaction pathways which will be presented. It is planned to use calorimetric analysis of such selected species (e.g. DSC, TGA) for further evidence on the different kinds of decomposition.

[1] Hessel, V., Löb, P., Löwe, H.; "Development of reactors to enable chemistry rather than subduing chemistry around the reactor - potentials of microstructured reactors for organic synthesis", Curr. Org. Chem. 9, 8 (2005) 765-787.

[2] Hessel, V., Hofmann, C., Löb, P., Löhndorf, J., Löwe, H., Ziogas, A.; "Aqueous Kolbe-Schmitt synthesis using resorcinol in a microreactor laboratory rig under high-p,T conditions", Org. Proc. Res. Dev. 9, 4 (2005) 479-489.

[3] Hessel, V.; Hofmann, C.; Löb, P.; Löhndorf, J.; Löwe, H.; "Minimizing reaction times for the Kolbe-Schmitt synthesis

with resorcinol using high p,T-processing in a micro reactor setup", Chem. Eng. Trans. 6 (2005) 49-55. [4] Löwe, H., Hessel, V., Löb, P., Hubbard, S.; "Addition of secondary amines to alpha, beta unsaturated carbonyl compounds and nitriles by using microstructured reactors", Org. Proc. Res. Dev. (2005) submitted.

[5] Löb, P., Löwe, H., Hessel, V.; "Fluorinations, chlorinations and brominations of organic compounds in micro structured reactors", Journal of Fluorine Chemistry 125, 11 (2004) 1677-1694.

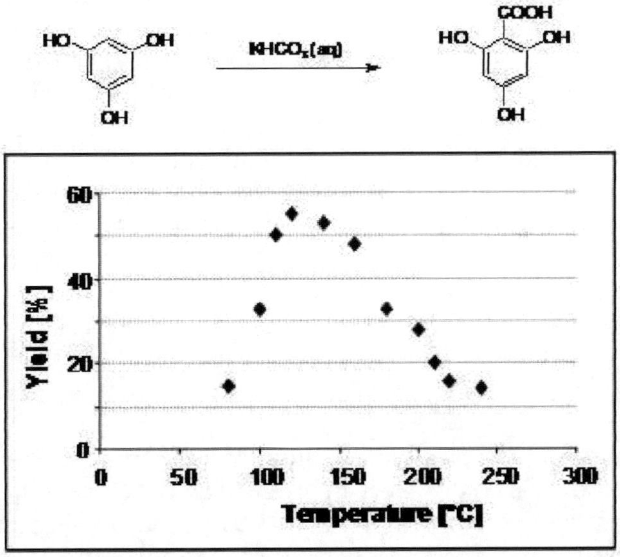

Catalyst-Trap Microreactor for Hydrogenation of a Pharmaceutical Intermediate

S. McGovern[1], H. Gadre[1], R.S. Besser[1], C.S. Pai[2], W. Mansfield[2], S. Pau[2]
[1]Chemical, BioMedical, & Materials Engineering Department, Stevens Institute of Technology, Castle Point on Hudson, Hoboken, NJ 07030, 201.216.5257, rbesser@stevens.edu
[2]New Jersey Nanotechnology Consortium (NJNC) at Lucent-Bell Labs, 600 Mountain Avenue, Murray Hill, NJ 07974

Abstract

A silicon microreactor has been developed to investigate gas-liquid-solid catalytic reactions. The reactor employs a three-channel "catalyst-trap" design, whereby solid catalyst is suspended in the liquid channel by an arrangement of posts. Such a device has advantages in that commercial catalysts are supported, and that pressure drop across the bed can be reduced by engineering the packing density. A model incorporating the transport and kinetic effects is developed to design the reactor. The reaction chosen is one in a family of reactions relevant to the pharmaceutical industry, the liquid-phase hydrogenation of o-nitroanisole to o-anisidine. The reaction is carried out across a range of gas and liquid flow rates that encompasses three distinct flow regimes, termed gas-dominated, liquid-dominated, and transitional. Experiments seek to assign a reaction conversion and selectivity to each point in this two-phase "flow map", then reconcile differences in performance with the characteristics of the respective flow regime. We observe the highest reaction conversion in the transitional flow regime, where competition between the two phases results in the generation of a large amount of gas-liquid interfacial area. The experimental conversion is greater than that predicted by the model, an effect attributable to the mass transfer enhancement induced by transitional flow. Reaction time within the microchannel is sufficiently small so that selectivity towards production of o-anisidine is nearly 100% across all flow regimes. In this work we relate our map of conversion to the flow behavior and reactor geometry, and we discuss steps for further exploring the mass transfer characteristics of the transitional flow regime. This reactor architecture may be useful for catalyst evaluation through rapid screening, or in large numbers as an alternative to macro-scale production reactors.

Introduction

The term "microreactor" is broadly used to describe devices ranging from tens of microns to several centimeters in size. Microreactor technology possesses significant advantages over conventional macro-scale reactors. Because of their small size, microreactors have inherently large surface-to-volume ratios, allowing for superior mass and heat transport. Surface-to-volume ratios of 20,000 m^2/m^3 or more are not uncommon, compared to 1,000 m^2/m^3 for a conventional reactor. In the case of catalytic reactions, where competition exists between the rate of diffusion to the catalyst sites and the rate of reaction, microreactors are able to virtually eliminate mass transport resistance, making them an extremely useful tool for isolating reaction kinetics. Excellent heat transfer properties ensure a uniform temperature throughout the reactor and prevent the formation of hotspots in the case of an exothermic reaction. In a stirred-tank reactor, for example, hotspots can lead to undesired changes in local concentration or pH. Low residence time and ease of heat removal also make microreactors more suitable for flammable service, where the potential for explosion or fire is greatly reduced.

In this paper, we consider the use of a microreactor for gas-liquid-solid catalytic reactions, where the rate of diffusion of gas into the liquid phase can also limit the observed reaction rate. Thus, we would like to be able to address both intra-particle and inter-particle diffusion. The former can be accomplished with reasonable certainty simply by using small-sized catalyst particles. In general, catalyst particles with a diameter smaller than 100 μm (1 x 10^{-04} m) exhibit sufficiently low transport resistance to enable their use with most reactions [9]. The latter is a function of gas-liquid interfacial area and driving force for transport, both of which depend on the gas-liquid flow regime.

For our purposes, three traditional flow regimes are adequate to describe the two-phase behavior, although other papers present more precise delineations [7]. Bubble flow is characterized by liquid as the continuous phase, with bubbles of gas dispersed into a fully-wetted channel. As the ratio of gas-to-liquid velocity increases, the Taylor (slug) flow regime emerges, in which alternating segments of gas and liquid occupy the channel. The relative lengths of each segment are constant for a given set of inlet conditions. As gas-to-liquid velocity further increases, annular (for a capillary) or trickle (for a wide reaction channel) flow occurs. In annular or trickle flow, gas is the continuous phase, and the liquid flows as a thin film along the channel walls [7]. Figure 1 illustrates the three flow regimes, where u_G and u_L are the gas and liquid superficial velocities, respectively.

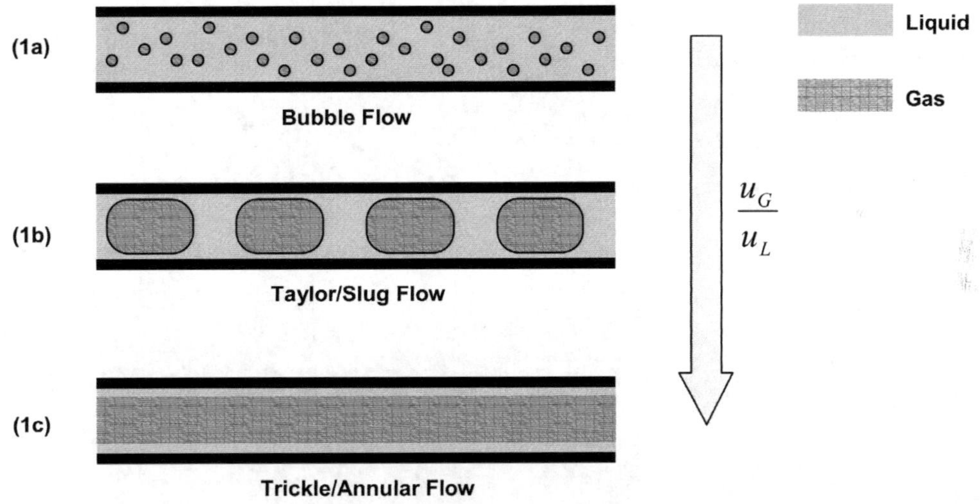

Figure 1- Illustration of Gas/Liquid Flow Regimes

In the context of a gas-liquid-solid catalytic reaction with reaction occurring in the liquid phase, two characteristics are critical in evaluating the advantages and disadvantages of a particular flow regime. First, we desire good mass transport between gas and liquid, which depends on the driving force for mass transport and the gas-liquid interfacial area created. Second, we desire a high liquid-solid interfacial area to effectively utilize the catalyst for rate of production. The highest liquid-solid interfacial area will be achieved when liquid is the continuous phase.

The slug flow regime in a capillary is typically associated with the best gas-liquid mass transport because of the no-slip condition at the channel walls. The high velocity gradient within the liquid slug provides turbulence, so a constant refreshing of the gas-liquid interface occurs. This provides a high driving force for diffusion at the interface, and good mixing within the liquid. Despite good mass transport properties, slug flow does not maximize the liquid-solid interfacial area because gas still occupies a significant portion of the channel. Bubble flow, in contrast, exhibits both a high liquid-solid and gas-liquid interfacial area. Because liquid is the continuous phase with only small pockets of gas, essentially all of the catalyst is in contact with liquid [6]. Thus we can generalize that slug flow will possess the best gas-liquid mass transport, owing to both interfacial area and driving force, and that bubble flow, although less turbulent, will have the highest liquid-solid interfacial area. On the basis of these mass transfer arguments, we speculate that the best reactor performance will fall in either the bubble or slug flow regimes. Likewise, we suppose that trickle flow, with the lowest liquid-solid and gas-liquid interfacial areas, will be the least effective for this particular reaction. We proceed to design the device and subsequent experiments with this hypothesis in mind.

Reactor Modeling, Design, and Fabrication

The reactor is designed for operation across the spectrum of flow regimes, while making use of the advantages of microchemical systems to relieve various transport resistances. The design is pictured below in Figure 2.

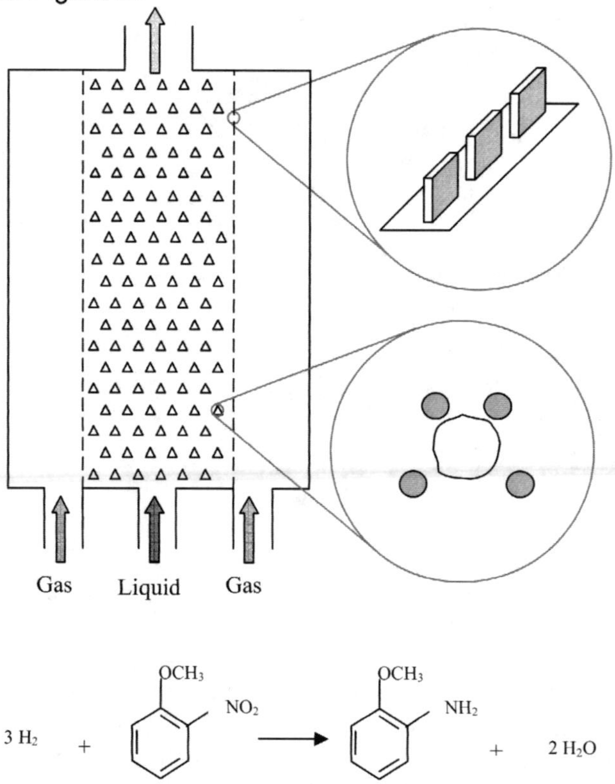

Figure 2- Proposed Reactor Design

The reactor will be used for the liquid-phase hydrogenation of o-nitroanisole to o-anisidine, with methanol as an inert solvent. The catalyst is 5% Pd by weight on carbon. Hydrogen gas enters along the two outside channels, and is allowed to diffuse into the liquid channel through a slotted wall, shown at the top inset. The reaction occurs in the liquid channel, where an arrangement of posts, or catalyst traps, holds the catalyst particles in place (lower inset). Each trap is a trapezoidal arrangement of four posts spanning the depth of the channel. The posts are spaced to “catch” catalyst particles and hold them in place, so one can imagine the liquid sees the particles stacked single-file on top of one another as it travels down the channel. The reaction is therefore a three-step process: (1) hydrogen must diffuse through the slots and dissolve in the bulk liquid; (2) the liquid must disperse to the catalyst particles and diffuse within the particles to the reaction sites; (3) reaction occurs at the catalyst sites.

To design the reactor, we model the bulk liquid-phase concentration of all species throughout the reactor. The liquid channel is treated as a plug flow reactor with a continuous source of hydrogen available at any axial position. The slots can be sufficiently sized to allow the gas to overcome liquid surface tension and pass into the liquid channel. Hydrogen concentration in the liquid phase is therefore limited by diffusion across the gas-liquid interface. The rate of change of hydrogen concentration has two contributing terms:

$$\frac{dc_A}{dt} = [\text{Rate of diffusion from gas into liquid}] + [\text{Rate of reaction}]$$

$$\frac{dc_A}{dz} = \frac{\varepsilon}{u_L} \cdot \left[k_L \left(\frac{P_A}{H_A} - c_A \right) a_h - \frac{3 w_c k c_A c_B}{(1 + K_A c_A)(1 + K_B c_B)} \right] \tag{1}$$

For the first term, the concentration of hydrogen at the gas-liquid interface is calculated from the Henry’s Law constant, H_A, for hydrogen in solutions of nitroanisole and methanol [2]. The difference between the interfacial and bulk liquid concentrations represents the driving force for diffusion. The parameter a_h is the gas-liquid interfacial area per unit volume, and is estimated from packed tower literature (strippers/absorbers) [10]. Uncertainty may exist in the application of these macroscale correlations to the microreactor geometry, but in the absence of alternatives we will use them for design purposes. The second term is the reaction rate expression [1]. For each remaining species, equation (1) will contain only the reaction term with the respective stoichiometric coefficient. The four expressions for rate of change of concentration are integrated numerically over the channel length from 0 to *z*. Figure 3 below shows the final liquid channel dimensions.

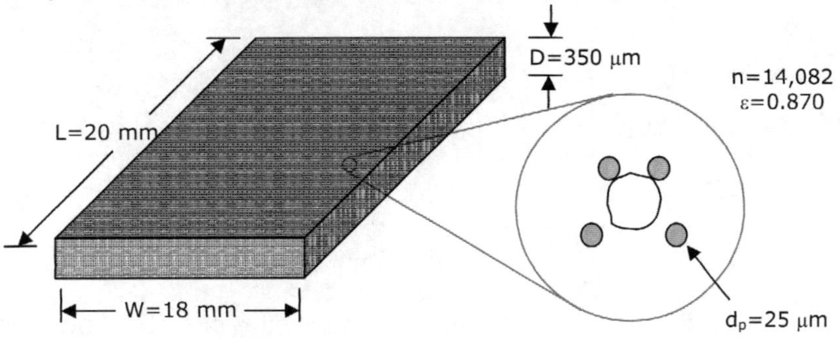

Figure 3- Design Parameters for Liquid Channel

The above dimensions are chosen to give an appreciable conversion across our target range of experimental conditions. The channel contains approximately 14,000 catalyst traps that, when loaded, give a void fraction of 0.870. The posts are 25 μm in diameter and are arranged as a trapezoid to hold a 40-50 μm particle. Traps are spaced 75 μm apart (closest edge-to-edge distance) to allow catalyst to pass through during loading.

The reactors are fabricated at NJNC-Bell Labs in Murray Hill, NJ using standard silicon photolithography techniques. Inlet and outlet ports are etched on the underside of the reactors. After etching, the silicon wafer is diced into individual chips. In our case, we received sixteen chips of dimensions 31 x 28 mm (0.031 x 0.028 m) per wafer. The reactors are then sealed by anodically bonding a piece of Pyrex glass over the surface of the chip. Anodic bonding applies voltage at high temperature, typically 1000 VDC and 450-490 C, to displace ions from the surface of the glass. The depletion of ions makes the glass surface highly reactive, forming a strong bond with the silicon substrate. The intent of the glass-covered reactor is two-fold, first as an aid in catalyst loading, and second for observation of the gas-liquid flow behavior.

Experimental

A schematic of the experimental setup is shown below in Figure 4. For overhead image and video capture, the setup includes a CCD camera above the reactor. The reactor itself is clamped onto a 4.5" x 3" stainless steel block, through which channels are machined for gas and liquid passage.

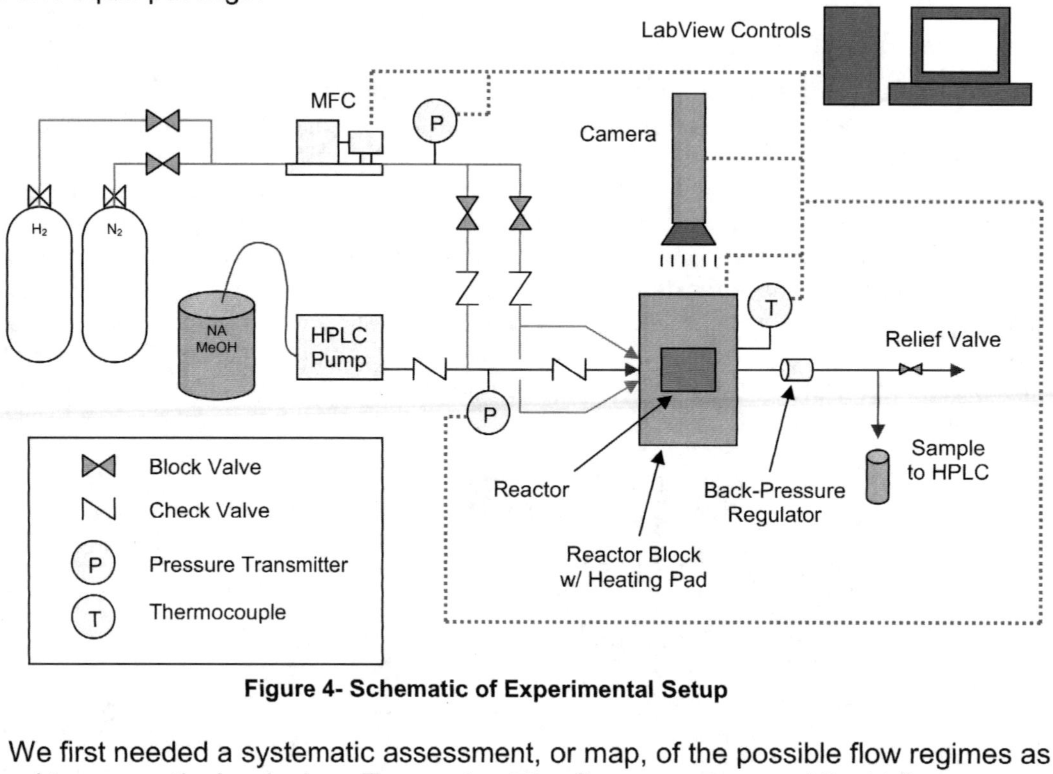

Figure 4- Schematic of Experimental Setup

We first needed a systematic assessment, or map, of the possible flow regimes as they pertained to our particular device. To construct the flow map, we used liquid flow rates ranging from 0.06 to 0.50 mL/min, and gas flow rates ranging from 2 to 20 sccm. We looked for differences in behavior with changing conditions and approximated borders between the three flow regimes. With the flow map established, we began measuring reaction conversion and selectivity at each point on the map. Catalyst was manually loaded by inverting the reactor and applying vacuum to draw the catalyst through the inlet port. Tapping the reactor against the countertop caused the catalyst to move down the channel and settle amongst the traps. Approximately 6 mg of catalyst were used for each run. For each set of conditions, we sampled the reactor outlet periodically over the course of approximately one hour. Samples were analyzed offline in an HPLC to measure the weight fractions of o-nitroanisole and o-anisidine, from which we calculated the conversion and selectivity. After each run, we regenerated catalyst and evaporated any residual liquid by heating to 300 C for ten minutes. Reactor conditions were 30 C and 25 psia for all experiments. The feed solution was approximately 16% nitroanisole by weight.

Results and Discussion

The reactor flow map is shown below in Figure 5:

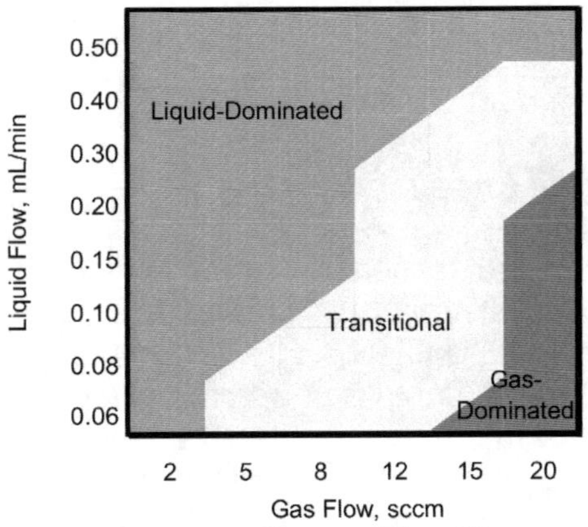

Figure 5- Reactor Flow Map

Because this particular microchannel is rectangular-shaped with a high width-to-depth ratio, it cannot be considered a capillary or cylindrical channel. Thus, the conventional nomenclature of bubble, slug, and trickle flow is not applicable to our geometry. Instead, "gas-dominated", "liquid-dominated", and "transitional" are more appropriate terms. Liquid- and gas-dominated flows are characterized by stable patterns in which the continuous phase impedes movement of the non-continuous phase throughout the channel. For example, in liquid-dominated flow, 80% or more of the channel volume is occupied by liquid, with several small pockets of gas. In gas-dominated flow, large gas pockets persist and force liquid to channel around them. Transitional flow is characterized by an unstable pattern in which gas and liquid compete for space within the channel. This competition causes a periodic refreshing of the gas-liquid pattern, so that in addition to significant interfacial area, high turbulence is exhibited within the channel. Pockets of gas, upon refreshing, often leave behind traps surrounded by a droplet of liquid. The small distance (~15 μm) between posts likely prevents gas from overcoming the liquid surface tension, and leaves behind "wetted" traps within the gas pocket, an effect we will discuss further when we present the reaction results. The presence of wetted traps is another indicator of transitional flow. It is important to note that the modified terminology is intended only to help visualize the correct flow behavior, and that liquid-dominated/transitional/gas-dominated are analogous to bubble/slug/trickle in the context of the mass transfer arguments presented in the Introduction section.

We obtained the conversion results in Figure 6 for the lower third of the flow map. Each chart shows the conversion plotted against time. Error bars represent plus/minus one standard deviation for each particular set of samples.

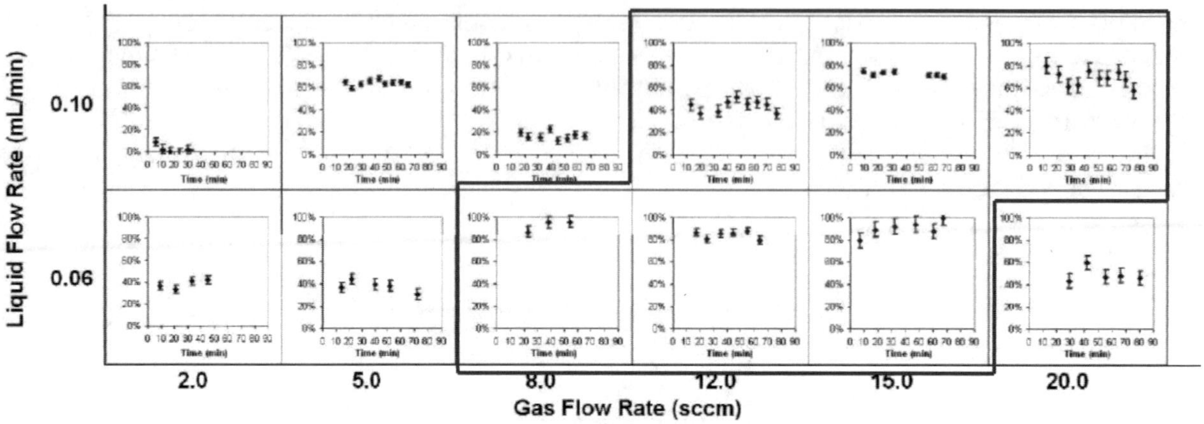

Figure 6- Conversion Results (30 C, 25 psia)

The highest conversion in Figure 6 occurs in the highlighted area, from approximately 8-15 sccm for 0.06 mL/min and 12-20 sccm for 0.10 mL/min. This range coincides with the transitional flow regime on Figure 5. Conditions outside the transitional region experienced significantly lower conversion. We also note that within a given flow regime, conversion is higher at 0.06 mL/min than at 0.10 mL/min, indicating the influence of liquid residence time, and giving some validity to the plug flow model. For all runs, the selectivity towards o-anisidine is nearly 100%, indicating that no side reactions occur. We can see from the data that over the period of approximately one hour, catalyst deactivation is negligible.

The conversion results are consistent with our premise that the best reactor performance would fall within the flow regime exhibiting either the best gas-liquid or liquid-solid mass transfer characteristics. In our case, this is transitional flow, with its continually refreshing pattern and high gas-liquid interfacial area. Though their mass transfer characteristics are similar, there is one outstanding difference between conventional slug flow and our observed transitional flow. In slug flow within a capillary, the refreshing occurs as a recirculation around the boundary of each slug. In general, slugs do not disturb the structure of adjacent slugs, and the pattern of alternating gas and liquid from one of the channel to the other is maintained. The refreshing in transitional flow affects the entire channel, as though the pattern is wiped clean and then redrawn in a different, usually random, way. Thus, it could be argued that the turbulence brought on by transitional flow enhances not only gas-liquid, but also liquid-solid mass transfer, making it clearly the preferred regime for the reaction.

The conversion predicted by the model ranges from 1.0 to 3.5% across our range of experimental conditions. The model does assume the reaction to be controlled by gas-liquid mass transfer, but even after adjusting the interfacial area parameter in equation (1), a_h, to large values, the experimental conversion is much higher. A possible explanation is the presence of wetted versus non-wetted catalyst traps. As we discussed above, when the gas pattern refreshes it often leaves behind traps encased in droplets of liquid. One can imagine that an array of several thousand traps covered by liquid droplets approximately 100 μm in diameter would create an extraordinary amount of gas-liquid interfacial area. If coupled with the proper refreshing gas-liquid pattern the traps would function as a network of mini-batch reactors, where refreshing periodically sweeps away the products and introduces new reactants. The droplets' small size would give very short diffusion distances for hydrogen, allowing for excellent transport of dissolved hydrogen throughout the liquid phase, and high reaction conversion. Figure 7 below shows a snapshot of the reactor in operation with areas of wetted and non-wetted traps.

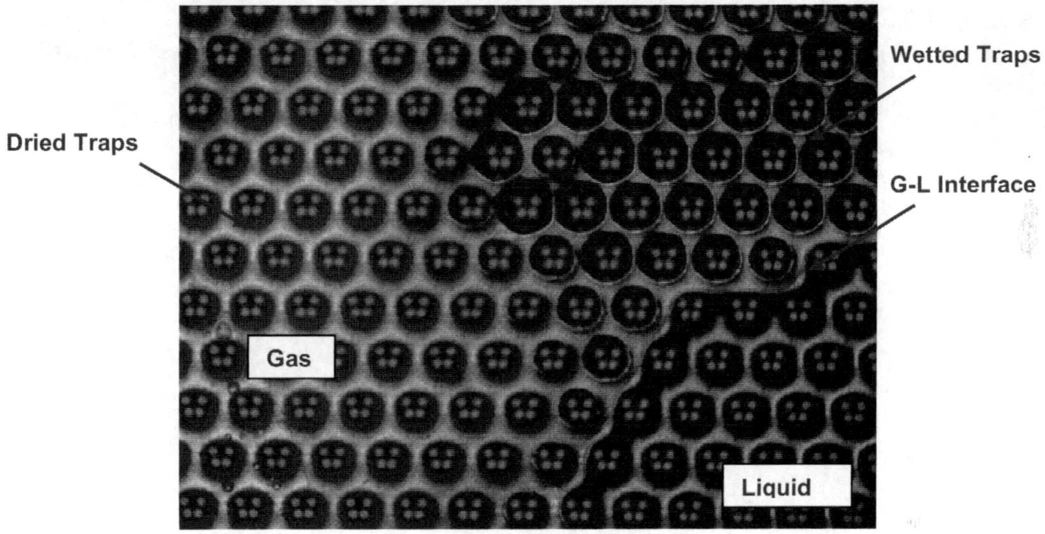

Figure 4- Reactor Photograph of Dried vs Wetted Traps

Conclusions and Future Work

After defining three gas-liquid flow regimes for the reactor, we began measuring conversion and selectivity at each point on the flow map. We found that the highest conversion falls within the transitional flow regime. The reaction is mass-transfer-controlled, as opposed to kinetically-controlled, and the turbulence induced by transitional flow provides the best gas-liquid and liquid-solid mass transfer. Experimental conversion is much higher than predicted by our kinetic model. The presence of wetted traps functioning as a network of small batch reactors is a possible cause for the discrepancy.

For future work, we should fill in conversion and selectivity for the remainder of the flow map. It is likely that 0.20-0.30 mL/min will be the maximum feasible liquid flow rate if the reaction becomes limited by residence time. As liquid flow rate is increased, the transitional flow regime will shift to the right in Figure 5, so the range of gas flow rates should be expanded to accommodate several gas-dominated points. The flow map should also be duplicated at a higher hydrogen partial pressure to observe the effect of hydrogen liquid-phase concentration. Finally, we can explore the effect of wetted traps by revising the model to include some prediction of interfacial area and residence time in a series of tiny batch reactors.

Literature Cited

1. R.V. Chaudhari, M.G. Parande, P.A. Ramachandran, and P.H. Brahme; I. ChemE Symposium Series No. 87.
2. P.H. Brahme, H.G. Vadgaonkar, P.S. Ozarde, and M.G. Parande; J. Chem. Eng. Data 1982, *27,* 461-462.
3. R.F. Guedes de Carvalho, A.M. Alves; AIChE Journal V. 45, Issue 12, Dec. 1999, 2495-2502.
4. A. Lawal, S. Tadepalli, R. Halder, H. Qiu, W.Y. Lee, R. Besser; AIChE Spring Meeting, Atlanta, GA, April 13, 2005.
5. E.H. Stitt; Chem. Eng. Journal 4025 (2002) 1-14.
6. J. Heiszwolf, M. Kreutzer, M. van den Eijnden, F. Kapteijn, J. Moulijn; Catalysis Today, 69 (2001) 51-55.
7. M. Kreutzer, F. Kapteijn, J. Moulijn, J. Heiszwolf; Chem. Eng. Science 60 (2005) 5895-5916.
8. M. Losey, R. Jackman, S. Firebaugh, M. Schmidt, K.F. Jensen; Journal of Microelectromachanical Systems, Vol. 11, No. 6, December 2002.
9. H.S. Fogler; *Elements of Chemical Reaction Engineering, 4th Ed.;* 2006
10. J.D. Seader and E.J. Henley; *Separation Process Principles;* 1998

Kinetic Study of Catalytic Hydrogenation of O-Nitroanisole to O-Anisidine in a Microchannel Reactor

Sunitha Tadepalli[1], Raghu Halder[1], Jale Muslehiddinoglu[2], Donald Kientzler[2], Adeniyi Lawal[1]
(1) Dept. of Chemical, Biomedical and Materials Engineering, Stevens Institute of Technology, New Jersey Center for Microchemical Systems, Castle Point, Hoboken, NJ 07030
(2) Process R&D, Bristol-Myers Squibb Company, One Squibb Drive, New Brunswick, NJ 08903

Abstract

Hydrogenation reactions are ubiquitous in the fine chemicals and pharmaceutical industries. The hydrogenation of o-nitroanisole to o-anisidine in methanol was selected as a model three-phase reaction for kinetics study in a packed-bed microchannel reactor using Pd metal supported on zeolite as a catalyst. The kinetics of three-phase hydrogenation reactions in conventional slurry or batch reactors is often limited by mass transfer of hydrogen through the liquid due to the limited solubility of hydrogen in organic substrates and solvents. These reactions are also highly exothermic, therefore adequate temperature control is of primary importance to reduce side reactions and to prevent thermal runaway conditions. The use of a microchannel reactor for such reactions provides improved mass and heat transfer rates which may ensure that the reaction operates close to intrinsic kinetics. In the present study, the rate data in the intrinsic kinetic regime were analyzed and found to be well represented by L-H type rate equations. Kinetic experiments were also conducted in a semi-batch reactor under similar reaction conditions as the microchannel reactor. The performance of the packed bed microchannel reactor was compared to that of the semi-batch reactor by evaluating the mass transfer resistance in each system.

1. Introduction

Hydrogenation reactions are versatile in the chemical industry accounting for 10-20% of all the reactions in the pharmaceutical industries. For example hydrogenation of nitro groups is important in the synthesis of drugs such as Viagra, Zyvox, and Agenerase [1]. In the pharmaceutical industry, most of the hydrogenation reactions are conducted in CSTR or semi-batch reactors utilizing large stirred tanks. These reactions are highly exothermic, with heats of reaction that exceed 500kJ/mole substrate. This heat is usually removed through a circulating cooling fluid operated by a centralized chiller which requires several hundred tons of refrigeration. This represents the biggest source of energy consumption in the slurry semi-batch process used for hydrogenation reactions in the pharmaceutical industry. Also, non-uniform temperature distribution in the batch reactors may lead to the formation of several byproduct(s) or intermediate(s) which can also react with one another, and further lead to violent decomposition of the substrate [2],[3] or thermal runaway conditions. Further, the reaction time in the batch reactors is much longer than kinetically needed because of the high heat and mass transfer resistances associated with the low specific surface area per unit volume in these reactors.

The use of a continuous microreactor for such reactions enables uniform temperature control throughout the reactor because of the high surface to volume ratio in microreactors. For most of the multiphase reactions, which are limited by mass and heat transfer effects, the microreactor offers improved overall reaction rate with higher yield and selectivity of the desired products when compared with conventional macroreactors. Further, the miniaturization of chemical reactors offers cleaner, safer and more energy efficient technology to the pharmaceutical industry which is constantly searching for safe, high-throughput production methods with reduction in energy consumption, liability risks (for certain products that may be extremely hazardous or potential targets of terrorism) and scale up issues that are often present in conventional reactors.

The aim of the present study is to evaluate the performance of the packed bed microreactor for the hydrogenation of o-nitroanisole to o-anisidine using Pd supported on zeolite catalyst. This study comprises the:

(1) Investigation of the kinetics of this reaction with detailed analysis of mass and heat transport effects at different operating conditions and development of rate equations that could be useful for the design of a multi-channel microreactor and,

(2) Comparison of the performance of the microreactor to that of the conventional semi-batch reactor.

2. Experimental

Figure 1 shows the experimental setup used for conducting the reaction. Two gas lines, one for hydrogen reactant (H_2) and the other for nitrogen (N_2) are shown in the system. The purpose of the nitrogen line is to flush the reactor system at the end of each experiment. The compressed gas flows through the Mass Flow Controller and is then mixed with the liquid stream using a tee connection. The liquid reactant solution (o-nitroanisole in methanol) is pumped using HPLC pump. The combined gas-liquid mixture then enters the reactor. The reactor is made of SS316L with an outer diameter of 1.5875mm and inner diameter of 0.775mm. The reactor is packed with catalyst and installed at the entrance and exit of the reactor are filters made of polypropylene wool to prevent the movement of the catalyst bed. The reactor is immersed in a constant temperature water bath to maintain isothermal conditions. The pressure values at the entrance and the exit are measured using pressure transducers (PI). From the reactor, the reaction mixture passes through a sampling loop where the product is collected for analysis. From the sampling loop, the reaction mixture passes through the back pressure regulator. The back pressure regulator is used to build up the desired pressure during an experimental run. From the back pressure regulator, the mixture is passed to a product receiver where the liquid is collected in a glass vessel and the gas phase is vented to the atmosphere. The product samples are analyzed using High Performance Liquid Chromatography (HPLC) from Shimadzu.

3. Results and discussion

3.1. Reaction Pathway

The catalytic hydrogenation of o-nitroanisole to o-anisidine involves a three-step reduction series reaction. First, the reduction of o-nitroanisole to 2-methoxynitrosobenzene, next to N-(2-methoxyphenyl) hydroxylamine, and, finally to o-anisidine. During the experiments only one intermediate was observed in the product stream along with the product, and reactant at all operating conditions. This intermediate was found to be 2-methoxynitrosobenzene by the Mass Spectrum Analysis. The presence of hydroxylamine derivative was not detected during analysis, indicating that it had only transient lifetime probably because of its instantaneous reaction with hydrogen to form o-anisidine, the rate of which far exceeds its formation from 2-methoxynitrosobenzene. Thus the hydrogenation of o-nitroanisole to o-anisidine was considered as a process consisting of two consecutive reactions, one leading to the formation of 2-methoxynitrosobenzene, and the other leading to the formation of o-anisidine as shown in fig. 2.

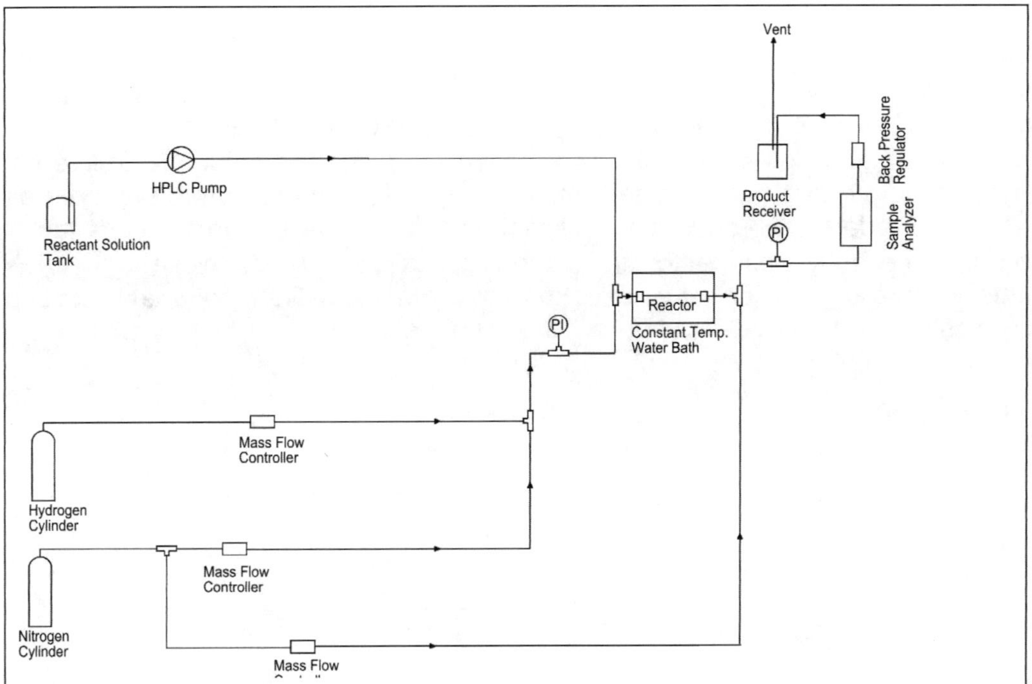

Fig. 1: Lab Set-up for hydrogenation of pharmaceutical intermediate

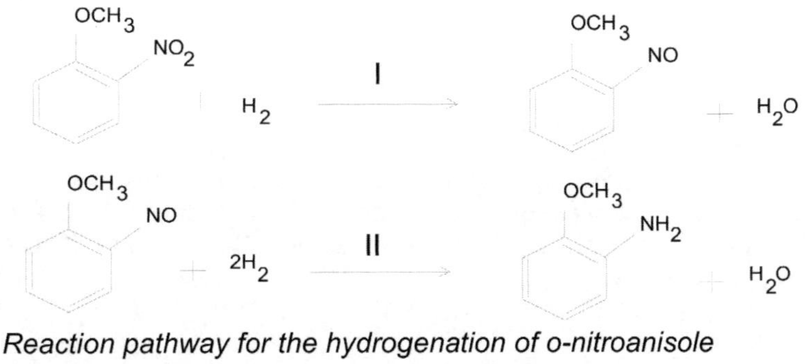

Fig. 2 Reaction pathway for the hydrogenation of o-nitroanisole

3.2. Kinetic Study

The main objective of the present study was to investigate the kinetics of the catalytic hydrogenation of o-nitroanisole to o-anisidine. This involves a number of competing mass, heat transfer and kinetic rate processes that contribute to the overall observed reaction rate. Optimization of any process requires that the contributing rate processes are understood individually, and that their impact on the total process rate is quantifiable.

3.2.1. Mass transfer limitations

The reaction is described as free from external mass transfer limitations or in order words operating in kinetic regime if the rates of all the external mass transfer processes mentioned above are higher than the rate of chemical reaction at the catalyst surface. If the effectiveness factor of the catalyst is close to unity, then internal mass transfer resistance is assumed negligible. To evaluate the possible mass transfer limitations, an experimental approach was used.

Internal Mass Transfer limitations:

The effect of particle size on the space-time yield (STY) and conversion at three different pressure levels and three different temperature levels are shown in figs. 3 & 4 respectively. From the data in figs. 3 & 4, it can be concluded that the influence of internal diffusion on the reaction rate can be neglected for the range of pressures and temperatures considered. To further validate this conclusion, calculation of the catalyst effectiveness factor η_{eff} was performed. The effectiveness factor under the reaction conditions is close to 1 ($\eta_{eff} \rightarrow 1$). Therefore, the calculations confirm the experimental data indicating that the hydrogen diffusion inside the catalyst pores does not affect the reaction rate.

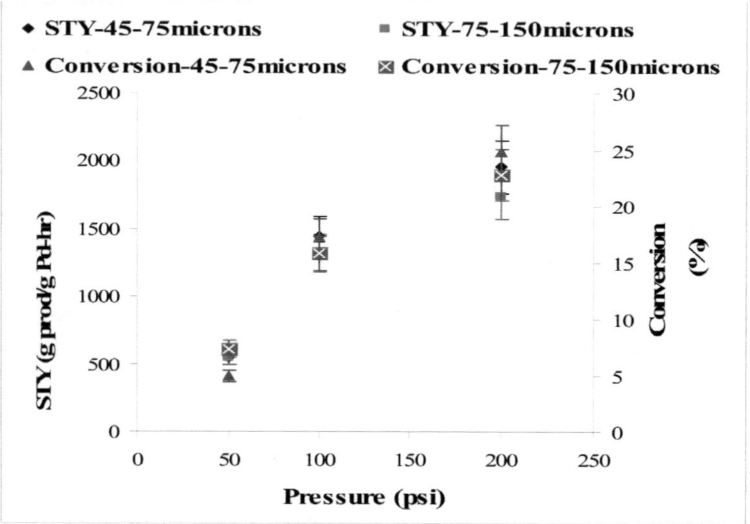

Fig. 3: Effect of particle size on STY and conversion at different pressures
(All runs were made with 23% (w/w) o-nitroanisole in methanol, liquid flow rate of 0.05 ml/min, H_2 flow rate = 5 sccm, 2.2 mg of 2% Pd/ Zeolite Catalyst and at 30°C)

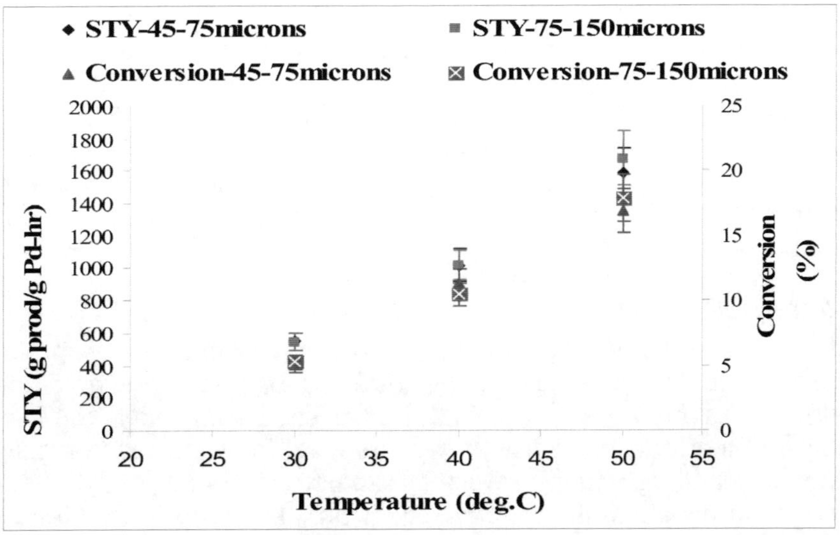

Fig. 4: Effect of particle size on STY and conversion at different temperature levels (All runs were made with 23% (w/w) o-nitroanisole in methanol, liquid flow rate of 0.05 ml/min, H_2 flow rate = 5 sccm, 2.2 mg of 2% Pd/Zeolite catalyst, at 50 psig pressure)

External Mass Transfer limitations:

Experiments were conducted to see the effect of overall flow velocity on the STY and determine the controlling regime in this gas-liquid-solid reaction. Figure 5 shows the results obtained from these experiments which were conducted at a constant residence time by varying the gas and liquid flow rates proportionately with the catalysts amounts so as to obtain constant residence time. The data clearly show that the reaction rate is independent of the flow velocity, which indicates the absence of bulk external mass transfer limitations for the reaction under the present operating conditions.

Therefore, the reaction is not mass transfer limited under the operating conditions utilized in the experiments.

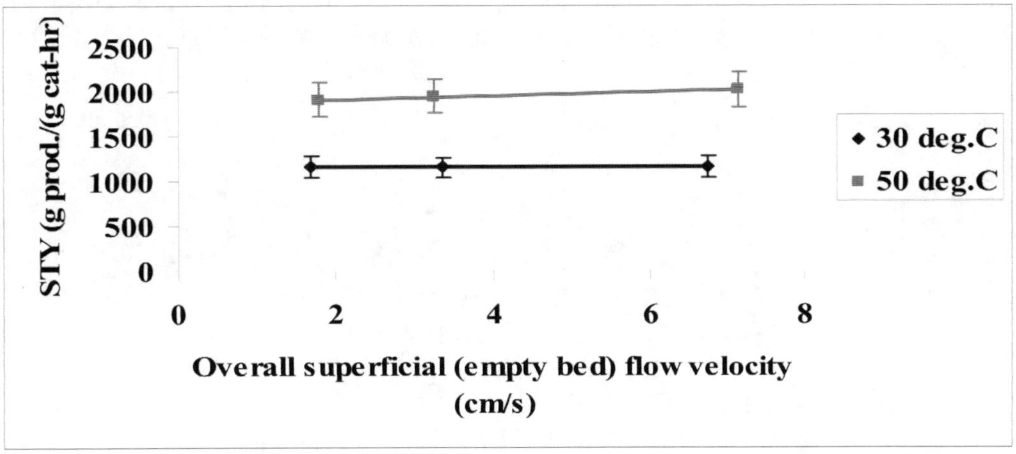

Fig. 5: Effect of overall flow velocity on reaction rate for the same residence time (All runs were made at reactor pressure of 100 psig and 23% (w/w) o-nitroanisole in methanol)

To further validate this conclusion, some calculations were made to compare the observed experimental rate with the maximum mass transfer rates using known correlations in the literature for mass transfer coefficients in fixed bed and trickle bed reactors. *It was observed that the actual mass transfer coefficient in the microreactor is much higher than the estimates obtained from these correlations.* Therefore, the calculations and the experimental data confirm that the reaction is not mass transfer limited under the operating conditions utilized in the experiments.

3.2.2. Heat transfer limitations

Apart from mass transfer limitations, heat transfer limitations can also have significant effect on the kinetics of the reaction. The intrusion of heat transfer resistances inside the reactor can lead to severe degradation in the catalyst's performance thus completely masking the true kinetics of the reaction. Efficient heat removal from the reactor is essential especially for hydrogenation reactions which are highly exothermic in nature. To ensure that the kinetic data obtained in an experimental reactor reflect only kinetic effects, Mears criteria [4-5] was used to determine the presence of heat transfer limitations in the microreactor. It was observed that heat transfer effects can be neglected in the microreactor

With all the above considerations, overall it can be concluded that under the present operating/processing conditions the reaction is controlled by surface kinetics and not significantly influenced by mass and heat transfer resistances.

3.2.3. Intrinsic kinetics

Experiments were conducted in the microreactor under different conditions to determine the kinetic rate expressions for both reactions. The conversion in the microreactor was limited to 10% so that a differential reactor can be assumed to find the dependence of the reaction rates on the reactant concentrations.

Dependence of rate of disappearance of o-nitroanisole on hydrogen concentration (or hydrogen pressure):

The effect of hydrogen pressure on the rate of disappearance of o-nitroanisole was investigated in a range of 50-300psig, at temperatures of 30, 40 and 50°C and o-nitroanisole concentration of 23% (w/w) in methanol. The results are presented in fig. 6.

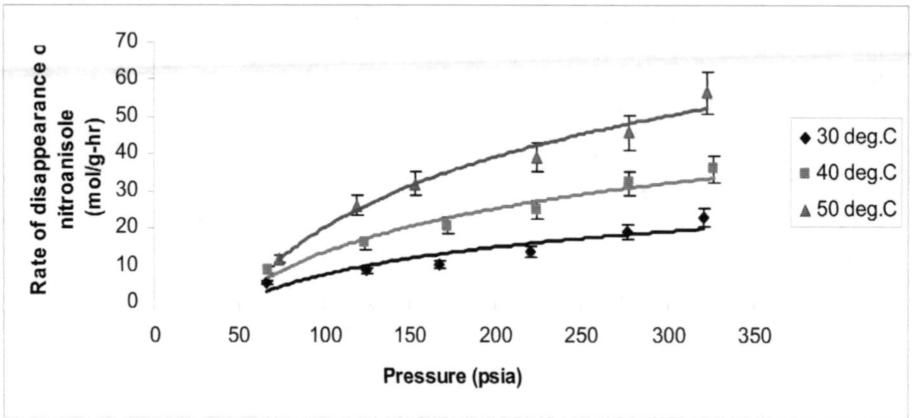

Fig. 6: Effect of H_2 pressure on the reaction rate of disappearance of o-nitroanisole

Dependence of rate of disappearance of o-nitroanisole on the concentration of o-nitroanisole:

The effect of o-nitroanisole concentration on the rate of disappearance of o-nitroanisole was investigated in the concentration range of 0.1 to 2 mol/L (2-30% (w/w) o-nitroanisole in methanol) at temperatures of 30, 40 and 50°C, and a hydrogen pressure of 100psig. The results are presented in fig. 7.

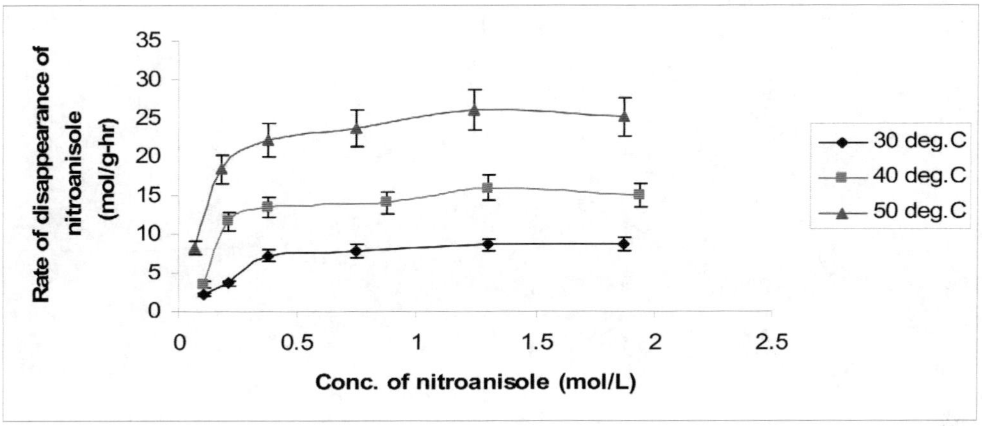

Fig. 7: Effect of o-nitroanisole concentration on the rate of disappearance of o-nitroanisole

Dependence of rate of formation of o-anisidine on hydrogen concentration (or hydrogen pressure):

The effect of hydrogen pressure on the rate of formation of o-anisidine was investigated in a range of 50-300psig at temperatures of 30, 40 and 50°C, and o-nitroanisole concentration of 23% (w/w) in methanol. The results are presented in fig. 8.

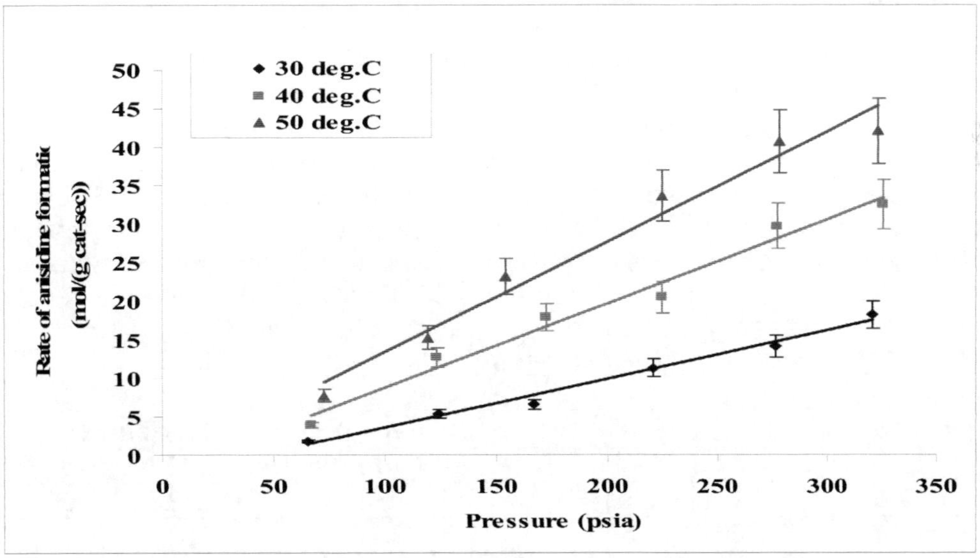

Fig. 8: Effect of H_2 pressure on the reaction rate of formation of o-anisidine

Dependence of rate of formation of o-anisidine on the concentration of o-nitroanisole:

The effect of o-nitroanisole concentration on the rate of formation of o-anisidine was investigated in a range of 0.1 to 2 mol/L (2-30% (w/w) o-nitroanisole in methanol), at temperatures of 30, 40 and 50°C, and a hydrogen pressure of 100psig. The results are presented in fig. 9.

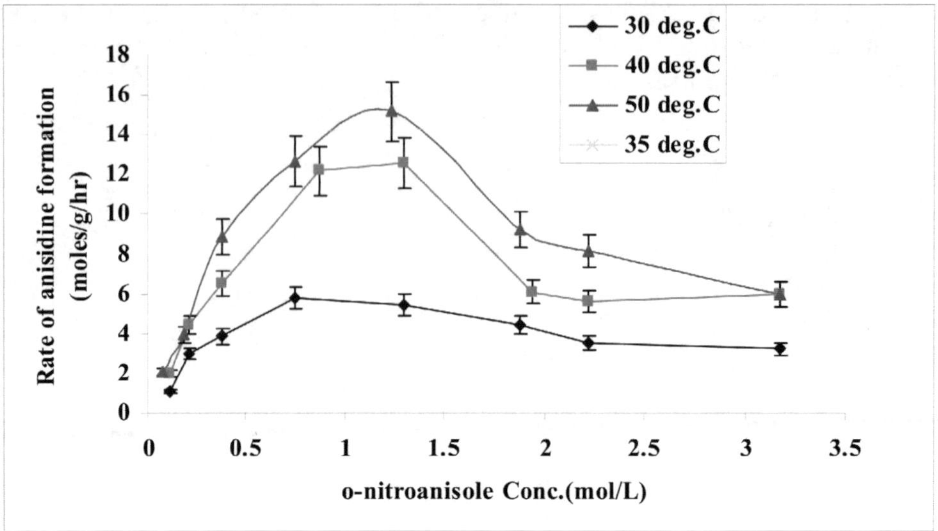

Fig. 9: Effect of o-nitroanisole concentration on the rate of formation of o-anisidine

Comparison of fig. 7 and fig. 9 indicates that the rate of o-nitroanisole disappearance is higher than the rate of formation of o-anisidine at the same conditions. The rate equations that best fitted the experimental data for reactions I and II are the following:

For reaction I (desorption of the product)

$$r_1 = k_I(C_{H2}C_N)/((1+(K_NC_N))(1+(K_{H2I}C_{H2}))) \quad (1)$$

For reaction II (surface reaction controlling)

$$r_2=k_{II}(K_{H2II}C_{H2})(K_IC_I)/((1+(K_IC_I))(1+(K_{H2II}C_{H2}))) \quad (2)$$

where, k_I and k_{II} are the rate constants for reactions I and II respectively, K_N, K_{H2I}, K_I and K_{H2II} are equilibrium constants, C_{H2}, C_N and C_I are concentrations of hydrogen, o-nitroanisole and 2-nitrosomethoxybenzene respectively. The values of the kinetic constants obtained using Polymath for these equations are given in Table 1. The activation energies for reactions I and II were determined by using the rate constant values at different temperature levels. The values of the activation energies determined are 61.5kJ/mol and 40.6kJ/mol for reactions I and II respectively which compare favorably to the literature value of 43.83kJ/mol [6].

***Table 1:** Kinetic constants and regression coefficients for the reactions*

Reaction I				
Temp. (deg.C)	k_I (mol/g-hr)	K_N (L/mol)	K_{H2} (L/mol)	R^2
30	456.68±0.0094	2.31±6.46E-05	0.21±1.75E-04	0.97
40	1037.38± 0.014	3.14±5.45E-05	1.38±1.29E-04	0.984
50	2066.67±0.396	3.93± 9.2E-04	2.61± 0.002	0.965
Reaction II				
Temp. (deg.C)	k_{II} (mol/g-hr)	K_I (L/mol)	K_{H2} (L/mol)	R^2
30	112.33± 0.34	11.09±0.084	1.93±.0073	0.954
40	207.59±0.55	9.19±0.056	2.00± .0067	0.963
50	303.99±2.71	2.98± 0.037	4.095± .056	0.965

3.3. Comparison between microreactor and batch reactor

Experiments were conducted in a 25 ml 'PARR' batch reactor to compare the result obtained in the microreactor with that of the semi-batch reactor. For purpose of comparison, a simple mass transfer model approach was chosen where the overall reaction rate R is expressed as a function of the saturated solubility of hydrogen in the liquid, $C_{H2,\,l}$, the intrinsic rate constant k, and the overall mass transfer coefficient, K_La as follows:

$$R = C_{H2,l}/(1/K_La+1/k) \quad (3)$$

Using this equation, the K_La value was calculated from the reaction rate and the intrinsic rate constant k for both the micro and the semi-batch reactors under selected operating conditions. The intrinsic rate constants for the reaction in the semi-batch reactor were obtained experimentally by determining the order of dependence of the rates of disappearance of nitroanisole and formation of anisidine on the concentrations of reactants. Table 2 shows the calculated K_La values at different reactor pressures for the micro and semi-batch reactors.

Table 2: *Comparison of mass transfer coefficients between microreactor and semi-batch reactor*

	K_La (s^{-1})	
Reactor Pressure (psig)	Micro reactor	Semi-batch reactor
100	3.56	0.0037
150	2.86	0.0065
200	3.04	0.0036
250	6.01	0.0022

Table 2 shows that the mass transfer coefficient values (K_La) in a microreactor are three orders of magnitude higher than the values in a semi-batch reactor. Hence the performance of the microreactor is much better than the semi-batch reactor with respect to mass transfer.

4. Conclusions

In the present study the kinetics of 2%Pd/Zeolite catalyzed hydrogenation of o-nitroanisole to anisidine was investigated in both a microreactor and a semi-batch reactor. In the microreactor, the conversion of o-nitroanisole in all the kinetic experiments was less than 10% to assume a differential reactor. Furthermore, reaction conditions were selected such that external and internal mass and heat transfer effects were negligible, and the reaction dominated by intrinsic kinetics. The rate of formation of o-anisidine was found to be lower than the rate of o-nitroanisole disappearance indicating that the intermediate 2-methoxy nitrosobenzene was present in significant amount in the reactor at low conversions. On the basis of experimental data obtained in the kinetic regime, Langmuir-Hinshelwood type rate equations were proposed for both reactions, and the kinetic parameters were evaluated. From the temperature dependence of the kinetic constants, the activation energies for both the first and second reactions were found to be 61.5 and 40.6 kJ/mole respectively. Further, it was confirmed through calculations that the actual mass transfer coefficient in the microreactor is much higher than the estimates obtained from standard correlations in conventional reactors. Verification of this observation was also done by conducting kinetic experiments in a semi batch reactor. The mass transfer coefficient in the microreactor was observed to be three orders of magnitude higher than that in the semi-batch reactor.

5. Acknowledgements

This work is funded by the DOE-ITP under contract DE-FC36-02GO13156. The authors gratefully acknowledge the support.

6. References

[1] Hoogenraad, M., J. B. Linden and A. A. Smith. (2004). Accelerated Process Development of Pharmaceuticals: Selective Catalytic Hydrogenations of Nitro Compounds Containing Other Functionalities, *Organic Process Research & Development, 8,* pp: 469-476

[2] Stoessel, F. (1993). Experimental study of thermal hazards during the hydrogenation of aromatic nitro compounds, *Journal of loss Prevention in the Process Industries, 6,* pp: 79-85

[3] Blaser, H. U., U. Siegrist and H. Steiner (2001) *Fine Chemicals through Heterogeneous Catalysis*, Wiley/VCH, Weinheim, Chapter 8.5, pp: 389

[4] Mears, D.E. (1971). Test for transport limitations in experimental catalytic reactors, *Industrial and Engineering Chemistry Process Design and Development, 10,* pp: 541-547

[5] Mears, D. E. (1971). Diagnostic criteria for heat transport limitations in fixed bed reactors, *Journal of Catalysis 20*, pp: 127-131

[6] Chaudhari, R.V., M.G. Parande, P.A. Ramachandran and P.H. Brahme (1983). Modelling of a batch slurry reactor for hydrogenation of o-nitroanisole to o-anisidine, *Institute of Chemical Engineers Symposium Series No. 87,* pp: 205-213

Microchemical engineering - a Powerful Tool for Process Intensification

O. Stange, A. Azzawi, F. Schael
Ehrfeld Mikrotechnik BTS GmbH, Mikroforum Ring 1, 55234 Wendelsheim, Germany

H.-E. Gasche, M. Grünewald, L. Mleczko
Bayer Technology Services GmbH, 51368 Leverkusen, Germany

Thomas Daszkowski, Eric Boonstra, Peter Ryan, Shaibal Roy
Bayer Technology Services, Americas
8500 West Bay Road, MS #52, Baytown, TX, 77520, USA

Miniaturization is setting the trend: within the electronic industry microtechnology has reached our daily life. Cell phones, MP3-player and notebooks could only become true through microtechnology. Also, miniaturization has found its way into the chemical-pharmaceutical industry. Although glass apparatuses in chemical labs for product and process development and big stirred vessels in the production are still being extensively used, miniaturization and transition to flexible production methods will soon introduce a change in these areas. Process intensification through miniaturization by microchemical engineering could have exceptional advantages for the chemical-pharmaceutical industry.

Microchemical engineering opens new pathways for the development of economical, innovative and intensified processes. In the chemical and pharmaceutical industry, this technology plays an increasingly important role in research, development and production because of its decisive advantages over conventional technology. For example: microstructured devices have up to three orders of magnitude higher surface to volume ratio compared to conventional apparatuses. This results in intensified heat transfer. Microreactors are an excellent solution for extremely fast and highly exothermic or endothermic reactions. These reactions can be much better controlled in a microreactor and the possibility of hot spot for homogenous or heterogeneous reaction will be minimized. In conventional reactors strong temperature profiles could influence the product spectrum, and in some case even the catalyst can be irreversibly damaged. Microreaction technology therefore can be used to perform hard to control reactions, which otherwise cannot be performed in a conventional reactor.

Due to very high surface to volume ratio, the mass transfer limited processes can also take advantage of microchemical engineering. In micromixers the mixing velocity is an order of magnitude higher then in conventional mixers, which results in smaller mixing sections, a few millimeters or even less. Another advantage of micro devices is the option to control and adjust the process parameters much better. This leads not only to a better yield, but also to a higher level of safety. Additionally costs for downstream processing can be reduced significantly: for example by a solvent free production, production of reduced waste material during start up or shut down, change of product or potential production interference because of the small hold ups in continuos operating microreaction plants etc.

From Lab to Production

Using microtechnology, the development of products or continuous processes, e.g. transfer from batch to continuous production could be accelerated significantly. Additionally, new synthesis routes are possible, which can not be realized in a conventional apparatuses in a cost effective manner, such as highly exothermic reactions. For this kind of applications microreactor technology, such as the Modular Microreaction System of Ehrfeld Mikrotechnik BTS GmbH (EMB) is most suitable (Figure 1).

Figure 1: Modules of the Modular Microreaction System Toolbox from Ehrfeld Mikrotechnik BTS GmbH (EMB)

EMB Modular Micro Reaction Systems consist of 40 different types of modules, including mixers, heat exchangers, reactors, sensors, actuators and fluidic and electrical interfaces. All these modules are fully or partially equipped with microstructures. These modules cover essential unit operations within the reaction technology. Additionally the system is equipped for measuring pressure, temperature and flow rates. Also analytic modules are available, which allow us to determine the quality of the product e.g. using NIR- or UV/VIS-spectroscopy. The individual microreaction modules can be freely arranged on a solid base plate, enabling the fast and flexible setup of a wide variety of microreaction plants. Various configurations can be employed for single- and multi-step reactions (homogeneous and also heterogeneous), depending on the reaction type applied. All fluid-wetted parts are made of stainless steel; sealing and further internals are realized in FFPM or PTFE. The Modular Micro Reaction system is also available in hastelloy C and some other materials.

With minimal effort, a whole lab or pilot plant can be installed with the Modular Microreaction System. First of all, the modules for the process have to be placed on a ground plate made of aluminum. Each row of modules is fixed to the base plate by clamping devices; for sealing purpose, plates with exchangeable sealing rings are placed between each module (Figure 2).

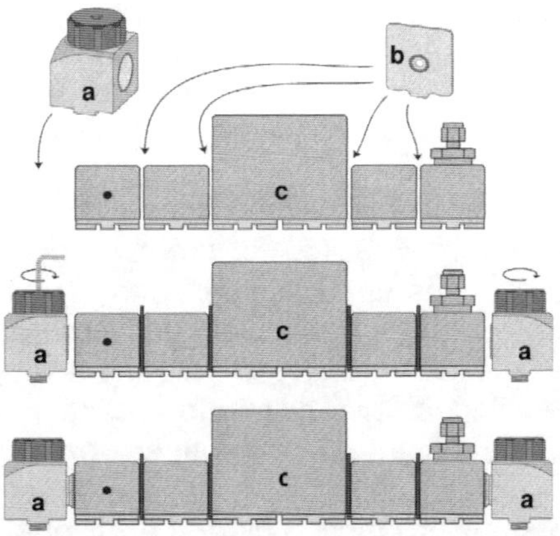

Figure 2: Assembly of a Modular Micro Reaction System (a: Clamping device, b: Sealing plate, c: Module).

The overall construction is extremely robust and there is no dead volume in the connections. The microreaction system allows operating parameters including a temperature range of –20 to 200 °C; operating pressures up to 100 bar (at room temperature); and rated throughputs of up to 25 L/h for low-viscosity media. Specially manufactured modules significantly extend this process range. Among the special items available are reactors for reaction temperatures from –80 °C to 500 °C. All modules can be totally disassembled for cleaning purpose.

In conjunction with highly flexible process control system and peripheral equipments, the Modular Microreaction System is a powerful development and production tool, which allows innovations in all areas of the chemical industry (Figure 3). Sustained development of modules and new module materials will widen the field of application for the Modular Microreaction System.

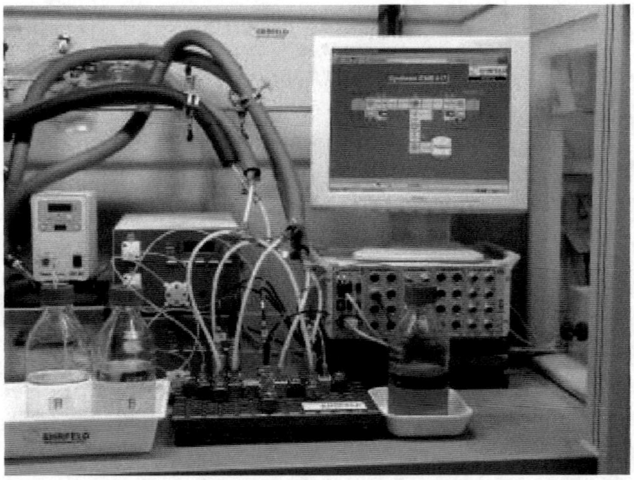

Figure 3: Lab configuration of a microreaction plant

As an example, let us have a closer look at one group of modules: the silt plat mixers. These mixers operate based on the multilamination principle. They can be applied to almost every mixing and emulsion processes through a wide range of mixing plates. For instance the new LH 2-Mixer and the LH 25-Mixer can cover flow rate from less than a milliliter per min up to several tens of liters per hour. This could enable a lab unit to operate as a small production unit without significant changes.

Modular Microreaction System as a flexible pilot plant has been demonstrated by Bayer Technology Services GmbH (BTS) for small production. Also, if the holdup of the unit is small, production rates up to several tons is possible. In addition to liquid, gas-liquid and gas phase reactions, the Modular Microreaction System can also be used to produce micro- and nanoparticles. For example, the valve-assisted mixer (Figure 4), which was developed for precipitation reactions, can be used to produce pigments and catalysts. The advantages of suspension handling and the production of solid particles with specified properties lead to process innovations, which have already been installed in small production plants. Additionally some reactors, such as the cascade mixer allow the handling of slurries.

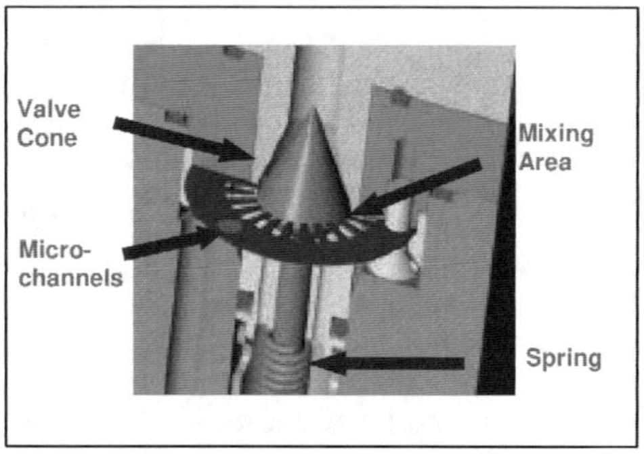

Figure 4: Structure of the valve-assisted mixer

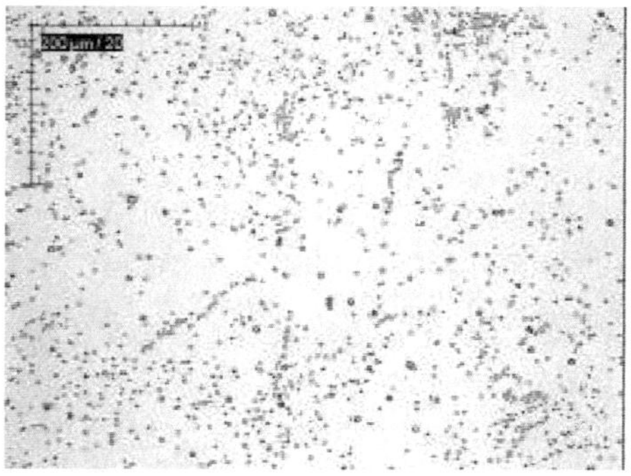

Figure 5: Particles produced by the valve-assisted mixer

The scale-up from lab to bulk production can be achieved by parallel connection of microtechnology plants or by scale-up of the modules by multiplying the microstructures. One example of bigger modules is the LH 1000-mixer, which can handle flow rates of several thousand liters with a very low pressure drop. This mixer contains 20000 slits instead of 300 slits as in a lab module. This cubic shaped slit plate mixer with edge length of around 10 cm reaches the same mixing efficiency as the small modules, but the production rate is several thousand tons per year. This indicates that microreactor technology could be a viable alternative to conventional production units.

For successful process development, a combined knowledge of microtechnology concepts along with conventional plant design process, reaction engineering and process chemistry is essential. This holistic approach is the key to innovative, microtechnology-based solutions for efficient processes and high-performance plants. This will be shown in this presentation for selected applications and the advantages of process intensification will be discussed.

180h A Novel Microfabricated Device Capable of Continuous-Flow Separations Utilizing Vapor-Liquid Equilibria

Stephen Cypes, Symyx Technologies Inc, 3100 Central Expressway, Santa Clara, CA 95051

Microfluidic devices involving single-phase operations, as well as those involving solid-gas and solid-liquid interactions, have been extensively studied both theoretically as well as experimentally. However, there are noticeably fewer published experimental results involving gas-liquid multiphase systems. The comparatively fewer number of published studies is not due to a lack of applicable unit operations within the chemical industry that would benefit from microfluidic technology, but rather due to the inherent challenges of constructing and operating a device capable of manipulating these two phases at length-scales in which surface tension forces dominate.

This presentation will describe a microfluidic device capable of a wide-range of unit operations involving gas-liquid multiphase flow, specifically those involving vapor-liquid equilibrium. Robustness of this device is achieved through careful design of the microcomponents, as well as via a computer-control scheme ensuring complete vapor-liquid flow control under a wide range of operating conditions. Although this device would be applicable to unit operations across widely diverse industries, examples in this specific presentation will focus on operations within the pharmaceutical industry traditionally performed as batch processes, now enabled as continuous-flow processes using the microfluidic device.

Specifically, the time-cycle consuming unit operation used extensively within the pharmaceutical industry known as a “solvent-switch” (a constant-volume batch distillation) will be demonstrated as a continuous-flow operation. Advantages of the microfluidic device in this specific example include the potential for reduced batch time-cycle, as well as increased product yield in cases where the molecule of interest is heat-sensitive (due to greatly reduced residence times at elevated temperatures). Examples of this microfluidic device with respect to other unit operations may also be presented.

American Institute of Chemical Engineers
3 Park Avenue
New York, New York 10016-5991

ISBN: 978-1-60423-525-8